Communications in Computer and Information Science 1657

Yixiang Chen · Songmao Zhang (Eds.)

Artificial Intelligence Logic and Applications

The 2nd International Conference, AILA 2022 Shanghai, China, August 26–28, 2022 Proceedings

Springer

Editors
Yixiang Chen
East China Normal University
Shanghai, China

Songmao Zhang
Chinese Academy of Sciences
Beijing, China

ISSN 1865-0929 ISSN 1865-0937 (electronic)
Communications in Computer and Information Science
ISBN 978-981-19-7509-7 ISBN 978-981-19-7510-3 (eBook)
https://doi.org/10.1007/978-981-19-7510-3

This Springer imprint is published by the registered company Springer Nature Singapore Pte Ltd.
The registered company address is: 152 Beach Road, #21-01/04 Gateway East, Singapore 189721, Singapore

Preface

This volume contains papers from the 2nd International Conference on Artificial Intelligence Logic and Applications (AILA 2022), establishing what will hopefully be a long-running series of conferences dedicated to logical formalisms and approaches to artificial intelligence (AI). Hosted by the Chinese Association for Artificial Intelligence (CAAI), AILA 2022 was organized by both the Technical Committee for Artificial Intelligence Logic under CAAI and the East China Normal University. The conference was held as a virtual online event during August 26–28, 2022.

Logic has been a foundation stone for symbolic knowledge representation and reasoning ever since the beginning of AI research in the 1950s. Besides, AI applications often make use of logical approaches, including decision making, fraud detection, cybernetics, precision medicine, and many more. With the prevalence of machine learning and deep learning, combining logic-related structures is becoming a common view so as to take advantage of the diverse paradigms. The AILA conference series aims to provide an opportunity and forum for researchers to share and discuss their novel ideas, original research achievements, and practical experiences in a broad range of artificial intelligence logic and applications. The first AILA was held in 2019 as a special session within the IEEE 14th International Conference on Intelligent Systems and Knowledge Engineering (ISKE 2019). AILA 2022 was therefore organized as a full-fledged event for the first time. We received a total of 27 submissions, and each paper was peer reviewed by three to four reviewers from our Program Committee with the help of subreviewers designated by the proceedings editors. Based on the scores received and confidence levels of the reviewers, 20 submissions with a final score higher than 0 were accepted as a full-length papers for publication in this proceedings. They are classified into three categories: program logic, fuzzy logic, and applications. Moreover, we were honored to have three prestigious scholars giving keynote speeches at the conference: Guo-Qiang Zhang (University of Texas Houston, USA), Sanjiang Li (University of Technology Sydney, Australia), and Meng Sun (Peking University, China). The abstracts of their talks are included in this proceedings.

AILA 2022 would not have been possible without the contribution and efforts of a dedicated scientific community. We sincerely appreciate members of our Program Committee and all the external reviewers for providing comprehensive and timely reviews. Moreover, we would like to express gratitude to the conference chair, Ruqian Lu (Chinese Academy of Sciences, China). The organization committee from the East China Normal University provided extensive support for the conference, and we especially thank Qin Li and Xuecheng Hou. The conference management system EasyChair was used to handle the submissions and conduct the reviewing and decision-making processes. We thank Springer for their trust and for publishing the proceedings of AILA 2022.

August 2022

Yixiang Chen
Songmao Zhang

Organization

Conference Chair

Ruqian Lu	Academy of Mathematics and Systems Science, Chinese Academy of Sciences, China

Program Committee Co-chairs

Yixiang Chen	East China Normal University, China
Songmao Zhang	Academy of Mathematics and Systems Science, Chinese Academy of Sciences, China

Program Committee

Cungen Cao	Institute of Computing Technology, Chinese Academy of Sciences, China
Shuwei Chen	Southwest Jiaotong University, China
Shifei Ding	China University of Mining and Technology, China
Lluis Godo	Artificial Intelligence Research Institute, CSIC, Spain
Qin Feng	Jiangxi Normal University, China
Xiaolong Jin	Institute of Computing Technology, Chinese Academy of Sciences, China
Qin Li	East China Normal University, China
Huawen Liu	Shandong University, China
Lin Liu	Tsinghua University, China
Yanfang Ma	Huaibei Normal University, China
Wenji Mao	Institute of Automation, Chinese Academy of Sciences, China
Dantong Ouyang	Jilin University, China
Haiyu Pan	Guilin University of Electronic Technology, China
Meikang Qiu	Texas A&M University Commerce, USA
Joerg Siekmann	German Research Center for Artificial Intelligence (DFKI), Germany
Yiming Tang	Hefei University of Technology, China
Hengyang Wu	Shanghai Polytechnic University, China
Maonian Wu	Huzhou University, China
Zhongdong Wu	Lanzhou Jiaotong University, China

Juanying Xie	Shaanxi Normal University, China
Min Zhang (1)	East China Normal University, China
Min Zhang (2)	East China Normal University, China
Hongjun Zhou	Shaanxi Normal University, China
Li Zou	Shandong Jianzhu University, China

Organization Committee Chair

Qin Li	East China Normal University, China

Organization Committee

Yang Yang	East China Normal University, China
Tingting Hu	East China Normal University, China
Li Ma	East China Normal University, China
Xuecheng Hou	East China Normal University, China
Xueyi Chen	East China Normal University, China
Xinyu Chen	East China Normal University, China

External Reviewers

Chaoqun Fei	Academy of Mathematics and Systems Science, Chinese Academy of Sciences, China
Zhenzhen Gu	Free University of Bozen-Bolzano, Italy
Pengfei He	Shaanxi Normal University, China
Xikun Huang	Academy of Mathematics and Systems Science, Chinese Academy of Sciences, China
Weizhuo Li	Nanjing University of Posts and Communications, China
Yangyang Li	Academy of Mathematics and Systems Science, Chinese Academy of Sciences, China
Hongwei Tao	Zhengzhou University of Light Industry, China
Na Wang	Shanghai Polytechnic University, China
Shuhan Zhang	The People's Insurance Company (Group) of China Limited, China

Keynotes

Temporal Cohort Logic

Guo-Qiang Zhang

Texas Institute for Restorative Neurotechnologies,
University of Texas, Houston, Texas, USA
Guo-Qiang.Zhang@uth.tmc.edu

Abstract. We introduce a new logic, called Temporal Cohort Logic (TCL), for cohort specification and discovery in clinical and population health research. TCL is created to fill a conceptual gap in formalizing temporal reasoning in biomedicine, in a similar role that temporal logics play for computer science and its applications. We provide formal syntax and semantics for TCL and illustrate the various logical constructs using examples related to human health. We then demonstrate possible further developments along the standard lines of logical enquiry about logical implication and equivalence, proof systems, soundness, completeness, expressiveness, decidability and computational complexity. Relationships and distinctions with existing temporal logic frameworks are discussed. Applications in electronic health record (EHR) and in neurophysiological data resource are provided. Our approach differs from existing temporal logics, in that we explicitly capture Allen's interval algebra as modal operators in a language of temporal logic (rather than addressing it purely in the semantic space). This has two major implications. First, it provides a formal logical framework for reasoning about time in biomedicine, allowing general (i.e., higher levels of abstraction) investigation into the properties of this framework independent of a specific query language or a database system. Second, it puts our approach in the context of logical developments in computer science (from the 70's to date), allowing the translation of existing results into the setting of TCL and its variants or subsystems so as to illuminate the opportunities and computational challenges involved in temporal reasoning for biomedicine.

Qualitative Spatial and Temporal Reasoning

Sanjiang Li

Centre for Quantum Computation and Intelligent Systems (QCIS),
Faculty of Engineering and Information Technology,
University of Technology Sydney, Australia
Sanjiang.Li@uts.edu.au

Abstract. Spatial and temporal information is pervasive and increasingly involved in our everyday life. Many tasks in the real or virtual world require sophisticated spatial and temporal reasoning abilities. Qualitative Spatial and Temporal Reasoning (QSTR) has the potential to resolve the conflict between the abundance of spatial/temporal data and the scarcity of useful, human-comprehensible knowledge. The QSTR research aims to design (i) human-comprehensible and cognitively plausible spatial and temporal predicates (or query languages); and (ii) efficient algorithms for consistency checking (or query preprocessing). For intelligent systems, the ability to understand the qualitative, even vague, (textual or speech) information collected from either human beings or the Web is critical. This talk will introduce core notions and techniques developed in QSTR in the past three decades. I will focus on introducing Allen's famous interval algebra and two well-known spatial relation models - the topological RCC8 algebra and the Cardinal Direction Calculus (CDC).

A Unifying Logic for Neural Networks

Meng Sun

School of Mathematical Sciences, Peking University, China
sunm@pku.edu.cn

Abstract. Neural networks are increasingly used in safety-critical applications such as medical diagnosis and autonomous driving, which calls for the need for formal specification of their behaviors to guarantee their trustworthiness. In this work, we use matching logic - a unifying logic to specify and reason about programs and computing systems - to axiomatically define dynamic propagation and temporal operations in neural networks and to formally specify common properties about neural networks. As instances, we use matching logic to formalize a variety of neural networks, including generic feed-forward neural networks with different activation functions, convolutional neural networks and recurrent neural networks. We define their formal semantics and several common properties in matching logic. This way, we obtain a unifying logical framework for specifying neural networks and their properties.

Contents

Program Logic

Fuzzy Logic

Applications

Program Logic

Finite Quantified Linear Temporal Logic and Its Satisfiability Checking

Yu Chen, Xiaoyu Zhang, and Jianwen Li(✉)

East China Normal University, Shanghai, China
{51205902034,52215902001}@stu.ecnu.edu.cn,
jwli@sei.ecnu.edu.cn

Abstract. In this paper, we present Finite Quantified Linear Temporal Logic (FQLTL), a new formal specification language which extends Linear Temporal Logic (LTL) with quantifiers over finite domains. Explicitly, FQLTL leverages quantifiers and predicates to constrain the domains in the system and utilizes temporal operators from LTL to specify properties with time sequences. Compared to LTL, FQLTL is more suitable and accessible to describe the specification with both restricted domains and temporal properties, which can be applied to the scenarios such as railway transition systems. In addition, this paper proposes a methodology to check FQLTL satisfiability, releasing the corresponding checker for potential users to further use. Towards experiments, we show that by applying the logic to the railway transit system, most of the safety specifications can be formalized and several inconsistent specifications are reported through our implemented satisfiability checker.

1 Introduction

In recent years, with the development of formal verification techniques and tools, formal methods [2] have been introduced in a variety of large-scale verification scenarios such as railway transit, autonomous driving, and aerospace [4,8].

Model checking [3] is an efficient approach to formal verification. Given an abstract model M from a system and a specification ϕ described in a formal language for the properties to be verified, model checking automatically determines whether the model M can satisfy ϕ, i.e., whether $M \models \phi$ holds. Linear Temporal Logic is a frequently-used property specification language for model checking [18,24,25]. In this paper, we focus on languages for formal-verification applications based on realistic scenarios and define a new specification language FQLTL (Finite Quantified Linear Temporal Logic), to specify properties with both domain restrictions and time constraints.

FQLTL is an extended logic of LTL [19], which introduces quantifiers and predicates to represent domains and relations on the basis of LTL temporal operators. LTL is a very popular logic language today and can be used in various fields in computer science. Due to different purposes, some variants of LTL have been proposed successively, such as LTL on finite traces (LTL_f) [6,7,27], which

Y. Chen and S. Zhang (Eds.): AILA 2022, CCIS 1657, pp. 3–18, 2022.
https://doi.org/10.1007/978-981-19-7510-3_1

is widely used in the fields of AI, as well as Mission-time linear temporal logic (MLTL) [15] and signal temporal logic (STL) [1], which are applicable to the real-time systems such as aerospace.

In previous works, researchers have paid a lot of effort to combine Linear Temporal Logic with quantifiers. For example, [17] proposed a combination of the two, which is named QLTL, for the verification of probabilistic model systems, while [5] proposed Variable Quantifier LTL (VLTL) for the field of data science. The focus of both works is to consider the domains to be infinite. However, the new logic proposed in this paper is an extended LTL based on finite domains. Although [11] also proposed a quantifier LTL over finite domains (FO-LTL), the atoms are formulas in first-order predicate expressions rather than the traditional Boolean variables. However, we argue that the proposal of our new logic FQLTL is still necessary, as FO-LTL is too complicated to provide the practical reasoning technique, e.g., satisfiability checking, for industry usage. Meanwhile, FQLTL is simpler but still robust enough to be applied to a wide spectrum of scenarios, and what's more, it is much easier to present a reasoning framework for FQLTL, which is considered as one of the main contributions of this paper.

Table 1. Requirements and the corresponding formal description of FQLTL.

Requirements	FQLTL description
If a switch is placed reverse, the switch is unlocked	*ALL switch* { *placed_reverse(switch)* → *unlocked(switch)* }
If a track, which contains switches, is released, then it is logically clear in the previous cycle	*ALL track* { *SOME switch* { *in_track(switch,track)* & *released(track)* → *PRE logically_clear(track)*}}

Table 1 shows two examples of security requirements in the railway transit system. The first row of requirements can be represented by first-order logic, but not by temporal logic. The second row of requirements contains the temporal property *previous cycle*, which should be described LTL as *released* → PRE *logically_clear* while the relation *contains switches* cannot be formalized in LTL. So neither LTL nor first-order logic can describe such requirements. However, the requirements are well-specified by using FQLTL in the right column of the table.

The paper first presents the syntax and semantic definition of the FQLTL language, then presents the theoretic foundations to check the satisfiability of formulas written by this logic (FQLTL-SAT). The motivation is based on the fact that FQLTL is an extension of LTL, as well as the domains in considering, are finite, so eliminating the quantifiers can reduce an FQLTL formula to its equivalent LTL formula. As a result, FQLTL-SAT is reducible to LTL-SAT, which can utilize the existing LTL solvers, e.g., Aalta [13], to check the satisfiability of the reduced LTL. In the process of reducing FQLTL to LTL, we need to map an entity class to specific entity instances, and a single FQLTL property of a class of entities may yield multiple instances of LTL properties, among which there may be some that have already been reduced to validity or unsatisfiability. The process of mapping a class to specific instances is the process of eliminating quantifiers and determining the domain, which performs vacuity checking and completed the Boolean-level satisfiability check, only the undetermined results will be instantiated and input to the LTL satisfiability checker [12,14,21] to check the satisfiability at the temporal level.

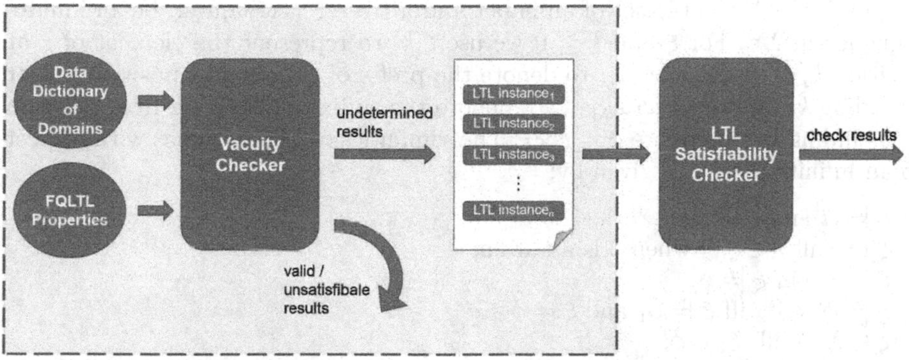

Fig. 1. The framework to implement FQLTL satisfiability checking.

Figure 1 shows the whole procedure of FQLTL satisfiability checking: the information about the domains in the system is abstracted into a data dictionary, and the properties to be checked are described in the FQLTL language. In the procedure, the vacuity checker gets the relevant domain information and the input FQLTL formula to check its satisfiability at Boolean level, ignore the valid and unsatisfiable results and keep the undetermined results as LTL instances. Then these instances are sent to the LTL satisfiability checker for the temporal-level checking and get the check results.

As far as we know, although there are many tools and algorithms for LTL satisfiability checking [10,13,15,22,23], our work is the first to propose a truly feasible solution for satisfiability checking for quantifier linear temporal logic.

The rest of the paper is organized as follows. Section 2 introduces the preliminaries; Sect. 3 gives the definition of syntax and semantics of FQLTL, and discuss the FQLTL satisfiablility problem. Then Sect. 4 gives the implementation details of the instantiation algorithm; Sect. 5 gives the experimental data of the work done in practical applications; Finally, Sect. 6 concludes the paper.

2 Preliminaries

2.1 Linear Temporal Logic

Let AP be a set of atomic properties. The syntax of LTL formulas is defined by:

$$\phi ::= tt \mid ff \mid p \mid \neg\phi \mid \phi \wedge \phi \mid \phi \vee \phi \mid \phi U \phi \mid \phi R \phi \mid X\phi$$

where tt, ff denote *true* and *false* respectively. $p \in AP$, ϕ is an LTL formula, X (Next), U (Until), and R (Release) are temporal operators. A literal is an atom $p \in AP$ or its negation $\neg p$. U and R are dual operators, i.e. $\phi_1 U \phi_2 \equiv \neg(\neg\phi_1 R \neg\phi_2)$. Boolean operators such as $\rightarrow$ and $\leftrightarrow$ can be represented by the combination $(\neg, \vee)$ or $(\neg, \wedge)$. Furthermore, we use the usual abbreviations $F\ a = tt\ U\ a$ and $G\ a = ff\ R\ a$.

Let $\Sigma = 2^{AP}$ be the set of alphabet and a trace $\xi = \omega_0\omega_1\omega_2...$ be an infinite sequence in Σ^ω. For ξ and $k \geq 0$ we use $\xi[k]$ to represent the element of ξ at position k, $\xi^k = \omega_0\omega_1...\omega_{k-1}$ to denote the prefix of ξ ending at position k (not including k), and $\xi_k = \omega_k\omega_{k+1}...$ to denote the suffix of ξ starting from position k (including k). Therefore, $\xi = \xi^k\xi_k$. The semantics of LTL formulas with respect to an infinite trace ξ is given by:

- $\xi \models tt$ and $\xi \not\models ff$;
- $\xi \models p$ iff $p \in \xi[0]$ when p is an atom;
- $\xi \models \neg\phi$ iff $\xi \not\models \phi$;
- $\xi \models \phi_1 \wedge \phi_2$ iff $\xi \models \phi_1$ and $\xi \models \phi_2$;
- $\xi \models X\ \phi$ iff $\xi_1 \models \phi$;
- $\xi \models \phi_1 U \phi_2$ iff there exists $i \geq 0$ such that $\xi_i \models \phi_2$ and for all $0 \leq j < i$, $\xi_j \models \phi_1$;
- $\xi \models \phi_1 R \phi_2$ iff either $\xi_1 \models \phi_2$ for all $i \geq 0$, or there exists $i \geq 0$ with $\xi_i \models \phi_1 \wedge \phi_2$ and $\xi_j \models \phi_2$ for all $0 \leq j < i$.

The above is the definition of standard LTL. Past-time LTL extends LTL by introducing the past operators such as PRE and S(SINCE), which are the temporal duals of the future operators and allow us to express statements on the past time instants. The PRE operator refers to the previous time instant. At any non-initial time, PRE ϕ is true if and only if ϕ holds at the previous time instant. For S operator, ϕ Sψ is true iff ψ holds somewhere in the past and ϕ is true from then up to now.

The satisfiability problem of LTL formulas are formalized as follows:

Definition 1 (LTL-SAT). *An LTL formula ϕ is satisfiable iff there exists an infinite trace ξ such that $\xi \models \phi$.*

Theorem 1 (*[26]*). *The complexity to solve the LTL satisfiability problem is in PSPACE-complete.*

There are already extensive studies on LTL satisfiability (LTL-SAT), e.g., [12,14,21,23], and several LTL checkers such as Aalta [13] are released for public. In this paper, we reduce the satisfiablity of our new logic FQLTL to that of LTL and directly utilize the off-the-shelf LTL satisfiability-checking tools for FQLTL-SAT.

2.2 First-Order Logic

While propositional logic deals with simple declarative propositions, First-Order Logic additionally covers predicates and quantification. The syntax of first-order logic is defined relative to a signature σ, which consists of a set of constant symbols, a set of function symbols and a set of predicate symbols. Each function and predicate symbol has an *arity* $k > 0$. We will often refer to predicates as *relations*. Typically we use letters c, d to denote constant symbols, f, g to denote function symbols and P, Q, R to denote predicate symbols. Note that the elements of a signature are symbols; only later will we interpret them as concrete functions or relations. Independent of the signature σ we also have an infinite set of variables $x_0, x_1, x_2, \ldots$. Given a signature σ, the set of σ-terms is defined by the following inductive process:

- Each variable is a term;
- Each constant symbol is a term;
- If $t_1, \ldots, t_k$ are terms and f is a k-ary function symbol then $f(t_1, \ldots, t_k)$ is a term.

Next the set of formulas is defined inductively as follows:

1. Given terms $t_1, \ldots, t_k$ and a k-ary predicate symbol P then $P(t_1, \ldots, t_k)$ is a formula;
2. For each formula ϕ, $\neg\phi$ is a formula;
3. For each pair of formulas ϕ_1, ϕ_2, $(\phi_1 \wedge \phi_2)$ and $(\phi_1 \vee \phi_2)$ are both formulas;
4. If ϕ is a formula and x is a variable then $\exists x\ \phi$ and $\forall x\ \phi$ are both formulas.

Atomic formulas are those constructed according to the first rule above. The symbol $\exists$ is called the *existential quantifier*. The symbol $\forall$ is called the *universal quantifier*. A general first-order formula is built up from atomic formulas using the Boolean connectives and the two quantifiers.

We will consider one important variant of first-order logic as described above, namely first-order logic with equality. This variant admits equality as built-in binary relation symbol. Thus, regardless of the signature, we admit $t_1 = t_2$ as an atomic formula for all terms t_1 and t_2.

Given a signature σ, a σ-structure $\mathcal{A}$ consists of:

- a non-empty set $U_\mathcal{A}$ called the universe of the structure;
- for each k-ary predicate symbol P in σ, a k-ary relation $P_\mathcal{A} \subseteq \underbrace{U_\mathcal{A} \times \cdots \times U_\mathcal{A}}_{k}$;
- for each k-ary function symbol f in σ, a k-ary relation $f_\mathcal{A} : \underbrace{U_\mathcal{A} \times \cdots \times U_\mathcal{A}}_{k} \to U_\mathcal{A}$;
- for each constant symbol c, an element $c_\mathcal{A}$ of $U_\mathcal{A}$;
- for each variable x an element $x_\mathcal{A}$ of $U_\mathcal{A}$.

The difference between constant symbols and variables is that the interpretation of variables can be overwritten. Given a structure $\mathcal{A}$, variable x, and

$a \in U_{\mathcal{A}}$, we define the structure $\mathcal{A}_{[x \mapsto a]}$ to be exactly the same as $\mathcal{A}$ except that $x\mathcal{A}_{[x \mapsto a]} = a$.

We define the value $\mathcal{A}[\![t]\!]$ of each term t as an element of the universe $U_{\mathcal{A}}$ inductively as follows:

- For a constant symbol c we define $\mathcal{A}[\![c]\!] \stackrel{\text{def}}{=} c_{\mathcal{A}}$;
- For a variable x we define $\mathcal{A}[\![x]\!] \stackrel{\text{def}}{=} x_{\mathcal{A}}$;
- For a term $f(t_1, ..., t_k)$, where f is a k-ary function symbol and $t_1, ..., t_k$ are terms, we define $\mathcal{A}[\![f(t_1, ..., t_k)]\!] \stackrel{\text{def}}{=} f_{\mathcal{A}}(\mathcal{A}[\![t_1]\!], ..., \mathcal{A}[\![t_k]\!])$.

We define the satisfaction relation $\mathcal{A} \vDash \phi$ between a σ-structure $\mathcal{A}$ and σ-formula ϕ by induction over the structure of formulas.

- $\mathcal{A} \vDash P(t_1, ..., t_k)$ iff $(\mathcal{A}[\![t_1]\!], \ldots, \mathcal{A}[\![t_k]\!]) \in P_{\mathcal{A}}$;
- $\mathcal{A} \vDash \phi_1 \wedge \phi_2$ iff $\mathcal{A} \vDash \phi_1$ and $\mathcal{A} \vDash \phi_2$;
- $\mathcal{A} \vDash \phi_1 \vee \phi_2$ iff $\mathcal{A} \vDash \phi_1$ or $\mathcal{A} \vDash \phi_2$;
- $\mathcal{A} \vDash \neg\phi_1$ iff $\mathcal{A} \nvDash \phi_1$;
- $\mathcal{A} \vDash \exists x\ \phi_1$ iff there exists $a \in U_{\mathcal{A}}$ such that $\mathcal{A}_{[x \mapsto a]} \vDash \phi_1$;
- $\mathcal{A} \vDash \forall x\ \phi_1$ iff $\mathcal{A}_{[x \mapsto a]} \vDash \phi_1$ for all $a \in U_{\mathcal{A}}$;
- $\mathcal{A} \vDash (t_1 = t_2)$ iff $\mathcal{A}[\![t_1]\!] = \mathcal{A}[\![t_2]\!]$.

3 FQLTL Language

3.1 Syntax and Semantics

LTL was introduced into computer science in the 1970s and is now used in various fields such as software verification [9], program synthesis [16], databases [28], and artificial intelligence [6,7]. LTL uses temporal operators to express the behavioral constraints and the temporal relationships between events that need to be satisfied by a system at each moment in the past, present, and future.

However, in some specifications, we need to limit the domain of the entities described by the properties. For example, in a multi-device system, device A need to satisfy property P_1, LTL can only determine whether P_1 is satisfied, but not in the restricted domain “device A”; or in a more complex scenario, all subdevices B of device A need to satisfy property P_2, which cannot be described by LTL. These situations need to determine both “domain” and “time”, which requires us to introduce a domain-related logic based on temporal logic.

Based on this, we combine first-order logic and linear temporal logic to define a new language, FQLTL, which can express relational and temporal properties, to restrict the finite domain of devices described in LTL specification. FQLTL is accessible to describe a system with multiple, interrelated devices, the syntax and semantics of which are defined as follows.

Definition 2 (Syntax of QLTL formulas). *A legal FQLTL formula ϕ has the following syntax:*

$$\begin{aligned}\phi ::= &\ t \mid (\phi)\\ &\neg\phi \mid \phi \wedge \phi \mid \phi \vee \phi \mid \phi \rightarrow \phi \mid\\ &P(t_1, ..., t_k) \mid t_1 = t_2 \mid \mathit{ALL}\ x \cdot \phi \mid \mathit{SOME}\ x \cdot \phi \mid\\ &\mathit{PRE}\ \phi \mid X\ \phi \mid \phi\ U\ \phi \mid \phi\ S\ \phi;\end{aligned}$$

In the above, $t, t_1, ..., t_k$ are the *terms* and P is the predicate symbol as defined in First-Order Logic. ALL is the universal quantifier, which is a syntax sugar of $\forall$. Meanwhile, SOME is the existential quantifier, which is a syntax sugar of $\exists$. PRE, X, U, and S are all temporal operators, where PRE is the previous period operator and X means the next period; U is the Until operator, and S is the Since (past) operator. In particular, we use $\phi_1\ R\ \phi_2$ to denote $\neg(\neg\phi_1 U \neg\phi_2)$, i.e., R is the dual operator of U; and we use the usual abbreviations: $G\ \phi = ff\ R\ \phi$, and $F\ \phi = tt\ U\ \phi$.

Definition 3 (Semantics of QLTL formulas). *Let ξ be an infinite trace, σ be a signature and $\mathcal{A}$ be the corresponding σ-structure such that the universe of $\mathcal{A}$, i.e., $U_{\mathcal{A}}$, is a finite set. Then the semantics of FQLTL formulas are interpreted over the tuple $\langle \xi, \mathcal{A}, i\rangle$ such that:*

- $\langle \xi, \mathcal{A}, i\rangle \vDash t$ *iff* $\mathcal{A}[\![t]\!] = true$*;*
- $\langle \xi, \mathcal{A}, i\rangle \vDash P(t_1, ..., t_k)$ *iff* $(\mathcal{A}[\![t_1]\!], \ldots, \mathcal{A}[\![t_k]\!]) \in P_{\mathcal{A}}$*;*
- $\langle \xi, \mathcal{A}, i\rangle \vDash (\phi)$ *iff* $\langle \xi, \mathcal{A}, i\rangle \vDash \phi$*;*
- $\langle \xi, \mathcal{A}, i\rangle \vDash \phi_1 \wedge \phi_2$ *iff* $\langle \xi, \mathcal{A}, i\rangle \vDash \phi_1$ *and* $\langle \xi, \mathcal{A}, i\rangle \vDash \phi_2$*;*
- $\langle \xi, \mathcal{A}, i\rangle \vDash \phi_1 \vee \phi_2$ *iff* $\langle \xi, \mathcal{A}, i\rangle \vDash \phi_1$ *or* $\langle \xi, \mathcal{A}, i\rangle \vDash \phi_2$*;*
- $\langle \xi, \mathcal{A}, i\rangle \vDash \neg\phi$ *iff* $\langle \xi, \mathcal{A}, i\rangle \nvDash \phi$*;*
- $\langle \xi, \mathcal{A}, i\rangle \vDash \mathit{ALL}\ x \cdot \phi$ *iff* $\forall\ a \in U_{\mathcal{A}}, \langle \xi, \mathcal{A}_{[x \mapsto a]}, i\rangle \vDash \phi$*;*
- $\langle \xi, \mathcal{A}, i\rangle \vDash \mathit{SOME}\ x \cdot \phi$ *iff* $\exists a \in U_{\mathcal{A}}, \langle \xi, \mathcal{A}_{[x \mapsto a]}, i\rangle \vDash \phi$*;*
- $\mathcal{A} \vDash (t_1 = t_2)$ *iff* $\mathcal{A}[\![t_1]\!] = \mathcal{A}[\![t_2]\!]$*;*
- $\langle \xi, \mathcal{A}, i\rangle \vDash \mathit{PRE}\ \phi$ *iff* $i > 0$ *and* $\langle \xi, \mathcal{A}, i-1\rangle \vDash \phi$*;*
- $\langle \xi, \mathcal{A}, i\rangle \vDash X\ \phi$ *iff* $i \geq 0$ *and*$\langle \xi, \mathcal{A}, i+1\rangle \vDash \phi$*;*
- $\langle \xi, \mathcal{A}, i\rangle \vDash \phi U \psi$ *iff there exists* $j \geq i$ *such that* $\langle \xi, \mathcal{A}, j\rangle \vDash \psi$ *and for all* $i \leq k < j, \langle \xi, \mathcal{A}, k\rangle \vDash \phi$*;*
- $\langle \xi, \mathcal{A}, i\rangle \vDash \phi S \psi$ *iff there exists* $j \leq i$ *such that* $\langle \xi, \mathcal{A}, j\rangle \vDash \psi$ *and for all* $j < k \leq i, \langle \xi, \mathcal{A}, k\rangle \vDash \phi$*.*

The tuple $\langle \xi, \mathcal{A}, i\rangle \models \phi$ means ϕ holds in $\langle \xi, \mathcal{A}\rangle$ at step i. In particular, we define $\langle \xi, \mathcal{A}\rangle \models \phi$ iff $\langle \xi, \mathcal{A}, 0\rangle \models \phi$. Two examples of using FQLTL formulas to specify the behaviours of railway transit systems are given as follows.

Example 1. If a switch device (sw) is in the normal position, it will not be detected as being in the reverse position.

$$\begin{aligned}&\mathit{formula\text{-}001} :=\\ &\quad \mathsf{ALL}\ sw \cdot (\\ &\qquad\qquad G(is\_normal(sw) \rightarrow \neg is\_reverse(sw))\\ &\qquad\qquad)\end{aligned}$$

In Example 1, ALL sw requires that the specification has to be satisfied for all switch objects. The inner formula $G(is\_normal(sw) \rightarrow \neg is\_reverse(sw))$ represents that, if the object is in the normal position, it will not be detected as being in the reverse position.

Example 2. If a track device (tr) releases, the switch device (sw) in the same area with tr must have already released in the previous period.

$$
\begin{aligned}
&formula\text{-}002 := \\
&\quad \mathsf{ALL}\ tr \cdot (\\
&\mathsf{SOME}\ sw \cdot (\\
&\qquad in\_area(tr, sw) \wedge released(tr) \rightarrow \mathsf{PRE}\ released(sw) \\
&\qquad))
\end{aligned}
$$

In Example 2, ALL tr means that this specification must be satisfied for all objects of the track device, while SOME sw means that it can be satisfied only for the objects of the switch devices which are in the same area with a given track object. *in_area* is a predicate that determines whether a switch is in the same area with a track, i.e., it is used to constraint the domain of switches. PRE is a temporal operator that refers to the last period. The sub-formula "$in\_area(tr, sw) \wedge released(tr) \rightarrow \mathsf{PRE}\ released(sw)$" is the expression of "if a switch device is in the same area with a track device and the track is released, then the switch must have been released in the previous period".

3.2 FQLTL Satisfiability

In this section, we discuss the satisfiability problem of FQLTL formulas. Formally, the problem is defined as follows.

Definition 4 (QLTL -SAT). *An FQLTL formula ϕ is satisfiable iff there exists an infinite sequence ξ and a σ-structure $\mathcal{A}$ of ϕ such that $\langle \xi, \mathcal{A} \rangle \models \phi$.*

Since in FQLTL, the universe $U_{\mathcal{A}}$ of $\mathcal{A}$ is restricted to be finite, the motivation comes up straightforward that the quantifiers can be eliminated for further processing. Let's re-consider Example 1 and assume that there are three switch devices $\{sw_1, sw_2, sw_3\}$ in the domain. Then the formula can be transformed into three sub-formulas (1), (2) and (3) below after eliminating the quantifier ALL sw.

$$G(is\_normal(sw_1) \rightarrow \neg is\_reverse(sw_1)) \tag{1}$$

$$G(is\_normal(sw_2) \rightarrow \neg is\_reverse(sw_2)) \tag{2}$$

$$G(is\_normal(sw_3) \rightarrow \neg is\_reverse(sw_3)) \tag{3}$$

The above example has no relational predicates, so only quantifiers need to be considered. If the formula is restricted by a relation that only sw_1 and sw_2 satisfy, the result of elimination will be only the subformulas (1) and (2) above. The above elimination process is called *instantiation*, which is formally defined as below.

Definition 5 (QLTL Instantiation). *For an FQLTL formula ϕ with the signature σ. Let $\mathcal{A}$ be its σ-structure and $U_{\mathcal{A}}$ be the universe in $\mathcal{A}$. The instantiation of ϕ under $\mathcal{A}$, denoted as $I(\phi)$, is an LTL formula such that*

- $I(P(t_1, \ldots, t_k)) = tt$ *iff* $(\mathcal{A}[\![t_1]\!], \ldots, \mathcal{A}[\![t_k]\!]) \in P_{\mathcal{A}}$; *Otherwise,* $I(P(t_1, \ldots, t_k)) = ff$;
- $I((\phi)) = I(\phi)$;
- $I(\phi_1 \wedge \phi_2) = I(\phi_1) \wedge I(\phi_2)$;
- $I(\phi_1 \vee \phi_2) = I(\phi_1) \vee I(\phi_2)$;
- $I(\neg\phi) = \neg I(\phi)$;
- $I(\mathsf{PRE}\ \phi) = \mathsf{PRE}\ I(\phi)$;
- $I(X\phi) = X\ I(\phi)$;
- $I(\phi_1\ S\ \phi_2) = I(\phi_1)\ S\ I(\phi_2)$;
- $I(\phi_1\ U\ \phi_2) = I(\phi_1)\ U\ I(\phi_2)$;
- $I(\mathsf{ALL}\ x \cdot \phi) = \bigwedge_{a \in U_{\mathcal{A}}} I(\phi_{[x \mapsto a]})$, *where* $\phi_{[x \mapsto a]}$ *is obtained from* ϕ *by replacing* x *to* a;
- $I(\mathsf{SOME}\ x \cdot \phi) = \bigvee_{a \in U_{\mathcal{A}}} I(\phi_{[x \mapsto a]})$, *where* $\phi_{[x \mapsto a]}$ *is obtained the same as above.*

Lemma 1. *For an FQLTL formula ϕ, the size of $I(\phi)$ is at most $|U_{\mathcal{A}}|^k \cdot |\phi|$, where k is the number of quantifier variables in ϕ. Moreover, $I(\phi)$ is semantically equivalent to ϕ, i.e., $\phi \equiv I(\phi)$.*

Proof. The proof can be done by induction over the types of ϕ. Since the proofs for the cases when ϕ is a non-quantifier formula are trivial, we focus on here the case when ϕ is an ALL/SOME formula. Assume now $\phi = \mathsf{ALL}\ x \cdot \psi$. Based on the inductive hypothesis, the size of $I(\psi_{[x->a]})$ for each $a \in U_{\mathcal{A}}$ is at most $|U_{\mathcal{A}}|^{k-1} \cdot |\psi|$. Then according to Definition 5, we have $I(\mathsf{ALL}\ x \cdot \psi) = \bigwedge_{a \subset U_{\mathcal{A}}} I(\phi_{[x \mapsto a]})$. So the size of $I(\mathsf{ALL}\ x \cdot \psi)$ is at most $|U_{\mathcal{A}}|^k \cdot |\psi|$, which is at most $|U_{\mathcal{A}}|^k \cdot |\phi|$, providing that $|\psi| < |\phi|$. The proof of the case when $\phi = \mathsf{SOME}\ x \cdot \psi$ is analogous.

To prove that $\phi \equiv I(\phi)$, it can be achieved also by induction over the types of ϕ. Since this part of proof is very similar to that of above, we omit the details here. □

Since Definition 5 reduces an FQLTL formula to the corresponding LTL formula, it inspires us to reduce FQLTL-SAT to LTL-SAT. The following theorem shows the complexity to check FQLTL-SAT.

Theorem 2. *For an FQLTL formula ϕ whose signature is σ and $U_{\mathcal{A}}$ is the universe of the formula's σ-structure $\mathcal{A}$, the complexity to determine ϕ's satisfiability is at most $|U_{\mathcal{A}}|^k \times 2^{O(n)}$, where k is the number of quantifier variables in $\mathcal{A}$ and n is the number of temporal operators in ϕ.*

Proof. First of all, we know that the cost of checking the satisfiability of an arbitrary LTL formula is at most $2^{O(n)}$, where n is the number of temporal operators in the formula [26]. Secondly, based on Lemma 1, the instantiation procedure can generate the LTL formulas with the size of at most $|U_{\mathcal{A}}|^k$. As a result, the total checking cost is at most $|U_{\mathcal{A}}|^k \times 2^{O(n)}$. □

4 Implementation

4.1 Three-Valued Logic

This section present a simplification based on the *three-valued logic*, which aims to remove the redundant tt or ff instances during instantiation.

When calculating the logical value of a FQLTL formula, the predicate has a clear evaluation of true or false, while the function and temporal operations are instantiated as part of the LTL formula, and their values can only be determined during the LTL satisfiability checking process. Thus, when doing the Boolean-level value calculation, the possible results are specified by false, true, and uncertain. For simplicity, we directly write 0, 1 and −1 respectively in this section.

For a FQLTL formula with a signature σ. Let $\mathcal{A}$ be its σ-structure and $U_{\mathcal{A}}$ be the universe in $\mathcal{A}$. There are four kinds of situations when calculating the logical value of the FQLTL formula.

- **No calculation.** Predicates and functions just return the logical value without any calculation. For a predicate $P(t_1, ...t_k)$, if $(\mathcal{A}[\![t_1]\!], \ldots, \mathcal{A}[\![t_k]\!]) \in P_{\mathcal{A}}$, the Boolean value is 1; otherwise, the value is 0. For a function, the value is always −1.
- **Normal logical calculation.** For all logical operations include NOT, AND, OR, and implication, we calculate the logical value according to the truth table in Table 2.

Table 2. Truth table of the three-valued logic.

p	q	¬ p	$p \wedge q$	$p \vee q$	$p \rightarrow q$
1	1	0	1	1	1
1	0	0	0	1	0
1	−1	0	−1	1	−1
0	1	1	0	1	1
0	0	1	0	0	1
0	−1	1	0	−1	1
−1	1	−1	−1	1	1
−1	0	−1	0	−1	−1
−1	−1	−1	−1	−1	−1

  Implication can be denoted by NOT and OR operations, $A \rightarrow B$ is logically equivalent to $\neg A \vee B$.
- **Reserved temporal calculation.** For the temporal operators, the results cannot be calculated directly at the Boolean level because of the temporal property, so it is necessary to reserve the operators to get the logical value during the LTL satisfiability checking. In this, the calculated result is not the final result, but the Boolean-level result. For unary temporal operation $X\phi$

and PRE ϕ, the reserved result is the logical value of ϕ ignoring the operator. For binary temporal operation $\phi S \psi$ and $\phi U \psi$, Table 3 shows the rules for the reserved calculation.

- **Quantifier calculation.** For formula in the form of ALL $x \cdot \phi$, if all objects $a \in U_{\mathcal{A}}$ satisfy ϕ, the logical value of the formula is 1; if there exists an object that does not satisfy, the value is 0; otherwise, the value is -1. For formula in the form of SOME $x \cdot \phi$, if all objects $a \in U_{\mathcal{A}}$ do not satisfy ϕ, the logical value of the formula is 0; if there exists an object that satisfies, the value is 1; otherwise, the value is -1.

Table 3. Logical calculation rules for U and S operators.

ϕ	ψ	$\phi U \psi$	$\phi S \psi$
1	1	1	1
1	0	0	0
1	−1	−1	−1
0	1	1	0
0	0	0	0
0	−1	−1	0
−1	1	1	−1
−1	0	0	0
−1	−1	−1	−1

Based on three-valued logic, we can check the vacuity for FQLTL formulas. In Example 1, if a switch itself is not in the normal position, then the left side of the implication $is\_normal(sw)$ will be false(0), the whole formula will be valid(1) and do not need to be checked later, so this switch object is removed during the calculation and not to be instantiated.

4.2 Implementation Process

Example 3. Taking formula-002 of Example 2 as an example, the process of instantiation is implemented as follows.

Suppose we have five tracks: $tr_1, tr_2, tr_3, tr_4, tr_5$, and four switches: sw_1, sw_2, sw_3, sw_4, and their relation are shown in the Table 4.

Assume that the variable that can be recognized by the satisfiability checker of the function *released* is the variable *RELEASED*. By providing the table as a data dictionary of domains, we instantiate the result as

$tr_1\text{-}RELEASED \rightarrow \mathsf{PRE}\ sw_1\text{-}RELEASED$

$tr_2\text{-}RELEASED \rightarrow \mathsf{PRE}\ sw_2\text{-}RELEASED \vee \mathsf{PRE}\ sw_3\text{-}RELEASED$

$tr_3\text{-}RELEASED \rightarrow \mathsf{PRE}\ sw_3\text{-}RELEASED \vee \mathsf{PRE}\ sw_4\text{-}RELEASED$

$tr_5\text{-}RELEASED \rightarrow \mathsf{PRE}\ sw_4\text{-}RELEASED$

Table 4. Device relation table.

tr	sw in the same area
tr_1	sw_1
tr_2	sw_2, sw_3
tr_3	sw_3, sw_4
tr_4	–
tr_5	sw_4

The result will be ORed up because tr_2 and tr_3 are associated with more than one switch; while tr_4 has no associated switch object, the logical value obtained from the calculation is 1, and the result is removed.

The above example finally gets all the results of the concrete object after replacing the device type with the object whose return value is not true or false and eliminating the quantifiers from the formula. That is, in the loop, the vacuity checking of validity (return true) and unsatisfiability (return false) has been completed, and the valid and unsatisfiable results are discarded, the undetermined objects are retained, i.e., instances of LTL are generated from FQLTL, to be checked in the next step.

5 Experiment and Case Study

5.1 Experiment

We apply the FQLTL language to a railway transit system to specify 50 safety requirements, we first instantiate these specified properties in FQLTL formula as described in this Sect. 4, afterwards we get 2031 LTL instances to be checked, and the total instantiation time consumption is about 200 s. Here, 10 of the detailed results are selected for demonstration in Table 5. The sets of devices involved are switches (total 74, 45 are singled and 29 are doubled), tracks (50), signals (50), routes (total 453, including 214 train routes).

The *device number* refers to the number of device types involved in the formula. The *total number* refers to the number of loop levels when instantiating the formula. It is written in the form of multiplication, each multiplier corresponds to the number of objects whose type is quantified in the formula. This result is also the number of pieces we send into the vacuity checker. The number of *to be checked* refers to the undetermined results that returns to the top-level quantifier, which is the number of LTL results we finally instantiated out. *Validity* refers to the number of results that return true to the top-level quantifier. The number of *unsatisfiability* is the result that is found to be false and discarded during intermediate operations.

It is meaningless to observe the value of *validity/unsatisfiability* and the *time cost* alone for that they are related to the structure of the formula. The ratio of *validity* and *unsatisfiability* sum to *the total* will be more significant, which is

Table 5. Instantiation datas of 10 formulas.

Formula	Device number	Total number	Validity/ unsatisfiability	To be checked	Pre-determination rate	Time cost
ϕ_1	1	74	0/0	74	–	0.005 s
ϕ_2	1	50	12/0	38	98.5%	0.05 s
ϕ_3	1	74	71/0	3	95.6%	0.003 s
ϕ_4	2	74*50	0/3650	50	98.6%	0.2 s
ϕ_5	2	74*50	47/3650	3	99.9%	0.5 s
ϕ_6	2	74*50	3626/0	74	98.0%	0.1 s
ϕ_7	2	50*214	12/12000	38	99.7%	0.08 s
ϕ_8	3	74*2*29	10/4218	64	98.5%	0.5 s
ϕ_9	3	50*214*214*453	92693/1037186673	34	99.9%	90 s
ϕ_{10}	3	74*50*74*29	16/7618242	58	95.9%	110 s

the pre-determination rate. This rate represents the performance of our vacuity checker to remove redundant instances.

In Table 5, except for ϕ_1, which does not contain a relation, other formulas have predicates to restrict the domain. It is not difficult to conclude that for each formula containing predicates, the vacuity checker can remove a large number of valid and unsatisfiable results, which is far more than the number of instantiated LTL results. The pre-determination rate of the formula is basically above 95%.

5.2 Case Study

We send the obtained LTL formulas to the satisfiability checker and get the check results. The checker reported several unsatisfiable and valid instances. Here we present some simplified cases to make them easier to study.

Case 1. The following is an instance containing two kinds of devices. The route rt_1 and the track tr_3 satisfy the relation in the origin FQLTL formula, so they can be instantiated out.

$$(\boldsymbol{G}\ rt_1\text{-}open) \wedge (tr_3\text{-}locked\ \boldsymbol{U}\ rt_1\text{-}open)$$

It is obvious that this formula is valid at temporal level.

Case 2. The following is an instance of the signal si_5.

$$((\boldsymbol{G}((\neg si_5\text{-}lxj) \vee (\neg si_5\text{-}zxj))) \wedge (((\boldsymbol{G}\ \boldsymbol{F}\ si_5\text{-}lxj) \wedge (\boldsymbol{G}\ \boldsymbol{F}\ si_5\text{-}zxj))\wedge$$
$$((\boldsymbol{G}(si_5\text{-}lxj \rightarrow \boldsymbol{X} si_5\text{-}lxj)) \wedge (\boldsymbol{G}(si_5\text{-}zxj \rightarrow \boldsymbol{X} si_5\text{-}zxj))))$$

The meaning of three subformulas that are ANDed up in the instance is:

- The si_5 signal is either not commanded LXJ up or not commanded ZXJ up.
- The si_5 will finally be commanded both LXJ up and ZXJ up.
- If the si_5 is commanded LXJ or ZXJ up, it will continue to be commanded it up in the next period.

It's also not hard to see that these requirements are in conflict with each other, which proves that the instantiated result is unsatisfiable at the temporal level.

Both of the above examples illustrate that the satisfiability check at the temporal level can be accomplished for instantiated LTL. Thus, at this step we complete the FQLTL-SAT on the basis of LTL-SAT.

6 Conclusion

In this paper, we proposes a new formal specification language FQLTL which extends LTL with quantifiers over finite domains. Based on FQLTL, the paper proposes a methodology for FQLTL satisfiability, and gives a formal definition of this elimination process, i.e., instantiation, then presents the implementation details. In practical applications, the language can formalize the safety specifications of the railway transit system. After applying the proposed methodology to these specifications and analyzing experimental data, it is found that the vacuity checking greatly reduces the size of the result formulas. Sending the instantiated formulas to the LTL satisfiability checker, we also completed the satisfiability check of the FQLTL.

Acknowledgment. We thank anonymous reviewers for their helpful comments. This work is supported by Chinese National Key Research and Development Program (Grant No. 2020AAA0107800), Shanghai Trusted Industry Internet Software Collaborative Innovation Center, Shanghai Pujiang Talent Plan (Grant No. 19511103602) and National Natural Science Foundation of China (Grant No. 62002118 and U21B2015).

References

1. Bae, K., Lee, J.: Bounded model checking of signal temporal logic properties using syntactic separation. In: Proceedings of the ACM on Programming Languages 3(POPL), pp. 1–30 (2019)
2. Bjørner, D., Havelund, K.: 40 years of formal methods. In: Jones, C., Pihlajasaari, P., Sun, J. (eds.) FM 2014. LNCS, vol. 8442, pp. 42–61. Springer, Cham (2014). https://doi.org/10.1007/978-3-319-06410-9_4
3. Clarke, E.M.: The birth of model checking. In: 25 Years of Model Checking-History, Achievements, Perspectives (2008)
4. Feng, J., et al.: FREPA: an automated and formal approach to requirement modeling and analysis in aircraft control domain. In: ESEC/FSE 2020: 28th ACM Joint European Software Engineering Conference and Symposium on the Foundations of Software Engineering, pp. 1376–1386 (2020)
5. Song, F., Wu, Z.: Extending temporal logics with data variable quantifications. In: 34th International Conference on Foundation of Software Technology and Theoretical Computer Science (FSTTCS 2014). Schloss Dagstuhl-Leibniz-Zentrum fuer Informatik (2014)
6. Giacomo, G.D., Vardi, M.Y.: Linear temporal logic and linear dynamic logic on finite traces. AAAI Press (2013)
7. Giacomo, G.D., Vardi, M.Y.: Synthesis for LTL and LDL on finite traces. AAAI Press (2015)

8. Hird, G.R.: Formal methods in software engineering. In: 9th IEEE/AIAA/NASA Conference on Digital Avionics Systems (2002)
9. Holzmann, G.J.: The model checker - spin. IEEE Trans. Softw. Eng. **23**, 279–295 (1997)
10. Hustadt, U., Konev, B.: TRP++2.0: a temporal resolution prover. In: International Conference on Automated Deduction-cade (2002)
11. Kuperberg, D., Brunel, J., Chemouil, D.: On finite domains in first-order linear temporal logic. In: Artho, C., Legay, A., Peled, D. (eds.) ATVA 2016. LNCS, vol. 9938, pp. 211–226. Springer, Cham (2016). https://doi.org/10.1007/978-3-319-46520-3_14
12. Li, J., Zhu, S., Pu, G., Vardi, M.Y.: SAT-based explicit LTL reasoning. In: Piterman, N. (ed.) HVC 2015. LNCS, vol. 9434, pp. 209–224. Springer, Cham (2015). https://doi.org/10.1007/978-3-319-26287-1_13
13. Li, J., Yao, Y., Pu, G., Zhang, L., He, J.: Aalta: an LTL satisfiability checker over infinite/finite traces. In: The 22nd ACM SIGSOFT International Symposium, pp. 731–734 (2014)
14. Li, J., Zhang, L., Pu, G., Vardi, M.Y., He, J.: LTL satisfiability checking revisited. In: 2013 20th International Symposium on Temporal Representation and Reasoning (TIME), pp. 91–98 (2013)
15. Li, J., Zhu, S., Pu, G., Zhang, L., Vardi, M.Y.: SAT-based explicit LTL reasoning and its application to satisfiability checking. Form. Methods Syst. Des. **54**(2), 164–190 (2019). https://doi.org/10.1007/s10703-018-00326-5
16. Michaud, T., Colange, M.: Reactive synthesis from LTL specification with spot. In: Proceedings of the 7th Workshop on Synthesis (2018)
17. Piribauer, J., Baier, C.B.N.: Quantified linear temporal logic over probabilistic systems with an application to vacuity checking. In: International Conference on Concurrency Theory (2021)
18. Pnueli, A.: The temporal logic in programs. In: Proceedings of the 18th Annual IEEE Symposium on Foundations of Computer Science, pp. 46–57 (1977)
19. Pnueli, A.: The temporal logic of programs. In: 18th Annual Symposium on Foundations of Computer Science (sfcs 1977), pp. 46–57 (1977). https://doi.org/10.1109/SFCS.1977.32
20. Putnam, H.: Three-valued logic. The Logico-Algebraic Approach to Quantum Mechanics (1975)
21. Rozier, K.Y., Vardi, M.Y.: LTL satisfiability checking. In: Bošnački, D., Edelkamp, S. (eds.) SPIN 2007. LNCS, vol. 4595, pp. 149–167. Springer, Heidelberg (2007). https://doi.org/10.1007/978-3-540-73370-6_11
22. Schwendimann, S.: A new one-pass Tableau calculus for **PLTL**. In: de Swart, H. (ed.) TABLEAUX 1998. LNCS (LNAI), vol. 1397, pp. 277–291. Springer, Heidelberg (1998). https://doi.org/10.1007/3-540-69778-0_28
23. Suda, M., Weidenbach, C.: A PLTL-prover based on labelled superposition with partial model guidance. In: Gramlich, B., Miller, D., Sattler, U. (eds.) IJCAR 2012. LNCS (LNAI), vol. 7364, pp. 537–543. Springer, Heidelberg (2012). https://doi.org/10.1007/978-3-642-31365-3_42
24. Vardi, M.Y.: Branching vs. linear time: final showdown. In: Margaria, T., Yi, W. (eds.) TACAS 2001. LNCS, vol. 2031, pp. 1–22. Springer, Heidelberg (2001). https://doi.org/10.1007/3-540-45319-9_1
25. Vardi, M.Y.: From church and prior to PSL. In: Grumberg, O., Veith, H. (eds.) 25 Years of Model Checking. LNCS, vol. 5000, pp. 150–171. Springer, Heidelberg (2008). https://doi.org/10.1007/978-3-540-69850-0_10

26. Vardi, M.Y.: An automata-theoretic approach to linear temporal logic. In: Moller, F., Birtwistle, G. (eds.) Logics for Concurrency. LNCS, vol. 1043, pp. 238–266. Springer, Heidelberg (1996). https://doi.org/10.1007/3-540-60915-6_6
27. Zhu, S., Giacomo, G.D., Pu, G., Vardi, M.: LTLf synthesis with fairness and stability assumptions. In: Proceedings of the AAAI Conference on Artificial Intelligence, vol. 34, No. 03, pp. 3088–3095 (2019)
28. Zhu, W.: Big data on linear temporal logic formulas. In: 2021 IEEE 4th Advanced Information Management, Communicates, Electronic and Automation Control Conference (IMCEC), vol. 4, pp. 544–547 (2021)

Quantitative BAN Logic Based on Belief Degree

Kaixuan Li[1,2], Hengyang Wu[3], Jinyi Xu[1,2], and Yixiang Chen[1,2(✉)]

[1] Software Engineering Institute, East China Normal University, Shanghai, China
{52205902008,52184501010}@stu.ecnu.edu.cn
[2] National Engineering Research Center of Trustworthy Embedded Software, Shanghai, China
yxchen@sei.ecnu.edu.cn
[3] School of Computer and Information Engineering, Shanghai Polytechnic University, Shanghai, China
wuhy@sspu.edu.cn

Abstract. Authentication protocols are the basis for secure communication in many distributed systems but are highly prone to errors in their design, preventing them from working properly. It is therefore necessary to analyze an authentication protocol to determine whether the designed protocol meets the requirements. Much attention has been paid to mathematical logic to analyze cryptographic protocols, particularly the logic proposed by Burrows, Abadi, and Needham (BAN logic). This logic has been successful in identifying weaknesses in various examples of authentication protocols. In this paper, we give a concept of "belief" for BAN logic based on the idea of possibility computation and further propose a quantitative BAN logic. It is also applied to the formal analysis and computation of a Radio Frequency Identification (RFID) authentication protocol to show how it works. The quantitative results on belief show that the proposed quantitative approach of BAN logic based on belief can more objectively reflect the security property of the authentication protocol.

Keywords: Belief logic · BAN logic · Quantitative logic · Possibility computation · Authentication protocols · Security and privacy

1 Introduction

Radio Frequency Identification (RFID) technology is a technology that automatically identifies objects in an open environment and is widely used in payment systems, supply chain management, and product anti-counterfeiting due to its low cost and ease of deployment [1]. As low-cost RFID becomes increasingly popular, RFID technology is widely used to ensure the low cost of tags in the system, but the low cost of tags limits their computing power and adds to the problem of ensuring privacy and security when working [2]. The security issues it brings have attracted widespread attention from academia and industry [3–7].

Y. Chen and S. Zhang (Eds.): AILA 2022, CCIS 1657, pp. 19–34, 2022.
https://doi.org/10.1007/978-981-19-7510-3_2

A typical RFID system consists of three legal entities: the tag, the tag reader, and the verifier (referred to as a back-end database or the server) [8–10]. The tag is usually embedded into an object, and when the tag enters the reading range of the tag reader, the tag and the tag reader verify each other's identities, and after verification, the tag transmits its content to the reader, which reads the tag content and passes the content information to the back-end database, which is shown in Fig. 1.

As is shown in Fig. 1, the connection channel between the tag reader and the back-end database is usually perceived as a secure channel. The communication between the tag and the tag reader is through the wireless channel, which is thought insecure and vulnerable to various security attacks [11], mainly passive attacks, active attacks, man-in-the-middle attacks, and asynchronous attacks. Therefore, the research on secure communication between tags and tag readers has received extensive attention. An authentication protocol is an interactive protocol executed between a tag and a tag reader. The owner of the tag (e.g. objects or owners) uses the tag to prove his identity to the reader, and the reader provides feedback on the authentication result. The RFID authentication protocols need to ensure the security and privacy of the communication. During the rapid development of RFID technology, various RFID authentication protocols that are based on different technologies and algorithms have been proposed to protect the security and privacy of data from RFID systems [1–3,5–10,12]. However, these techniques or algorithms are not enough to resist malicious attacks. Therefore, formal analysis and verification technologies are introduced to analyse the security and privacy of RFID authentication protocols [13–19] for they are built on a solid foundation of mathematics, e.g. first-order logic, set theory, graph theory, and category theory. BAN logic is widely used for its ease of use in the authentication of re-authentication protocols. Burrows, Abadi, and Needham propose a logic for analysing and verifying authentication protocols [20] in 1990, which is named BAN logic for it is simple but powerful in the analysis of authentication protocols. However, when using BAN logic to validate authentication protocols, it is often necessary to make initial assumptions about the protocol. This is dependent on the knowledge base and subjective experience of the expert, which can vary from expert to expert; and the specific environmental conditions in which the RFID authentication protocol is applied are unknown, such as the number, type and attack capability of potential attack devices in the surrounding area, the communication quality of the communication environment in which it is used, etc. Therefore, the verification process is fraught with uncer-

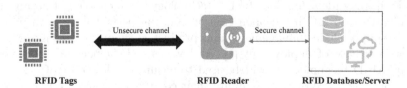

Fig. 1. A typical RFID system.

tainty. Furthermore, BAN logic is a belief logic. As far as beliefs are concerned, this again depends on the subjective experience of different subjects. Therefore, the mere use of binary logic in the original BAN logic is not precise enough and needs to be further quantified. Possibility computation [21] is a concept in possibility theory [22] that can be used to describe and measure the possibilities under uncertainty scenarios. We consider some scenarios that can be handled by possibility computation as follows.

The real-world environment in which RFID protocols are used may have malicious attackers. In the process of communication between the Server (S) and the Tag (T), supposing S sends a message M_1 to T, in the process of using the original BAN logic, the initial assumption of the protocol is that S will receive the message M_1 from T. In fact, if a malicious attacker obtains the key between S and T, he can tamper with M_1 or further forge a new message M_2 to be sent to S. So this protocol assumption above would be too idealistic. To describe this scenario more precisely, by introducing the idea of possibility computation, by considering the strength of the key between S and T and the attacker's capabilities, S can be described as believing that the message M_1 (is secure) with a possibility of δ, in which the $\delta \in [0, 1]$.

This paper focuses on BAN Logic which is broadly used to describe and analyze RFID authentication protocol, proposing a concept named "belief" from possibility computation to precisely quantify the uncertainty in the verification scenario. Then quantitative reasoning rules of BAN logic based on belief for the security property are further proposed. And we test the soundness and viability of our quantitative approach on an authentication protocol, the TSMCA PUF by Liang et al. [23]. The experimental results show that the proposed quantitative BAN logic can objectively reflect the security of TSMCA PUF.

In summary, our primary contributions are:

- We propose a novel quantitative approach for BAN logic based on possibility computation, named belief degree, in which we also develop quantitative reasoning rules corresponding to the original framework in BAN logic to quantify the belief.
- We present a case study to show how our quantitative approach works on an RFID authentication protocol [23], by analysing and calculating the original proof goals using quantitative BAN logic.

The remainder of this paper is organized as follows. Section 2 introduces related works. Section 3 provides the formalism to BAN logic. Section 4 presents our quantitative approach to BAN logic based on belief. Section 5 discusses the case study on the RFID authentication protocol (the TSMCA PUF [23]) in which we apply our novel approach to analyze its security property by the proposed quantitative BAN logic, with the process of computation and results presented afterwards. Section 6 concludes this paper and presents some future directions.

2 Related Work

Although RFID technology is widely used in various aspects of life, such as logistics management, intelligent healthcare [7,12], and product safety tracing, due to its fast identification speed, low cost and long lifespan, the security and privacy issues of RFID technology in applications are still a major concern. Formal analysis in RFID authentication protocols is widely studied in academia [13–15,17,18,23–25]. Burrows, Abadi, and Needham propose BAN logic which is simple but powerful in the analysis of authentication protocols. And David Basin et al. provide an overview of the main applications of model checking in security protocol analysis and explain the central concepts involved in the analysis of security protocols in [26], in which they take the Needham-Schroeder protocol as an example o show the impact of model checking. Vaudenay [19] proposes privacy models for RFID in 2007, in which he gives a strong definition of security and privacy of RFID authentication protocols and adversaries with different attacking abilities are portrayed. In 2019, Liang et al. [23] propose a double-PUF based RFID protocol and analyse the security with BAN logic, and the improved protocol on it presented by Li et al. [24] in 2021 adds the hash operations when the messages are transmitted and introduce a time threshold during the negotiation between the tag and the server's pseudo-random number generator seed. It is proved secure by analysing and verifying using BAN logic and Vaudenay model.

Although BAN logic has been successful in finding the vulnerability and weakness in authentication protocols for its simplicity and convenience, there are some limitations on it [27,28]. Boyd et al. illustrate the limitation of BAN logic on two authentication protocols [27] and give a formalisation of BAN logic in the [28].

3 The Formalism of BAN Logic

Belief logic [29] reasons about beliefs of principals (people, computers, and so on) on the security properties of the communication channels [28], e.g. [20,30]. BAN logic, a kind of belief logic, is proposed by Burrows, Abadi, and Needham [20] in 1990. This logic operates on an abstract level and filters the redundant information of security protocols like implementation errors or inappropriate use. And BAN logic is widely used for the analysis of security protocols for its simplicity. Therefore, BAN logic is widely used to analyze the security of authentication protocols which focus on exchanging messages between principals in the protocols.

To validate an authentication protocol using BAN logic, we must first establish models of the principals and their initial beliefs in the protocol. Each message exchanged during the operation of the authentication protocol is then idealized (referred to as the protocol idealization process). We must also be able to express these beliefs and messages by using the logic formulae in BAN logic reasoning system.

Commonly, a logic reasoning system consists of basic notations (or basic symbol representations), formulae, and reasoning rules. We will introduce the basic notations, logic formulae, and reasoning rules as follows in this section.

3.1 Basic Notations and Logic Formulae in BAN Logic

The basic notation of BAN logic and some symbol descriptions in this paper are shown in Table 1. P and Q in the table represent two principals in the authentication protocol, X and Y for messages, and K for the encryption key.

Table 1. Formulae and instructions

Predicates or symbols	Instructions
1. $P \models X$	P believes message X
2. $P \models \#X$	P believes message X is fresh
3. $P \triangleleft X$	P receives message X
4. $Q \mid\!\sim X$	Q sent message X
5. $Q \Rightarrow X$	Q can control X
6. $\{X\}_K$	X is encrypted by the key K
7. $P \overset{K}{\leftrightarrow} Q$	P and Q share the key K
8. $h(X)$	X is encrypted by the hash operator
9. $X \oplus Y$	X XOR Y
10. $X; Y$	X cascades Y

The logic formulae composed of the basic notations in BAN logic is represented in the form of belief logic. BAN logic can be viewed as a predicate logic constructed on several sorts of objects: principals, encryption keys, messages, and formulae (also called statements) [28].

Predicates construction are used to interpret organised objects into logical statements with truth values. The 1–5 in Table 1 are the predicates in BAN logic, and the others are the symbols that would be used in this paper.

3.2 Reasoning Rules in BAN Logic

For a logical system, proof reasoning mechanisms are vital. There are four reasoning rules in BAN logic as follows.

The reasoning rules are expressed using the Gentzen-style Representation. The rules are divided into two parts in which the upper part of the expression represents the condition and the lower part for the conclusion.

For example,

$$\frac{A, B}{C}$$

where the symbol "," is the boolean conjunctive operator with the following operational semantics: statement A, B is true if and only if both A and B are true which can be represented in Backus Naur Form (BNF) [31] version if necessary, as

$$A \wedge B \rightarrow C$$

Specific rules are strictly followed when performing BAN logic analysis as follows [20].

- Rule 1 (The message-meaning rule). It explains how to derive beliefs about the origin of messages.

$$\frac{P \models P \overset{K}{\leftrightarrow} Q, \quad P \lhd \{X\}_K}{P \models Q \mid\sim X}$$

That is, if the principal P believes that the key K is shared between P and Q, and P received a message X which is encrypted by the K, then P believes that the message X is sent by the principal Q.
- Rule 2 (The nonce-verification rule). This rule expresses the check that a message is recent and, hence, that the sender still believes in it.

$$\frac{P \models \#(X), \quad P \models Q \mid\sim X}{P \models Q \models X}$$

That is, if the principal P believes that the message X is fresh (which means X could have been uttered only recently) and that Q once sent X, then the principal P believes that Q believes the message X.
- Rule 3 (The jurisdiction rule). It states that if P believes that Q has jurisdiction over X(i.e., P believes that Q controls X) then P trusts Q on the truth of X:

$$\frac{P \models Q \Rightarrow X, \quad P \models Q \models X}{P \models X}$$

- Rule 4 (The freshness rule). If one part of a formula is fresh, then the entire formula must also be fresh:

$$\frac{P \models \#X}{P \models \#(X, Y)}$$

That is, if the principal P believes that the message X is fresh, then P believes that the message (X, Y), a combination of X and Y is fresh.

There are a few other reasoning rules, e.g.

$$\frac{P \lhd (X, Y)}{P \lhd X}$$

that is, if P received the combined message (X, Y) of the message X and Y, then it is clear that P received the message X (the same as the message Y).

We shall not list them one by one.

4 Quantitative BAN Logic Based on Belief

In this section, we will introduce the proposed quantitative BAN logic to calculate the belief when analyzing and verifying the security and privacy of authentication protocols.

To be sure, since the real number interval $[0, 1]$ has a richer mathematical content than the binary set $\{0, 1\}$, it is possible to model qualitative methods within different forms of quantitative intervals [32].

In BAN logic, we can only know whether a formula can be believed, that is, the formula of BAN logic is mapped to binary logic i.e., $\{0.1\}$. However, binary logic cannot describe well in practical scenarios, where uncertainty exists everywhere. Moreover, when the logic finds a bug in a protocol, everyone believes that it is a bug. However, when the logic finds proof of correctness, people seem to have trouble believing that it is proof [27,28]. Therefore, according to the basic notations and original reasoning rules in BAN logic, we further introduce the belief degree δ to quantify the belief of the set of formulae Φ in BAN logic by proposing a quantitative approach for BAN logic. In other words,

$$\Delta : \Phi \to [0, 1]$$

And the δ is denoted as the belief degree (in other words, the confidence) of formula ϕ ($\phi \in \Phi$), then we define the relations among the formula ϕ, the belief degree of ϕ and the formulae set Φ below:

$$\delta = \Delta(\phi)$$

Since BAN logic belongs to the category of belief logic, we map the formula in BAN logic to $[0, 1]$ to express the belief degree with the above equation.

Before introducing the quantitative reasoning rules in the proposed quantitative BAN logic, we need to explain the corresponding quantitative formulae as follows.

Quantifying the belief degree of $P \models X$ in BAN logic is defined in the following notations

$$P \models_{\delta} X$$

that is, the principal P believes the message X is secure with a belief degree of δ. The above representation in quantitative BAN logic can also be expressed in terms of $\delta = \Delta(P \models X)$. For the sake of convenience, we will use the former notation when expressing the quantitative reasoning rules.

Similarly, the quantitative formula following

$$P \models_{\delta} \#X$$

states that P believes message X is fresh with a belief of δ, which can be represented as $\delta = \Delta(P \models \#X)$, if necessary. And, because we take into account the quantification of uncertainty during protocol verification, we further quantify the scenarios of message combinations that will occur in the protocol, which were not originally available in the BAN logic. The combination of X and Y includes these three types as follows.

Using Y as a key to encrypt message X:

$$\{X\}_Y$$

Note that there exist some cases where the combination of messages is used to encrypted, like performing a bitwise XOR operation on two messages, etc. Therefore, we quantify the rules for this scenario. Considering this case where the combination of X and Y actually enhances the security relative to the two messages before they were combined. Therefore, the belief (of P in $\{X\}_Y$) should also be no less than the belief in X and Y at this point. To compute the principal P's belief in the combination $\{X\}_Y$:

$$\Delta(P \models \{X\}_Y) = max(\Delta(P \models X), \Delta(P \models Y))$$

For instance, in the TSMCA PUF protocol, if we know the principal P's belief to M_2 is, if we know that principal P has a belief in M_2 of 0.6 and a belief in GW of 0.8, then we can obtain P's belief in $\{GW\}_{M_2}$ is 0.8.

Message X and message Y in XOR operation:

$$X \oplus Y$$

Similarly, in this case, $X \oplus Y$ is obtained by a binary operation by bit, so the combination of the messages becomes more secure than before. Therefore, the belief of P to $X \oplus Y$ is supposed to be no less than the both in X and Y. To compute the principal P's belief in the combination $X; Y$:

$$\Delta(P \models (X \oplus Y)) = max(\Delta(P \models X), \Delta(P \models Y))$$

Message X and message Y in cascade operation, i.e., message X is followed by message Y:

$$X; Y$$

Considering in this case, $X; Y$ is obtained by a simple concatenation (similar to a string operation), so the combination does not improve its security. Therefore, the belief of P to $X; Y$ is supposed to be no greater than the both in X and Y. To compute the principal P's belief in the combination $X; Y$:

$$\Delta(P \models (X; Y)) = min(\Delta(P \models X), \Delta(P \models Y))$$

For instance, if we know that P has a belief in M_1 of 0.5 and a belief in M_2 of 0.6, then we can obtain P's belief in $(M_1; M_2)$ is 0.5.

Similarly, we can convert the basic symbols and representations of the original BAN logic. The rest of the symbol representations remain unchanged, like $P \triangleleft X$ still indicates that P receives a message X.

It also holds for

$$\frac{P \triangleleft (X, Y)}{P \triangleleft X}$$

in quantitative BAN logic. That is, if principal P received a combination of messages like (X, Y), then we can conclude that P received each part.

According to the original reasoning rules of BAN logic in Sect. 3, we propose quantitative reasoning rules of BAN logic based on belief as follows.

- Quantitative Rule 1 (Quantitative message-meaning rule). Similar to Rule 1 in Sect. 3, it explains how to derive belief about the origin of messages.

$$\frac{P \models_{\delta} P \overset{K}{\leftrightarrow} Q, \quad P \triangleleft \{X\}_K}{P \models_{\delta} Q \mid\sim X} \tag{1}$$

  that is, according to Rule 1 of BAN logic, if the principal P believes that the key K is shared between P and Q, and P received a message X which is encrypted by the K, then P believes that the message X is sent by the principal Q.
- Quantitative Rule 2 (Quantitative nonce-verification rule). Corresponding to Rule 2 in Sect. 3, it is noted that this rule states that if P believes that the fresh message X is secure with a belief as δ and that Q once sent X with a belief as δ', then P believes that Q believes the message X with a belief as $\min(\delta, \delta')$.

$$\frac{P \models_{\delta} \#(X), \quad P \models_{\delta'} Q \mid\sim X}{P \models_{\min(\delta,\delta')} Q \models X} \tag{2}$$

  Considering the security property of authentication protocols, $P \models \#(X)$ and $P \models Q \mid\sim X$ are preconditions for $P \models Q \models X$, so the belief of $P \models Q \models X$ depends on and is no greater than the lower of δ and δ' according to the **Barrel Principle**. Note that, this differs from the computation of the belief in messages combination (X, Y) mentioned earlier.
- Quantitative Rule 3 (Quantitative jurisdiction rule). It states that if P believes that Q has jurisdiction over X with a degree δ (i.e., P believes that Q controls X) and that Q sent the message X with a belief δ', then P trusts Q on the truth of X with a belief $\min(\delta, \delta')$:

$$\frac{P \models_{\delta} Q \Rightarrow X, \quad P \models_{\delta'} Q \models X}{P \models_{\min(\delta,\delta')} X} \tag{3}$$

  Similarly, the belief of $P \models X$ is expected to be no greater than the lower of δ and δ'. That is,

$$\Delta(P \models X) = \min(\Delta(P \models Q \Rightarrow X), \Delta(P \models Q \models X))$$

- Quantitative Rule 4 (Quantitative freshness rule). Note that, it indicates that if P believes the fresh message X with a belief as δ and that for Y as δ', then P believes that the message (X, Y) is fresh and P believes the fresh message with a degree as $\max(\delta, \delta')$.

$$\frac{P \models_{\delta} \#X, \quad P \models_{\delta'} \#Y}{P \models_{\max(\delta,\delta')} \#(X, Y)} \tag{4}$$

that is, the belief of that P believes message (X,Y) is fresh is

$$\Delta(P \models \#(X,Y)) = \max(\Delta(P \models \#X), \Delta(P \models \#Y))$$

Specially, if the message X is fresh but the message Y is not, we have the quantitative rule as follows.

$$\frac{P \models_{\delta} \#X}{P \models_{\delta} \#(X,Y)}$$

- Quantitative Rule 5 (Quantitative hash rule). If P believes the message X with a belief δ, then P believes the message $h(X)$ (hash operations has done for the message X) with a belief δ because we consider Hash encryption to be irreversible and therefore secure enough to maintain the security of X.

$$\frac{P \models_{\delta} X}{P \models_{\sqrt{\delta}} h(X)} \tag{5}$$

That is, for the principal P, the belief to $h(X)$ is greater than to X, because the message is encrypted, and more secure than before. i.e.,

$$\Delta(P \models h(X)) = \sqrt{\Delta(P \models X)}$$

5 Protocol Analysis Using Quantitative BAN Logic

This section shows how our quantitative BAN logic operates on authentication protocols. And we choose an RFID authentication protocol the TSMCA PUF [23] which is analyzed and proved in BAN logic in their paper, to reflect the soundness and validity of our quantitative reasoning rules.

To ensure a general result of the experiment, we randomly generated the initial beliefs of the protocol assumptions using a normal distribution $X \sim N(0.8, 0.05)$ by MATLAB in which we exclude the data not less than 1.

The transmission channel between the reader and the back-end database is secure and has a high performance. The channel between the reader and the tag is wireless and hence insecure. The transmission between the reader and the tag is initiated by the reader. Once the required information has been exchanged, the reader and the tag declare the protocol finished [11]. When analyzing the security of authentication protocols, it is common to perceive the reader and the back-end as one principal **Server**, so that there are two principals after abstracting from the RFID authentication protocols, and the same is true of the two protocols we will use. We will use T for the tag and S for the server in the following.

5.1 The TSMCA PUF Protocol

The protocol idealization and assumptions of the TSMCA PUF[1] protocol are shown in Table 2 and Table 3.

Table 2. Protocol idealization of the TSMCA PUF

Messages	Instructions
1. $S \to T : h(T_A) \oplus T_A$	S sends message $h(T_A) \oplus T_A$ to T
2. $T \to S : h(T_A) \oplus T_A \oplus T_B; T_B$	T sends a message $h(T_A) \oplus T_A \oplus T_B; T_B$ to S
3. $T \to S : \{GW\}_{M_2}$	T sends a message $\{GW\}_{M_2}$ to S
4. $S \to T : \{GW_2'\}_{PW}$	S sends a message $\{GW_2'\}_{PW}$ to T

Table 3. Protocol initial assumptions, beliefs and instructions of the TSMCA PUF

Assumption	Belief	Instructions
$A_1 : S \models \#(h(T_A) \oplus T_A)$	0.76	S believes $h(T_A) \oplus T_A$ is fresh with a belief 0.76
$A_2 : S \models T \Rightarrow h(T_A) \oplus T_A \oplus T_B$	0.62	S believes that T can control this message with a belief 0.62
$A_3 : S \models S \overset{h(T_A)\oplus T_A}{\longleftrightarrow} T$	0.88	S believes that S and T share information $h(T_A) \oplus T_A$ with a belief 0.88
$A_4 : S \models S \overset{M_2}{\longleftrightarrow} T$	0.71	S believes that S and T share M_2 with a belief 0.71
$A_5 : T \models S \overset{PW}{\longleftrightarrow} T$	0.81	T believes that S and T share PW with a belief 0.81
$A_6 : S \models T \Rightarrow \{GW\}_{M_2}$	0.75	S believes that T can control $\{GW\}_{M_2}$ with a belief 0.75
$A_7 : T \models S \Rightarrow \{GW_2'\}_{PW}$	0.83	T believes that S can control $\{GW_2'\}_{PW}$ with a belief 0.83
$A_8 : S \models \#M_2$	0.74	S believes message M_2 is fresh with a belief 0.74
$A_9 : T \models \#PW$	0.85	T believes message PW is fresh with a belief 0.85

Objectives to calculate by quantitative BAN logic in the TSMCA PUF protocol:

- **Goal A**: $S \models h(M_2)$.
- **Goal B**: $S \models \{GW\}_{M_2}$.
- **Goal C**: $T \models \{GW_2'\}_{PW}$.

The computation of Goal A is carried out in six steps by using our proposed quantified BAN logic as follows.

Computation of Goal A.

(i) According to the assumption A_1 and the reasoning rule (4), we can get

$$\frac{S \models_{0.76} \#(h(T_A) \oplus T_A)}{S \models_{0.76} \#(h(T_A) \oplus T_A \oplus T_B)} \tag{6}$$

(ii) According to the protocol idealisation 2 in Table 2 with the reasoning rule in Sect. 3 which still holds in our quantitative BAN logic, we have

$$\frac{S \lhd (h(T_A) \oplus T_A \oplus T_B; T_B)}{S \lhd h(T_A) \oplus T_A \oplus T_B} \tag{7}$$

(iii) Given the quantitative reasoning rule (1), With the assumption A_3 and (7), we have

$$\frac{S \models_{0.88} S \overset{h(T_A)\oplus T_A}{\longleftrightarrow} T, \quad S \lhd h(T_A) \oplus T_A \oplus T_B}{S \models_{0.88} T \mid\sim h(T_A) \oplus T_A \oplus T_B} \tag{8}$$

[1] Since $M_2 = h(T_A) \oplus T_A \oplus T_B$, for the sake of convenience for representation, we use M_2 to replace $h(T_A) \oplus T_A \oplus T_B$ if needed as follows.

(iv) Taking the (6) and (8) by using the quantitative reasoning rule (3), we can easily obtain

$$\frac{S \models_{0.76} \#h(T_A) \oplus T_A \oplus T_B, S \models_{0.88} T \mid\sim h(T_A) \oplus T_A \oplus T_B}{S \models T \models h(T_A) \oplus T_A \oplus T_B \quad 0.76} \tag{9}$$

(v) Given the quantitative reasoning rule (3), assumption $A2$ and (9), we have

$$\frac{S \models_{0.62} T \Rightarrow h(T_A) \oplus T_A \oplus T_B, \quad S \models_{0.76} T \models M_2}{S \models_{0.62} M_2} \tag{10}$$

(vi) By the quantitative rule (5), we can obtain

$$\frac{S \models_{0.62} M_2}{S \models_{0.79} h(M_2)} \tag{11}$$

By adopting the five steps above, we can get the belief of goal A by using the proposed quantitative BAN logic, which is 0.79. In other words, the principal S believes that the message $h(M_2)$ is secure with a belief as 0.79. i.e., $\delta = \Delta(S \models h(T_A) \oplus T_A \oplus T_B) = 0.79$.

The belief of Goal B is carried out in four steps by using the proposed quantified BAN logic as follows.

Computation of Goal B.

(i) Given the quantitative reasoning rule (1), the assumption A_4, and the protocol idealization 3 in Table 2, we have

$$\frac{S \models_{0.71} S \overset{M_2}{\longleftrightarrow} T, \quad S \lhd \{GW\}_{M_2}}{S \models_{0.71} T \mid\sim \{GW\}_{M_2}} \tag{12}$$

(ii) Given the rule (4) and the assumption A_8, we can get

$$\frac{S \models_{0.74} \#M_2}{S \models_{0.74} \#\{GW\}_{M_2}} \tag{13}$$

(iii) By the rule (2), (12) and (13), we can obtain

$$\frac{S \models_{0.74} \#\{GW\}_{M_2}, \quad S \models_{0.71} T \mid\sim \{GW\}_{M_2}}{S \models_{0.71} T \models \{GW\}_{M_2}} \tag{14}$$

(iv) According to the reasoning rule (3), the assumption A_6 and (14), we can obtain the belief of goal B

$$\frac{S \models_{0.75} T \Rightarrow \{GW\}_{M_2}, \quad S \models_{0.71} T \models \{GW\}_{M_2}}{S \models_{0.71} \{GW\}_{M_2}} \tag{15}$$

We conclude the belief of goal B is 0.71. i.e., $\delta = \Delta(S \models \{GW\}_{M_2}) = 0.71$ It indicates that the principal S believes that the message $\{GW\}_{M_2}$ is secure with a belief 0.71.

The computation of Goal C is carried out in three steps by using our proposed quantified BAN logic as follows.

Computation of Goal C.

The proof and quantitative reasoning process of goal C (i.e., $T \models \{GW_2'\}_{PW}$) is in the following.

(i) By the quantitative reasoning rule (1), the assumption A_5, and the protocol idealization 4 in Table 2, we have

$$\frac{T \models_{0.81} S \overset{PW}{\longleftrightarrow} T, \quad T \triangleleft \{GW_2'\}_{PW}}{T \models_{0.81} S \mid\sim \{GW_2'\}_{PW}} \tag{16}$$

(ii) Given the rule (2), the assumption A_9 and (16), the (17) can be obtained as

$$\frac{T \models_{0.85} \#PW, \quad T \models_{0.81} S \mid\sim \{GW_2'\}_{PW}}{T \models S \models_{0.81} \{GW_2'\}_{PW}} \tag{17}$$

(iii) By the rule (3), the assumption A_7 and (17), we can obtain

$$\frac{T \models_{0.83} S \Rightarrow \left\{GW_2'\right\}_{PW}, \quad T \models_{0.81} S \models \{GW_2'\}_{PW}}{T \models_{0.81} \{GW_2'\}_{PW}} \tag{18}$$

In the three steps above, we get the belief of goal C by using the quantitative approach we present in Sect. 4, which is 0.81. i.e., $\delta = \Delta(T \models \{GW_2'\}_{PW}) = 0.81$.

It reflects the fact that the principal T(referred to as the Tag of an RFID system in the protocol) believes that the message $\{GW_2'\}_{PW}$ is secure with a belief up to 0.81.

Table 4. The results of protocol analysis in quantitative BAN logic

Protocol name	Results of δ	Mean value	Min value
TSMCA PUF	0.79	0.77	0.71
	0.71		
	0.81		

As is shown in Table 4, the beliefs of the three goals in the TSMCA PUF protocol are 0.79, 0.71 and 0.81 respectively. To reflect the overall level of the protocol proof goal, we obtain a mean value of 0.77. Considering the security of the protocol, we arrive at a minimum value of 0.71 for the protocol belief.

By quantifying the belief in this protocol proof goal, we can have a more specific estimate of the security of this protocol than before. That is, we can estimate the possibility of the protocol as secure to be 0.77 if considering the overall situation of the protocol's proof target, or the probability of the protocol as secure to be 0.71 if based on the idea of the barrel principle.

6 Conclusion

In this paper, We propose a quantitative approach of BAN logic based on belief by the idea of possibility computation, and present some quantitative computing rules on compositional operations of message flow. Then, we give a case study to show how our approach works when analysing and verifying security on the TSMCA PUF protocol. To the best of our knowledge, it is the first piece of work on quantitative BAN logic based on belief to study the security and privacy of RFID authentication protocols, by introducing the possibility computation.

We have done a comparison of the strength of security of two similar authentication protocols using quantitative logic, but space does not allow us to present it in this paper. Next, we will consider further giving a security grading of the protocols by calculating the belief level of the protocol proof targets.

Acknowledgement. This work is supported by the East China Normal University - Huawei Trustworthiness Innovation Center and the Shanghai Trusted Industry Internet Software Collaborative Innovation Center.

References

1. Li, C.-T., Weng, C.-Y., Lee, C.-C.: A secure RFID tag authentication protocol with privacy preserving in telecare medicine information system. J. Med. Syst. **39**(8), 1–8 (2015). https://doi.org/10.1007/s10916-015-0260-0
2. Chien, H.Y.: SASI: a new ultralightweight RFID authentication protocol providing strong authentication and strong integrity. IEEE Trans. Dependable Secure Comput. **4**(4), 337–340 (2007)
3. Yang, L., Han, J., Qi, Y., Liu, Y.: Identification-free batch authentication for RFID tags. In: The 18th IEEE International Conference on Network Protocols, pp. 154–163. IEEE (2010)
4. Tewari, A., Gupta, B.B.: Cryptanalysis of a novel ultra-lightweight mutual authentication protocol for IoT devices using RFID tags. J. Supercomput. **73**(3), 1085–1102 (2016). https://doi.org/10.1007/s11227-016-1849-x
5. Fan, K., Luo, Q., Li, H., Yang, Y.: Cloud-based lightweight RFID mutual authentication protocol. In: 2017 IEEE Second International Conference on Data Science in Cyberspace (DSC), pp. 333–338. IEEE (2017)
6. Fan, K., Luo, Q., Zhang, K., Yang, Y.: Cloud-based lightweight secure RFID mutual authentication protocol in IoT. Inf. Sci. **527**, 329–340 (2020)
7. Kang, J., Fan, K., Zhang, K., Cheng, X., Li, H., Yang, Y.: An ultra light weight and secure RFID batch authentication scheme for IoMT. Comput. Commun. **167**, 48–54 (2021)
8. Das, A.K., Goswami, A.: A secure and efficient uniqueness-and-anonymity-preserving remote user authentication scheme for connected health care. J. Med. Syst. **37**(3), 1–16 (2013). https://doi.org/10.1007/s10916-013-9948-1
9. Lee, C.C., Chen, C.T., Li, C.T., Wu, P.H.: A practical RFID authentication mechanism for digital television. Telecommun. Syst. **57**(3), 239–246 (2014). https://doi.org/10.1007/s11235-013-9844-5
10. Li, C., Lee, C., Weng, C., Fan, C.: A RFID-based macro-payment scheme with security and authentication for retailing services. ICIC Express Lett. **6**(12), 3163–3170 (2012)

11. Liu, Y., Ezerman, M., Wang, H.: Double verification protocol via secret sharing for low-cost RFID tags. Futur. Gener. Comput. Syst. **90**, 118–128 (2019)
12. Agrahari, A.K., Varma, S.: A provably secure RFID authentication protocol based on ECQV for the medical internet of things. Peer-to-Peer Netw. Appl. **14**(3), 1277–1289 (2021). https://doi.org/10.1007/s12083-020-01069-z
13. Clarke, E.M., Henzinger, T.A., Veith, H., Bloem, R. (Eds.): Handbook of Model Checking, vol. 10. Springer, Cham (2018). https://doi.org/10.1007/978-3-319-10575-8
14. Clarke, E.M.: Model checking. In: Ramesh, S., Sivakumar, G. (eds.) FSTTCS 1997. LNCS, vol. 1346, pp. 54–56. Springer, Heidelberg (1997). https://doi.org/10.1007/BFb0058022
15. Baier, C., Katoen, J.P.: Principles of Model Checking. MIT Press, Cambridge (2008)
16. Sihan, Q.: Formal analysis of authentication protocols. J. Softw. **7**, 107–114 (1996)
17. Woo-Sik, B.: Formal verification of an RFID authentication protocol based on hash function and secret code. Wireless Pers. Commun. **79**(4), 2595–2609 (2014). https://doi.org/10.1007/s11277-014-1745-8
18. Sohrabi-Bonab, Z., Alagheband, M.R., Aref, M.R.: Formal cryptanalysis of a CRC-based RFID authentication protocol. In: 2014 22nd Iranian Conference on Electrical Engineering (ICEE), pp. 1642–1647 (2014)
19. Vaudenay, S.: On privacy models for RFID. In: Kurosawa, K. (ed.) ASIACRYPT 2007. LNCS, vol. 4833, pp. 68–87. Springer, Heidelberg (2007). https://doi.org/10.1007/978-3-540-76900-2_5
20. Abadi, M., Tuttle, M.R.: A logic of authentication. In: ACM Transactions on Computer Systems, vol. 8, pp. 18–36. Citeseer (1990)
21. Chen, Y., Wu, H.: Domain semantics of possibility computations. Inf. Sci. **178**(12), 2661–2679 (2008)
22. De Cooman, G., Ruan, D., Kerre, E.: Foundations and applications of possibility theory. Advances in Fuzzy Systems Applications and Theory, vol. 8 (World Scientific 1995) (1995)
23. Liang, W., Xie, S., Long, J., Li, K.C., Zhang, D., Li, K.: A double PUF based RFID identity authentication protocol in service-centric internet of things environments. Inf. Sci. **503**, 129–147 (2019)
24. Li, T., Liu, Y.: A double PUF-based RFID authentication protocol. J. Comput. Res. Dev. **58**(8), 1801 (2021)
25. Ha, J.H., Moon, S.J., Zhou, J., Ha, J.C.: A new formal proof model for RFID location privacy. In: Jajodia, S., Lopez, J. (eds.) ESORICS 2008. LNCS, vol. 5283, pp. 267–281. Springer, Heidelberg (2008). https://doi.org/10.1007/978-3-540-88313-5_18
26. Basin, D., Cremers, C., Meadows, C.: Model checking security protocols. In: Handbook of Model Checking, pp. 727–762. Springer, Cham (2018). https://doi.org/10.1007/978-3-319-10575-8_22
27. Boyd, C., Mao, W.: On a limitation of BAN logic. In: Helleseth, T. (ed.) EUROCRYPT 1993. LNCS, vol. 765, pp. 240–247. Springer, Heidelberg (1994). https://doi.org/10.1007/3-540-48285-7_20
28. Mao, W., Boyd, C.: Towards formal analysis of security protocols. In: Proceedings Computer Security Foundations Workshop VI, pp. 147–158. IEEE (1993)
29. Blum, A.: A logic of belief. Notre Dame J. Form. Log. **17**(3), 344–348 (1976)
30. Gong, L., Needham, R.M., Yahalom, R.: Reasoning about belief in cryptographic protocols. In: IEEE Symposium on Security and Privacy, vol. 1990, pp. 234–248. Citeseer (1990)

31. Knuth, D.E.: Backus normal form vs. backus naur form. Commun. ACM **7**(12), 735–736 (1964)
32. Hawthorne, J., Makinson, D.: The quantitative/qualitative watershed for rules of uncertain inference. Stud. Logica. **86**(2), 247–297 (2007). https://doi.org/10.1007/s11225-007-9061-x

Predicate Logic Network: Vision Concept Formation

Bang Chen[1,3], Maonian Wu[1,3](✉), Bo Zheng[1,3], Shaojun Zhu[1,3], and Wei Peng[2]

[1] The Information Engineering College, Huzhou University, Huzhou, China
wmn@zjhu.edu.cn

[2] Department of Computer Science, Guizhou University, Guiyang, China

[3] Zhejiang Province Key Laboratory of Smart Management and Application of Modern Agricultural Resources, Huzhou, China
2021388105@stu.zjhu.edu.cn

Abstract. Although deep learning has shown good performance in many fields, it still lacks the most basic human intelligence, which we often called the ability to draw inferences about other cases from one instance. Therefore, how to empower model with logical reasoning ability has received much attention. Thus, we propose neural predicate networks, a model that combines deep learning methods with first-order logic. It converts visual tasks into first-order logic problems by deconstructing them into objects, concepts and relations. Then, achieve first-order logic differentiable by learning logical predicates as neural networks. Finally, the differentiable model can be trained by back propagation to simulate the formation of concepts in the human brain and solve the problem. Experimental results on two image concept classification datasets demonstrate the effectiveness and advantages of our approach.

Keywords: Neural network · Neural-symbolic · Cognitive AI

1 Introduction

In recent years, deep learning methods have demonstrated excellent capabilities in many fields as representatives of connectionism [1], especially when faced with tasks such as image and text translation, where models can gain learned experience from a large number of samples. However, a growing number of scholars have also identified the shortcomings of deep learning, and the story of Clever Hans is quite instructive for the field. In the early 20th century, a horse named Hans caused a great sensation, after a period of training by its owner Osten, Hans could master simple mathematical operations. However, it was found that Hans would be in confusion if no one present knew the answer, or if Hans could not see the questioner. It turned out that Hans did not analyze and decide the result based on the question, but by observing the body posture, facial expressions, etc. of the questioner and the audience. Training Hans to do math problems is like training a model in deep learning. Osten's constant training is equivalent to providing the training set, and he thinks he is teaching Hans math, but what Hans actually learns is how to stop

Y. Chen and S. Zhang (Eds.): AILA 2022, CCIS 1657, pp. 35–48, 2022.
https://doi.org/10.1007/978-981-19-7510-3_3

knocking his hooves when he observes Osten's subtle reactions. Similarly, deep learning models are highly dependent on and constrained by the quality of the training data set. In the training process of classical image classification models, scholars often need to solve the problem of overfitting by enhancing and flipping the images, or optimizing the model by using weight recession, dropout, and so on. Even so, we can still only consider the model as a black box [2].

Symbolic systems with logic as their core are the opposite of Neural Networks. Symbolism is more adept at dealing with conceptual knowledge at a higher level. It constructs the world through objects, properties, and relations to provide a formal reasoning process. Therefore, it is extremely interpretable and well suited as a complementary approach to deep learning. However, logical symbols are discrete and cannot handle the high-dimensional continuous features of neural networks such as vectors and matrices, so how to combine symbolic reasoning with neural networks to integrate the respective advantages of both paradigms has become a popular research topic in academia [3–5].

A good medium for linking logical worlds and neural networks is object-centered conceptual learning. Indeed, there are many concepts and memories that humans have difficulty expressing in verbal symbols but can grasp well, which we often refer to as "concepts". By learning basic concepts, infants can generalize their conceptual knowledge to objects that share the same concept but have not been seen before [6]. In the human brain, more than 80% of knowledge is visual concepts, which can form visual propositions, including scene structure and dynamic structure; visual propositions can also constitute visual narratives, and in general, visual concepts are one of the basic units of visual reasoning or causal inference.

As a result, we propose a new image classification architecture, the neural predicate network, which treats pictures and categories as objects and concepts and represents the relationship as a neural network module. The objects, concepts (attributes) and relations are used to construct first-order logical expressions and accomplish the classification task. To summarize, this work makes the following contributions:

a) We construct a neural predicate network, use the neural network to achieve first-order logical differentiability, and use a simple triangular concept dataset to confirm that the neural predicate network is capable of generalizing learned concepts to objects that have not been seen but have the same properties, as infants do.
b) Subsequently, we generalize the neural predicate network to the Chinese calligraphy dataset for multiple classification tasks and change the model architecture so that it can learn multiple concepts simultaneously. These contributions are valuable for embedding first-order logic in visual inference and are one of the precursors for visual inference on complex pictures with multiple objects or for building interpretable image classifications.

In Sect. 2, we introduce the relevant prior work on neural predicate networks; in Sect. 3, we detail the specific architecture of the model when facing binary and multi-classification tasks; in Sect. 4, we document our two experiments conducted and some interesting details; finally, we conclude in Sect. 5 and provide an outlook on future work.

2 Related Work

2.1 Neural Symbolic Artificial Intelligence

Historically, AI has transitioned from symbolism to connectionism, but the current advantages of symbolism are once again gaining attention from various research scholars. Neuro-symbolic integration is a field that combines classical symbolic knowledge with neural networks, with the expectation that models will provide both computational power and logical reasoning. Deep learning, on the other hand, is seen as a promising way to overcome the gap between symbols and sub-symbols [7–9]. In recent years, a number of scholars have attempted to use deep learning approaches to solve logic problems. For example, Johnson et al. [10] and Yi et al. [11] designed a deep network model to generate programs and perform visual reasoning. Yang et al. [12] proposed a neural logic inference system for knowledge bases based on first-order logic. Dong et al. [13] constructed a neural logic model for relational reasoning and decision making. However, all of these works presuppose a single architecture to handle different logical inputs, which shows good performance in the face of idiosyncratic problems but is slightly less flexible in the face of complex datasets that require generalization capabilities. In order to retain better generalization performance of the inference model, Shi et al. [14] proposed a logic integration network in which the three basic propositional logic symbols are considered as neural networks for fitting learning to serve as a good medium between neural networks and symbolism. Although this idea improves the flexibility of the model to some extent, it is limited by the inherent limitations of propositional logic to represent more complex statements. Therefore, there is a need to integrate this idea with first-order logic so that it can be applied to complex visual tasks.

2.2 Object-Centered Visual Concept Reasoning

How to perform visual concept learning is one of the current hot topics, and unlike traditional neural networks that excel in processing continuous data, logical systems always construct the world based on objects, attributes and relationships. Decomposing images into sets consisting of objects is a promising step to convert from low-level perceptual features to efficient abstract reasoning [15]. Concepts are a form of abstraction of attributes. Current main approaches for learning visual concepts are mainly by introducing more representations with additional knowledge base [16, 17] or by using end-to-end neural networks to jointly learn visual concepts and reasoning [18, 19]. Besides, there are also attempts to decouple concept learning from inference as a way to obtain better efficiency and generalizability [20, 21].

In this paper, we treat a concept as an atom in an inference system and consider an image as an object to decouple the model, i.e., to construct logical expressions using objects, concepts, and relations. Subsequently, a neural network is used to fit the predicate logic and use vectors to represent objects and concepts to finally realize a neurosymbolic image classification system where the model not only learns how to generate object representations, but also corrects concepts based on the objects it has seen, which will simulate the process of continuous formation of infant concepts.

3 Neural Predicate Network

In this part, we introduce the Neural Predicate Network (NPN), a model that treats the image classification task as a problem of solving first-order logical expressions. In order to transform a discrete mathematical representation into a continuous tensor form, any logical variable is represented as a vector and the logical relations are learned as a neural network module. In short, the model generates vectors that can represent objects with the help of a convolutional neural network, and stores the categories in the model as concept vectors, while using a neural network to fit the functions of logical predicates. Note that the generation of concept vectors is also trained in backpropagation, which is similar to the formation of children's ideas in constant change.

3.1 NPN Binary Classification Model

First, we introduce the model structure of NPN in a binary classification task, which is shown in Fig. 1. Suppose faced with a cat and dog classification task, the model would define the classification problem as: Does this picture belong to a cat? The problem can be represented using first-order logic. Assuming that the picture is considered as an object, the category as a concept, and the classification relation as a logical predicate belong(), then when category 0 is cat, the model considers the picture as category 0 as long as the picture object is subordinate to the concept cat, and vice versa for category 1. This can be expressed in the following first-order logical form:

$$\forall \text{obj}.\, belong(obj, cat) \rightarrow isclass(obj, 0) \tag{1}$$

$$\forall \text{obj}.\, \neg belong(obj, cat) \rightarrow isclass(obj, 1) \tag{2}$$

The model will then solve the first half of the expression to arrive at the answer. The solution process has two steps, perception and inference, respectively. First, a d-dimensional vector w_i is extracted as an object feature with the help of a convolutional neural network (obviously the dimensionality is an arbitrary hyperparameter, in the example the value is 64), and a d-dimensional vector w_j is generated to represent the concept (cat), then we feed both vectors to the logical predicate module belong(.,.) which is implemented by a multi-layer perceptron(MLP) with one hidden layer:

$$\text{belong}(w_i, w_j) = \text{H}_{a2}f(\text{H}_{a1}(w_i \oplus w_j) + b_a) \tag{3}$$

where $H_{a2} \in R^{d \times d}$, $H_{a1} \in R^{d \times 2d}$, $b_a \in R^d$ are the parameters of the module belong(.,.). $\oplus$ means vector concatenation. f(·) is the activation function, we use ReLU in our networks.

This module outputs a d-dimensional vector as the solution vector of the expression and calculates the similarity with the T-vector which represents True to derive the probability that the expression is true. The T-vector, also with dimension d, is randomly generated when the model is declared and remains constant during training and computation.

We calculate the likelihood of the expression being true based on the cosine similarity, while we multiply by a value and use the activation function Sigmoid to ensure that the value domain ∈ [0,1]. The specific formula is:

$$Sim(w_i, w_j) = sigmoid(\alpha \frac{w_i \cdot w_j}{\|w_i\| \|w_j\|}) \tag{4}$$

The loss function for NPN is similar to the classification task, where we use a single cross-entropy loss function computed for the binary classification task, where o is the image of each represented as an object and p is the similarity of the expression vector to the T vector:

$$L_{Binary} = -\sum_{o_i \in O} y_i \log(p_i) + (1 - y_i) \log(1 - p_i) \tag{5}$$

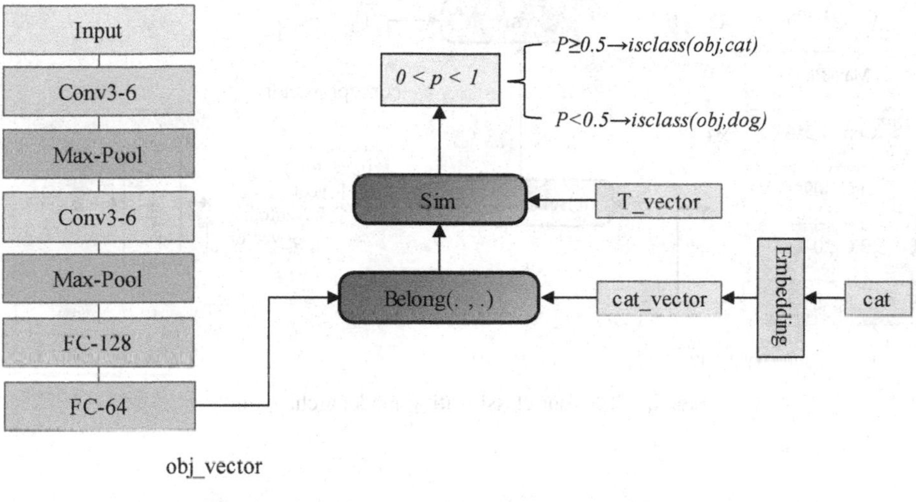

Fig. 1. NPN binary classification model architecture

3.2 NPN Multi-classification Model

This section presents the architecture of NPN when used for a multi-category task. Suppose that the number of categories is 4, the model structure is shown in Fig. 2. The perceptual part is the same as the binary classification task, i.e., a convolutional neural network is needed to generate object feature vectors representing the pictures, but in the inference part, the model needs to generate a concept vector for each category and the object vector needs to construct a first-order logical expression with each concept vector and solve it. Meanwhile, the concept with the largest p-value is calculated to be the predicted category.

In the multi-category task, each concept will have its loss value, and the overall loss of the model is the sum of the losses of each concept:

$$L_{Multi} = \sum_{i=0}^{n} L_i = -\sum_{i=0}^{n} \sum_{o_j \in O} y_{ij} \log(p_{ij}) + (1 - y_{ij}) \log(1 - p_{ij}) \tag{6}$$

It is worth noting that during the backpropagation process, the concept vectors are also updated, i.e., the concept vectors can be observed to know exactly what the model has learned, which can also be helpful for interpretability.

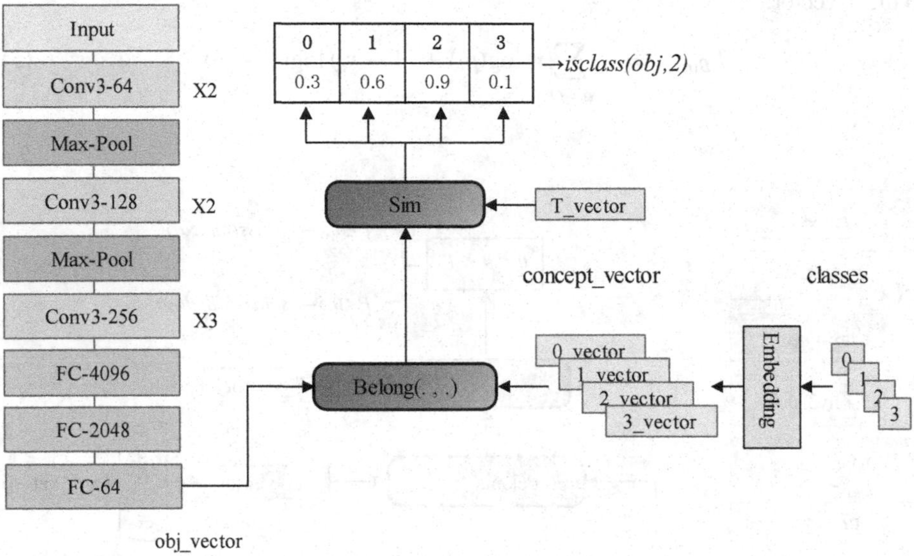

Fig. 2. NPN four classification model architecture

4 Experiments

In this part, the experimental results of the NPN model and the baseline model on the binary and multiclassification tasks are described. All the models, including baselines, are trained with Adam in weight_decay set to 0.02. The learning rate is 0.0001 and early-stopping is conducted according to the performance on the validation set. Models are trained at most 100 epochs. Vector sizes of the objects and the concepts are set to 64. In the binary classification, We run the experiments with ten different random seeds and report the average results. In the multiclassification task. We run the experiments with three different random seeds and report the average results.

4.1 Triangle Concept Identification

We first created a triangle concept recognition binary dataset, as shown in Fig. 3: in the training set, the positive examples are various triangles (but not right triangles), and the negative examples are other shapes. Unlike most publicly available datasets, the positive examples in the test set only contain right triangles that do not appear in the training set, i.e., the model needs to recognize objects which it has not seen before but have the same concept based on the concepts it has already learned. Such a task is very easy for children, and we expect to test whether the model has the ability to learn by example.

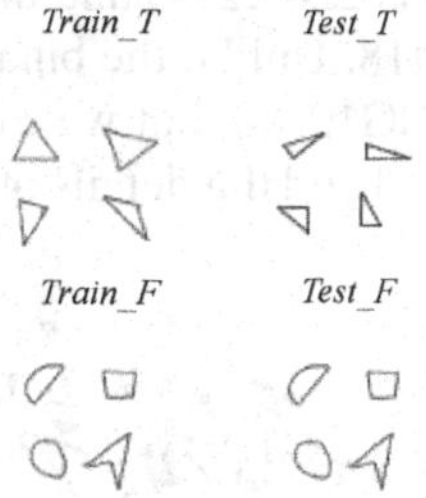

Fig. 3. Example of triangle concept dataset

In this dataset, the image size is 28X28, so LeNet5 and VGG5 were chosen as Baseline. The dataset includes 360 training sets, 120 validation sets and 120 test sets each, where the positive and negative ratio is 1:1. Table 1 shows the results of this experiment. It can be seen that all three models (including our NPN) show apparent overfitting, and the Precision metrics of all three models are higher than the Recall metrics, which indicates that the above models do show higher judgment errors when facing right triangles that have not been seen before. Of course, although NPN does not achieve 100% accuracy in the training set as the other two models, its effect in the test set is significantly better than the other two models, its results in the test set are significantly better than the other two models, with 6% and 2.66% improvement in accuracy and 7.17% and 2.2% improvement in recall, respectively, which indicates that the method of classification by learning concepts does play a certain effect.

Table 1. Experimental results of each model on the triangular concept dataset

	Train data			Test data		
	Acc	Recall	Precision	Acc	Recall	Precision
LeNet5	1	1	1	0.7733	0.7340	0.8567
VGG5	1	1	1	0.8067	0.7837	0.8500
NPN	0.9978	0.9978	0.9978	**0.8333**	**0.8057**	**0.8767**

4.2 Chinese Calligraphy Style Concept Identification

After the successful validation of NPN in the binary classification task, we choose the Chinese calligrapher style dataset to further test the model's ability of multi-concept learning. This dataset consists of Chinese characters in various calligraphic styles, and all the characters in the test set are not present in the training set, which will test the model's ability to solve new problems through existing knowledge[1]. We conducted experiments with four and eight classifications. In the four categories, the calligraphers were Xizhi Wang, Zhenqing Yan, Tingjian Huang, and Gongquan Liu. In the eight categories, additional fonts of Xun Ouyang, Huizong Song, Guiting Sun, and Fu Mi were added. The input image size is 224X224, and the baseline models are AlexNet8, VGG10, VGG13, VGG16, ResNet18. Unlike the binary classification task, we modify the convolution part of NPN to VGG10, so that we can compare them better. Example calligraphic fonts are shown in Fig. 4 and the details of the dataset are shown in Table 2.

Fig. 4. Example of the calligraphy style dataset

Table 2. Statistics of the Chinese calligraphy style dataset

		Calligrapher	Train data	Val data	Test data
8	4	Xizhi Wang	5393	1348	1348
		Zhenqing Yan	5405	1351	1351
		Tingjian Huang	5371	1343	1343
		Gongquan Liu	5410	1353	1353
		Xun Ouyang	2808	702	702
		Huizong Song	5410	1353	1353
		Guiting Sun	5001	1250	1250
		Fu Mi	5410	1353	1353

The overall performances on two dataset are shown on Table 3, The NPN model showed optimal performance on both the four- and eight-classification datasets, with correct rates 1.29% and 2.7% higher than the second place on the test set, respectively. Neither VGG13 nor VGG16 could converge on the training set, which may be due to the disappearance of gradients caused by the over-deepness of the neural network. ResNet18 solved the problem of over-deepness of the network, but showed a very serious overfitting problem. In contrast, the NPN model, although using VGG10 as a base to further increase the computational complexity of the model as in VGG13 and VGG16, is able to avoid overfitting while ensuring normal convergence of the model, which further validates the superiority of the object, concept, and relationship learning approach for image classification tasks.

Table 3. Experimental results of each model on the Chinese calligraphy style dataset

	Calligraphy style-4		Calligraphy style-8	
	TrainAcc	TestAcc	TrainAcc	TestAcc
AlexNet8	0.9824	0.9685	0.9727	0.9520
VGG10	0.9932	0.9757	0.9899	0.9588
VGG13	0.2499	0.2487	0.1369	0.1338
VGG16	0.2499	0.2484	0.1369	0.1338
ResNet18	0.9936	0.3300	0.9947	0.1666
NPN	**0.9966**	**0.9886**	**0.9977**	**0.9858**

Table 4. Experimental detail of each model on the Chinese calligraphy style test set

		Calligraphy style-4			Calligraphy style-8		
		Recall	Precision	F1	Recall	Precision	F1
1	AlexNet8	0.9807	0.9742	0.9774	0.9711	0.9513	0.9611
	VGG10	**0.9881**	0.9737	0.9808	0.9102	0.9639	0.9363
	ResNet18	0.2470	0.3310	0.2829	0.1573	0.1381	0.1471
	NPN	0.9874	**0.9852**	**0.9863**	**0.9814**	**0.9757**	**0.9785**
2	AlexNet8	**0.9955**	0.9911	0.9933	0.9800	0.9881	0.9840
	VGG10	0.9882	0.9874	0.9878	0.9845	0.9666	0.9755
	ResNet18	0.2036	0.2712	0.2326	0.2036	0.1554	0.1763
	NPN	0.9948	**0.9956**	**0.9952**	**0.9948**	**0.9933**	**0.9940**
3	AlexNet8	0.9702	0.9796	0.9749	0.9613	0.8984	0.9288
	VGG10	0.9798	0.9879	0.9838	0.9889	0.9595	0.9740
	ResNet18	0.6008	0.3648	0.4540	0.3246	0.2859	0.3040
	NPN	**0.9903**	**0.9911**	**0.9907**	**0.9948**	**0.9795**	**0.9871**
4	AlexNet8	0.9955	0.997	0.9962	0.9889	0.9933	0.9911
	VGG10	0.9852	0.9926	0.9889	0.9697	0.9791	0.9744
	ResNet18	0.2520	0.2932	0.2710	0.1574	0.2171	0.1825
	NPN	**0.9963**	**0.9985**	**0.9974**	**0.9985**	**0.9978**	**0.9981**
5	AlexNet8		0.9829	0.9857	0.9843		
	VGG10		0.9772	0.9581	0.9676		
	ResNet18		0.1211	0.1214	0.1212		

(continued)

Table 4. (*continued*)

		Calligraphy style-4			Calligraphy style-8		
		Recall	Precision	F1	Recall	Precision	F1
	NPN		**0.9957**	**0.9887**	**0.9922**		
6	AlexNet8		0.9993	0.9992	0.9992		
	VGG10		0.9985	1	0.9992		
	ResNet18		0.1375	0.3483	0.1972		
	NPN		**1**	**1**	**1**		
7	AlexNet8		0.9600	0.9554	0.9577		
	VGG10		0.9736	0.9419	0.9575		
	ResNet18		0.1472	0.1679	0.1569		
	NPN		**0.9760**	**0.9839**	**0.9799**		
8	AlexNet8		0.8825	0.9692	0.9238		
	VGG10		0.9217	0.9439	0.9327		
	ResNet18		0.2195	0.1553	0.1819		
	NPN		**0.9579**	**0.9774**	**0.9676**		

Table 4 shows the details of each model on the test set, including the results for the classifications of each category.

Subsequently, we conducted a more interesting experiment in which the training set of the Chinese calligraphy dataset was used as the test set, and the test set was used as the training set, expecting to test the ability of the model to infer universal knowledge from a small amount of data. In that experiment, we kept only the four models that could converge for comparison. The results are shown in Table 4. On the training set, there is not much difference in the results of each model compared with the previous experiment. However, in terms of test accuracy, all three models showed a decline except for ResNet18, which had poor results. However, the NPN model still demonstrates optimal performance and is the least affected, with accuracy rates on the test set higher than the second-ranked VGG model by 2.92% and 3.9%, respectively. The performance gap widens further compared to the previous experiment, which may be due to the better generalizability of the approach of using objects and concepts to store knowledge when learning from a small amount of data (Table 5).

In terms of parameter sensitivity, we tested the effect of different vector dimensions on the model performance, and we selected 32, 64, 96, 128, 160, and 192 for validation. The results are shown in Fig. 5, and it can be seen that the dimensionality has a small impact on the model performance on this dataset, which indicates that the NPN model is robust. The possible reason is that the dependence of the model on the vector dimensionality is not significant due to the low number of categories in the dataset.

Finally, as illustrated in Figs. 6 and 7, we also use the visualization technique of projecting high-dimensional to low-dimensional data to examine the link between picture

Table 5. Experimental results of each model on the Chinese calligraphy style reversal dataset

	Calligraphy style-4		Calligraphy style-8	
	TrainAcc	TestAcc	TrainAcc	TestAcc
AlexNet8	0.9805	0.9529	0.9707	0.9269
VGG10	0.9859	0.9537	0.9901	0.9336
ResNet18	0.9873	0.3077	0.9914	0.1792
NPN	**0.9912**	**0.9829**	**0.9927**	**0.9726**

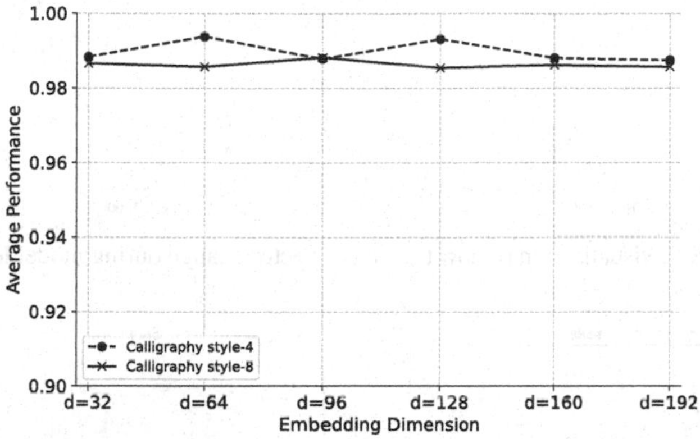

Fig. 5. Impact of vector dimensionality on model correctness

objects using vectors produced by the model during training. Figures 6 and 7 employ the PCA and SNE algorithms, respectively. We selected the cases of the 0th, 100th, 1000th, and 2000th batch of the training process (this is because the model has largely converged by the 2nd epoch, so we did not choose epoch as the observation unit) to be projected to the low-dimensional space for observation. As we can see, at the beginning, the object vector is chaotically present in the model, which means that the concept is not yet formed. As training progresses and the model observes more and more different data, concepts start to form gradually. Eventually, the concepts and the corresponding objects are clearly separated, which indicates that the model is developing in the way we expected.

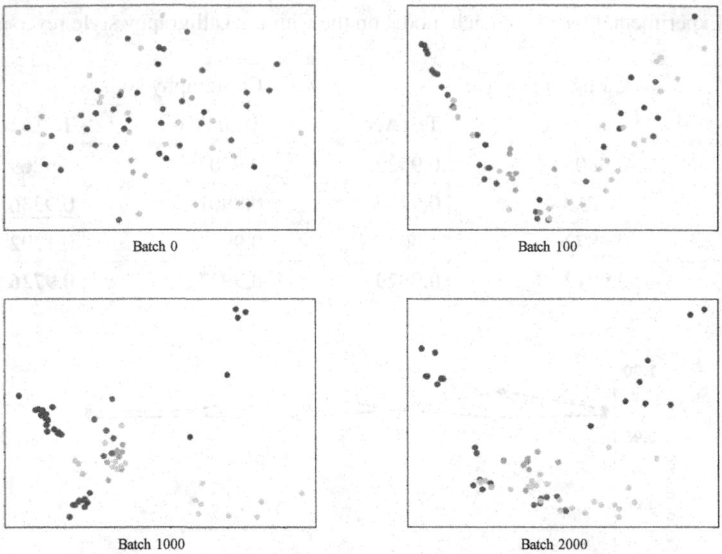

Fig. 6. PCA visualization of how the object vector change during model learning

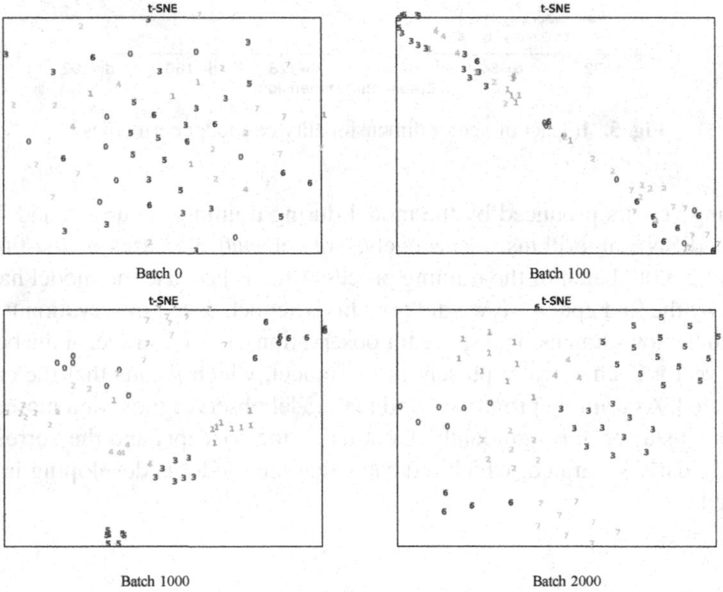

Fig. 7. t-SNE visualization of how the object vector change during model learning.

5 Conclusion and Future Work

We propose the Neural Predicate Network (NPN), a neural symbolic concept inference model. Specifically, NPN constructs first-order logical expressions by objects, concepts (attributes) and relations. Then, a neural network is used to fit logical predicates to achieve the integration of discrete space with continuous features and logical differentiability. Finally, NPN contributes to visual concept formation and completes the classification task. Experiments on two special datasets demonstrate that the visual classification approach through object, concept and relationship learning empowers the model to draw inferences about other cases from one instance and effectively suppresses the appearance of overfitting.

The approach is still in the preliminary stage of research, and the direction of future expansion is clear: that is, visual inference tasks. Visual inference tasks are well suited to object, concept (attribute) and relationship representation, especially when there are multiple complex objects in the visual scene, traditional target detection algorithms can easily fall into confusion or overfitting, while neural symbolic integration methods can solve this problem well by integrating first-order logic. Another possible direction is the interpretability of the visual task, we can understand what the model has learned by observing how the concept vectors change during training, that is benefiting from the fact that we learn concepts in the form of vectors. Beyond that, we can decompose a visual proposition or a visual narrative into objects and concepts as a way to interpret the mechanism of the model.

Acknowledgements. This work is supported in part by the Natural Science Found of China (61906066), and supported in part by the Zhejiang Provincial Education Department Scientific Research Project (Y202044192), Huzhou University Research and Innovation Fund (2022KYCX43).

References

1. LeCun, Y., Bengio, Y., Hinton, G.: Deep learning. Nature **521**(7553), 436–444 (2015)
2. Stammer, W., Schramowski, P., Kersting, K.: Right for the right concept: revising neuro-symbolic concepts by interacting with their explanations. In: Proceedings of the IEEE/CVF Conference on Computer Vision and Pattern Recognition, pp. 3619–3629 (2021)
3. De Raedt, L., Dumancic, S., Manhaeve, R., Marram, G.: From statistical relational to neuro-symbolic artificial intelligence. In: Proceedings of the 29th International Joint Conference on Artificial Intelligence (IJCAI 2020), pp. 4943–4950 (2021)
4. van Krieken, E., Acar, E., van Harmelen, F.: Analyzing differentiable fuzzy logic operators. Artif. Intell. **302**, 103602 (2022)
5. Bengio, Y.: From system 1 deep learning to system 2 deep learning. In: Thirty-third Conference on Neural Information Processing Systems (2019)
6. Dehaene, S.: How we learn: why brains learn better than any machine... for now. Penguin (2021)
7. Jiang, Z., Zheng, Y., Tang, H., Zhou, H.: Variational deep embedding: an unsupervised and generative approach to clustering. In: Proceedings of the 26th International Joint Conference on Artificial Intelligence, pp. 1965–1972 (2017)
8. Hohenecker, P., Lukasiewicz, T.: Ontology reasoning with deep neural networks. J. Artif. Intell. Res. **68**, 503–540 (2020)
9. Makni, B., Hendler, J.: Deep learning for noise-tolerant RDFS reasoning. Semantic Web **10**(5), 823–862 (2019)

10. Johnson, J., et al.: Inferring and executing programs for visual reasoning. In: Proceedings of the IEEE International Conference on Computer Vision, pp. 2989–2998 (2017)
11. Yi, K., Wu, J., Gan, C., Torralba, A., Kohli, P., Tenenbaum, J.B.: Neural-symbolic VQA: disentangling reasoning from vision and language understanding. In: Proceedings of the 32nd International Conference on Neural Information Processing Systems, pp. 1039–1050 (2018)
12. Yang, F., Yang, Z., Cohen, W.W.: Differentiable learning of logical rules for knowledge base reasoning. In: Proceedings of the 31st International Conference on Neural Information Processing Systems, pp. 2316–2325 (2017)
13. Dong, H., Mao, J., Lin, T., Wang, C., Li, L., Zhou, D.: Neural logic machines. In: International Conference on Learning Representations (2018)
14. Shi, S., Chen, H., Ma, W., Mao, J., Zhang, M., Zhang, Y.: Neural logic reasoning. In: Proceedings of the 29th ACM International Conference on Information & Knowledge Management, pp. 1365–1374 (2020)
15. Locatello, F., et al.: Object-centric learning with slot attention. Adv. Neural. Inf. Process. Syst. **33**, 11525–11538 (2020)
16. Wu, J., Tenenbaum, J.B., Kohli, P.: Neural scene de-rendering. In: Proceedings of the IEEE Conference on Computer Vision and Pattern Recognition, pp. 699–707 (2017)
17. Thoma, S., Rettinger, A., Both, F.: Towards holistic concept representations: embedding relational knowledge, visual attributes, and distributional word semantics. In: d'Amato, C., et al. (eds.) ISWC 2017. LNCS, vol. 10587, pp. 694–710. Springer, Cham (2017). https://doi.org/10.1007/978-3-319-68288-4_41
18. Mascharka, D., Tran, P., Soklaski, R., Majumdar, A.: Transparency by design: closing the gap between performance and interpretability in visual reasoning. In: Proceedings of the IEEE conference on computer vision and pattern recognition, pp. 4942–4950 (2018)
19. Hudson, D.A, Manning, C.D.: Compositional attention networks for machine reasoning. In: International Conference on Learning Representations (2018)
20. Mao, J., Gan, C., Kohli, P., Tenenbaum, J.B., Wu, J.: The neuro-symbolic concept learner: Interpreting scenes, words, and sentences from natural supervision. In: International Conference on Learning Representations (2019)
21. Yi, K., et al.: CLEVRER: collision events for video representation and reasoning. In: International Conference on Learning Representations (2019)

Approximate Simulation for Transition Systems with Regular Expressions

Xinyu Cui[1(✉)], Zhaokai Li[2], Yuting Chang[1], and Haiyu Pan[1]

[1] Guangxi Key Laboratory of Trusted Software, Guilin University of Electronic Technology, Guilin 541004, China
cuixy9704@163.com

[2] School of Computer Science and Engineering, North Minzu University, Yinchuan 750000, China

Abstract. Simulation is a well-established technique for verifying wheth-er the behaviors of one labeled transition system (LTS) can mimic all behaviors of another LTS. Transition systems with regular expressions (RE-TSs) are an extension of LTSs, which are used as semantic models in modal or temporal logics to solve model checking problems. This paper presents approximate simulation, an extension of simulation of an LTS by a RE-TS, by combining general simulation and metrics, and discusses its properties. First, the notion of approximate simulation is introduced. Then, we investigate properties and an equivalent formalism of approximate simulation. On the other hand, we propose two approaches of fixed point characterization for approximate simulation, and study the relationship between them.

Keywords: Approximate simulation · Regular expression · Metric · Graph pattern matching

1 Introduction

Analysis and verification of concurrent and reactive systems [1] is a well-established research field. Labeled transition systems (LTSs) [1,2] are typically used as models to describe the behaviors of concurrent and reactive systems. In order to compare the behaviors of LTSs, researchers proposed a variety of verification methods. Among them, simulations [1,3,4,30] have a wide range of applications in the analysis of LTSs.

Classical simulation verification techniques return a boolean answer that indicates whether one system can mimic all behaviors of another system. However, as pointed out in [8,11–13], these techniques are restrictive and not robust: Two systems either are simulated or are not simulated, regardless of how close the behaviors of two systems are. To overcome this limitation, the majority of existing works can be roughly grouped into two directions. One of them is based on the notion of metric, which assigns a non-negative real number to each pair

This research is supported by National Natural Science Foundation of China under Grant 62162014.

Y. Chen and S. Zhang (Eds.): AILA 2022, CCIS 1657, pp. 49–62, 2022.
https://doi.org/10.1007/978-981-19-7510-3_4

of states of systems (e.g., [7,8,13,18,25–27]). The other direction is to propose numerous approximate simulations (e.g., [10–12,14,28]), which characterize two almost similar states by a parameter δ.

In fact, in order to model the systems which are required to satisfy the requirements of different aspects, there are a large number of extensions of LTSs in existing literatures (e.g., [5–9,16,21,24]), such as transition systems with regular expressions (RE-TSs) (e.g., [15,17,19,20,22,23]), fuzzy transition systems (e.g., [31,33,34]) and probabilistic transition systems (e.g., [11,12]). RE-TSs have been used in classical modal and temporal logics as semantic models to express the properties of models of systems. For example, Bozzelli [17] investigated the model checking problem for interval temporal logic extended with regular expressions. Beer [20], Brazdil [23] and Mateescu [22] extended computation tree logic (CTL) by applying regular expressions so as to enhance the expression of CTL. From a different point of view, Fan [15] added regular expressions to pattern graphs, and used simulation to solve graph pattern matching which is a classical graph challenge, and considered as one of the most studied problems in the literature. It is regrettable that the simualtion of an LTS by a RE-TS has been ignored in the setting of approximate. To alleviate the aforementioned problem, we will use the notion of metrics to propose an approximate simulation, and study some properties about the approximate simulation.

The paper is organized as follows. Some preliminaries are given in Sect. 2. In Sect. 3, we give the notion of the approximate simualtion. We study some related properties about the approximate simualtion, and provide two fixed point characterizations in Sect. 4 and conclude the paper in Sect. 5.

2 Preliminaries

In this section, we recall some notations and definitions about regular expressions, metric spaces and transition systems with regular expressions.

We denote the sets of real numbers, non-negative reals, natural numbers and positive integers by $\mathbb{R}$, $\mathbb{R}^+$, $\mathbb{N}$, $\mathbb{Z}^+$, respectively. We use I to denote the set of indexes. Let Σ be a finite set. We denote the set of finite strings over Σ by Σ^*. We write $\mathcal{P}(\Sigma)$ for the power set of Σ. Let $\rho = a_1 \dots a_n$, $\sigma = b_1 \dots b_m \in \Sigma^*$ be two strings. Then, the concatenation of ρ and σ is the string $\rho\sigma = a_1 \dots a_n b_1 \dots b_m$. We also denote the i-th symbol of ρ and the length of ρ by ρ_i and $|\rho|$, respectively. Let $\mathcal{P}$, $\mathcal{Q} \subseteq \Sigma^*$. The concatenation of $\mathcal{P}$ and $\mathcal{Q}$ is $\mathcal{PQ} = \{\rho\sigma \in \Sigma^* : \rho \in \mathcal{P}, \sigma \in \mathcal{Q}\}$.

Regular expressions [29] ω over Σ are defined by the following grammar,

$$\omega ::= a|a^k|a^+|(\omega_1)|\omega_1\omega_2|\omega_1 + \omega_2,$$

where $a \in \Sigma$. The set of all regular expressions over Σ is written as $\varpi(\Sigma)$. The language $L(\omega) \subseteq \Sigma^*$ of a regular expression $\omega \in \varpi(\Sigma)$ is defined indutively by

(1) $L(a) = \{a\}$;
(2) $L(a^k) = \{\underbrace{a \cdots a}_{k}\}$;

(3) $L(a^+) = L(a) \cup L(a^2) \cup \cdots$;
(4) $L((\omega_1)) = L(\omega_1)$;
(5) $L(\omega_1\omega_2) = L(\omega_1)L(\omega_2)$;
(6) $L(\omega_1 + \omega_2) = L(\omega_1) \cup L(\omega_2)$.

We recall the definition of metrics taken from [32]. A function $d : \Sigma \times \Sigma \to \mathbb{R}$ is a metric over Σ if for all $x, y, z \in \Sigma$

(1) $d(x, y) \geq 0$, $d(x, y) = 0$ iff $x = y$;
(2) $d(x, y) = d(y, x)$;
(3) $d(x, z) \leq d(x, y) + d(y, z)$.

And, the pair (Σ, d) is called a metric space. When d is clear from context, we write Σ instead of (Σ, d).

Next, we review the definition of transition system with regular expressions [15,17,20,22,23]. A transition system with regular expressions (for short, RE-TS) is a tuple $\mathcal{RT} = (S, s_0, \varpi(\Sigma), \to)$ where S is a finite set of states, $s_0 \in S$ is a initial state, $\varpi(\Sigma)$ is the set of all regular expressions over Σ, and $\to \subseteq S \times \varpi(\Sigma) \times S$ is a set of transitions.

A labeled transition system (LTS) can be viewed as a special case of RE-TS, where $\varpi(\Sigma)$ is replaced with Σ. Let $\mathcal{RT} = (S, s_0, \varpi(\Sigma), \to)$ be a RE-TS. We write $s \xrightarrow{\omega} s'$ for $(s, \omega, s') \in \to$, where $s, s' \in S$. A trace σ is an infinite sequence of elements in $\varpi(\Sigma)$. For $j \geq 1$, let σ_j denote the jth element. A path from s in $\mathcal{RT}$ is an infinite sequence $\pi = s \xrightarrow{\omega_1} s' \xrightarrow{\omega_2} s'' \cdots$, and we denote by $\mathrm{tr}(\pi) = \omega_1\omega_2\cdots$ the trace induced by it. For $s \in S$, we denote by Path(s) the set of paths from s and by Trace$(s) = \{tr(\pi) : \pi \in \text{Path}(s)\}$ the set of traces from s. We use $\mathcal{L}(\mathcal{RT})$ to denote the languages of $\mathcal{RT}$, where $\mathcal{L}(\mathcal{RT}) = \text{Trace}(s_0)$. If $\mathcal{RT}$ is an LTS, $\to \subseteq S \times \Sigma \times S$ can be extended to $\to^* \subseteq S \times \Sigma^* \times S$. We write $s \xrightarrow{\rho}{}^* s'$ for $(s, \rho, s') \in \to^*$.

3 Approximate Simulation

In this section, we define an approximate simulation. Before formally defining approximate simulation, we will introduce the notion of exact simulation of an LTS by a RE-TS [15].

Definition 1. *Let* $\mathcal{RT}_1 = (S_1, s_{1,0}, \Sigma, \to_1)$ *be an LTS and* $\mathcal{RT}_2 = (S_2, s_{2,0}, \varpi(\Sigma), \to_2)$ *be a RE-TS. A relation* $R \subseteq S_1 \times S_2$ *is called a simulation if for any* $(s_1, s_2) \in R$ *and for each* $s_1 \xrightarrow{\rho}{}^* s_1'$, *there exists* $s_2 \xrightarrow{\omega} s_2'$ *such that* $\rho \in L(\omega)$ *and* $(s_1', s_2') \in R$. *We say that* $\mathcal{RT}_2$ *simulates* $\mathcal{RT}_1$, *denoted by* $\mathcal{RT}_1 \preceq \mathcal{RT}_2$, *if there exists a simulation* R *such that* $(s_{1,0}, s_{2,0}) \in R$.

Example 1. Consider an LTS $\mathcal{RT}_1 = (S_1, s_{1,0}, \Sigma, \to_1)$ and a RE-TS $\mathcal{RT}_2 = (S_2, s_{2,0}, \varpi(\Sigma), \to_2)$. The transition diagrams of $\mathcal{RT}_1$ and $\mathcal{RT}_2$ are depicted in Fig. 1. By Definition 2, we can find a relation $R = \{(s_{1,0}, s_{2,0}), (s_{1,1}, s_{2,1}), (s_{1,2}, s_{2,1}), (s_{1,3}, s_{2,2}), (s_{1,4}, s_{2,2})\}$ which is a simulation, and $(s_{1,0}, s_{2,0}) \in R$. Therefore, we can obtain $\mathcal{RT}_1 \preceq \mathcal{RT}_2$.

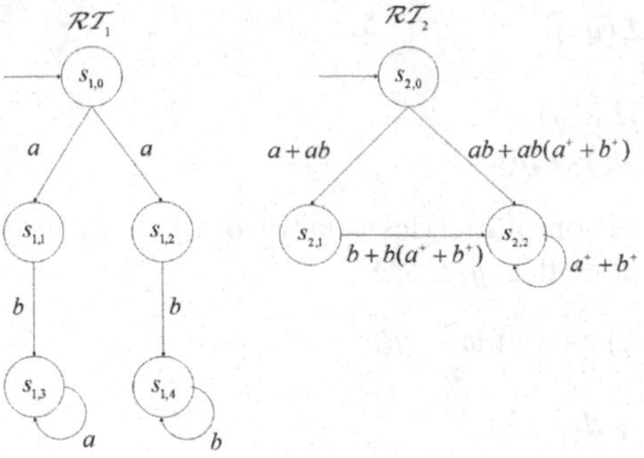

Fig. 1. An LTS and a RE-TS.

The following definitions and propositions will help us to define the approximate simulation.

The definition of string distance sd will be developed in a discounted version [7,8], in which distance of each step is decreased exponentially over time by a discounting factor $\alpha \in (0, 1]$.

Let (Σ, d) be a metric space, and $\alpha \in (0, 1]$. The string distance $sd : \Sigma^* \times \Sigma^* \to \mathbb{R} \cup \{+\infty\}$ is defined as

$$sd(\rho, \sigma) = \begin{cases} \max\limits_{1 \le i \le |\rho|} \alpha^{i-1} d(\rho_i, \sigma_i) & \text{if } |\rho| = |\sigma|, \\ +\infty & \text{otherwise,} \end{cases}$$

for all strings $\rho, \sigma \in \Sigma^*$.

Proposition 1. *Let d be a metric over Σ. Then, sd is a metric on the set Σ^*.*

Proof. By the definition of metric, we need to prove the follow properties:

(1) $sd(\rho, \sigma) \ge 0$ and $sd(\rho, \sigma) = 0$ iff $\rho = \sigma$,
(2) $sd(\rho, \sigma) = sd(\sigma, \rho)$,
(3) $sd(\rho, \varsigma) + sd(\varsigma, \sigma) \ge sd(\rho, \sigma)$

where $\rho, \sigma, \varsigma \in \Sigma^*$.

The properties (1) and (2) are obvious.

For property (3): We have to discuss the following four cases.

The cases of $|\rho| \ne |\sigma| \ne |\varsigma|$, $|\rho| = |\sigma| \ne |\varsigma|$, and $|\rho| \ne |\sigma| = |\varsigma|$ follow immediately from the definition of sd. We only consider the case of $|\rho| = |\sigma| = |\varsigma| = n$.

From the definition of sd, we know that

$$sd(\rho, \varsigma) + sd(\varsigma, \sigma) = \max_{1 \le i \le n} \alpha^{i-1} d(\rho_i, \varsigma_i) + \max_{1 \le j \le n} \alpha^{j-1} d(\varsigma_j, \sigma_j)$$

and $sd(\rho, \sigma) = \max\limits_{1 \le k \le n} \alpha^{k-1} d(\rho_k, \sigma_k)$. It is sufficient to show that there exists a $1 \le p \le n$ such that $\alpha^{p-1} d(\rho_p, \sigma_p) = sd(\rho, \sigma)$. Then, $d(\rho_p, \sigma_p) \le d(\rho_p, \varsigma_p) + d(\varsigma_p, \sigma_p)$.

Therefore,

$$\begin{aligned} sd(\rho,\sigma) &= \alpha^{p-1}d(\rho_p,\sigma_p) \\ &\le \alpha^{p-1}d(\rho_p,\varsigma_p) + \alpha^{p-1}d(\varsigma_p,\sigma_p) \\ &\le \max_{1\le i\le n}\alpha^{i-1}d(\rho_i,\varsigma_i) + \max_{1\le j\le n}\alpha^{j-1}d(\varsigma_j,\sigma_j) \\ &= sd(\rho,\varsigma) + sd(\varsigma,\sigma). \end{aligned}$$

Let (Σ, d) be a metric space. The distance between a string $\rho \in \Sigma^*$ and the language $L(\omega) \subseteq \Sigma^*$ of a regular expression $\omega \in \varpi(\Sigma)$ is defined as

$$d^*(\rho, L(\omega)) = \inf_{\sigma\in L(\omega)} sd(\rho,\sigma).$$

Proposition 2. *Let (Σ, d) be a metric space. Then, the following properties hold:*

(1) $d^*(\rho, L(\omega)) \ge 0$, for all $\rho \in \Sigma^*$ and $\omega \in \varpi(\Sigma)$.
(2) $\rho \in L(\omega)$ iff $d^*(\rho, L(\omega)) = 0$, where $\rho \in \Sigma^*$, $\omega \in \varpi(\Sigma)$.

Proof. For property (1): It follows immediately from Proposition 1.

For property (2): First, for the 'if' part, let $d^*(\rho, L(\omega)) = 0$. Without loss of generality, suppose $|\rho|=n$. Since Σ is a finite set, assume that there are m elements in Σ. According to the definition of sd, there are at most $nm + 1$ values. Then, there exists $\sigma \in L(\omega)$ such that $sd(\rho, \sigma) = 0$. So by Proposition 1 and the definition of sd, $\rho = \sigma$ holds. Hence $\rho \in L(\omega)$.

Second, for the 'only if' part, suppose that $\rho \in L(\omega)$. Then, there exists $\sigma \in L(\omega)$ such that $\rho = \sigma$. By Proposition 1 and the definition of sd, $sd(\rho, \sigma) = 0$ holds. Therefore, $\inf_{\sigma\in L(\omega)} sd(\rho,\sigma) = 0$, and $d^*(\rho, L(\omega)) = 0$.

By the above definitions and propositions, we will introduce the definition of approximate simulation.

Definition 2. *Let $\mathcal{RT}_1 = (S_1, s_{1,0}, \Sigma, \rightarrow_1)$ be an LTS, $\mathcal{RT}_2 = (S_2, s_{2,0}, \varpi(\Sigma), \rightarrow_2)$ be a RE-TS, (Σ, d) be a metric space, and $\delta \in \mathbb{R}^+$. A relation $R_\delta \subseteq S_1 \times S_2$ is called a δ-simulation if for any $(s_1, s_2) \in R$ and for each $s_1 \xrightarrow{\rho}{}^* s_1'$, there exists $s_2 \xrightarrow{\omega} s_2'$ such that $d^*(\rho, L(\omega)) \le \delta$ and $(s_1', s_2') \in R$. We say that $\mathcal{RT}_2$ δ-simulates $\mathcal{RT}_1$, denoted by $\mathcal{RT}_1 \preceq_\delta \mathcal{RT}_2$, if there exists a δ-simulation R_δ such that $(s_{1,0}, s_{2,0}) \in R_\delta$.*

We replace $s_{1,0} \xrightarrow{a} s_{1,1}$ by $s_{1,0} \xrightarrow{c} s_{1,1}$ in Fig. 1, where $d(a, c) = 0.5$. Then, according to Definition 2, we can obtain that there exists a 0.5-simulation $R_{0.5}$ such that $(s_{1,0}, s_{2,0}) \in R_{0.5}$. Thus, $\mathcal{RT}_1 \preceq_{0.5} \mathcal{RT}_2$.

In above definition, when $\delta = 0$, we recover the established definition of exact simualtion.

4 Related Properties

In this section, we will introduce some properties related with the approximate simulation.

Given a precision parameter δ, which used to measure the degree of simulation, the following lemma ensures that the set of δ-simulation has a maximal element, denoted by $R_\delta^{\max}$. Moreover, by using $R_\delta^{\max}$, we give an equivalent expression for δ-simulation.

Lemma 1. *Let* $\mathcal{RT}_1 = (S_1, s_{1,0}, \Sigma, \rightarrow_1)$ *be an LTS,* $\mathcal{RT}_2 = (S_2, s_{2,0}, \varpi(\Sigma), \rightarrow_2)$ *be a RE-TS,* (Σ, d) *be a metric space, and* $\delta \in \mathbb{R}^+$. *Suppose that* $\{R_\delta^i\}_{i\in I}$ *is a family of* δ*-simulation of* $\mathcal{RT}_1$ *by* $\mathcal{RT}_2$*, and* $R_\delta^{\max} = \bigcup_{i\in I} R_\delta^i$. *Then, the following properties hold:*

(1) $R_\delta^{\max}$ *is a* δ*-simulation of* $\mathcal{RT}_1$ *by* $\mathcal{RT}_2$.
(2) $\mathcal{RT}_1 \preceq_\delta \mathcal{RT}_2$ *iff* $(s_{1,0}, s_{2,0}) \in R_\delta^{\max}$.

Proof. For property (1): Consider any $(s_1, s_2) \in R_\delta^{\max} = \bigcup_{i\in I} R_\delta^i$. It is enough to show that there exists a $i \in I$ such that $(s_1, s_2) \in R_\delta^i$. By Definition 2, we can know that for each $s_1 \xrightarrow{\rho} {}^* s_1'$ in $\mathcal{RT}_1$, there exists $s_2 \xrightarrow{\omega} s_2'$ in $\mathcal{RT}_2$ such that $d^*(\rho, L(\omega)) \leq \delta$ and

$$(s_1', s_2') \in R_\delta^i \subseteq \bigcup_{i\in I} R_\delta^i = R_\delta^{\max}.$$

Hence, $R_\delta^{\max}$ is a δ-simulation of $\mathcal{RT}_1$ by $\mathcal{RT}_2$.

For property (2): It follows immediately from Definition 2 and property (1).

Proposition 3. *Let* $\mathcal{RT}_i = (S_i, s_{i,0}, \Sigma, \rightarrow_i)$, $i = 1, 2$, *be two LTSs,* $\mathcal{RT}_3 = (S_3, s_{3,0}, \varpi(\Sigma), \rightarrow_3)$ *be a RE-TS, and* (Σ, d) *be a metric space. Then, the following properties hold:*

(1) For all $\delta \geq 0$, $\mathcal{RT}_1 \preceq_\delta \mathcal{RT}_1$.
(2) For all $\delta \geq 0$, *if* $\mathcal{RT}_1 \preceq_\delta \mathcal{RT}_3$, *then for all* $\delta' > \delta$, $\mathcal{RT}_1 \preceq_{\delta'} \mathcal{RT}_3$.
(3) For all $\delta, \delta' \geq 0$, *if* $\mathcal{RT}_1 \preceq_\delta \mathcal{RT}_2$ *and* $\mathcal{RT}_2 \preceq_{\delta'} \mathcal{RT}_3$, *then* $\mathcal{RT}_1 \preceq_{\delta+\delta'} \mathcal{RT}_3$.

Proof. For property (1): Let $R = \{(s, s) \in S_1 \times S_1 : s \in S_1\}$. Then, for each $(s, s) \in R$ and $s \xrightarrow{a} {}^* s'$, it is obvious that $s \xrightarrow{a} s'$, $d^*(a, L(a)) = 0 \leq \delta$ and $(s', s') \in R$. Therefore, R is a δ-simulation by Definition 2. Thus $\mathcal{RT}_1 \preceq_\delta \mathcal{RT}_1$ because $(s_{1,0}, s_{1,0}) \in R$.

For property (2): Suppose that $\mathcal{RT}_1 \preceq_\delta \mathcal{RT}_3$ and $\delta' > \delta \geq 0$. From Lemma 1, we can know that $(s_{1,0}, s_{3,0}) \in R_\delta^{\max} = \bigcup_{i\in I} R_\delta^i$. Therefore, there exists $i \in I$ such that $(s_{1,0}, s_{3,0}) \in R_\delta^i$. By Definition 2, we have that for each $(s_1, s_3) \in R_\delta^i$ and $s_1 \xrightarrow{\rho} {}^* s_1'$, there exists $s_3 \xrightarrow{\omega} s_3'$ such that $d^*(\rho, L(\omega)) \leq \delta < \delta'$ and $(s_1', s_3') \in R_\delta^i$. Hence, R_δ^i is a δ'-simulation. Thus, $\mathcal{RT}_1 \preceq_{\delta'} \mathcal{RT}_3$.

For property (3): Let $R = \{(s_1, s_3) \in S_1 \times S_3 : \exists s_2 \in S_2, (s_1, s_2) \in R_\delta^{\max}$ and $(s_2, s_3) \in R_{\delta'}^{\max}\}$. For each $(s_1, s_3) \in R$, suppose that

$$s_1 \xrightarrow{a_1} {}^* s_1' \xrightarrow{a_2} {}^* \cdots \xrightarrow{a_n} {}^* s_1''.$$

We know from Lemma 1 that there exists

$$s_2 \xrightarrow{a'_1} s'_2 \xrightarrow{a'_2} \cdots \xrightarrow{a'_n} s''_2$$

such that $d^*(a_i, L(a'_i)) \leq \delta$ and $(s''_1, s''_2) \in R_\delta^{\max}$, where $a_i, a'_i \in \Sigma$ for every $i \in \mathbb{N}$. Let $\rho = a_1 a_2 \cdots a_n$ and $\rho' = a'_1 a'_2 \cdots a'_n$. By the definition of sd, we have $sd(\rho, \rho') \leq \delta$. From Lemma 1, we know that there exists $s_3 \xrightarrow{\omega} s'_3$ such that $d^*(\rho', L(\omega)) \leq \delta'$ and $(s''_2, s'_3) \in R_{\delta'}^{\max}$ for each $s_2 \xrightarrow{\rho'}{}^* s''_2$ and $(s_2, s_3) \in R_{\delta'}^{\max}$. Therefore, there exists $\sigma \in L(\omega)$ such that $sd(\rho', \sigma) \leq \delta'$. By Proposition 1,

$$sd(\rho, \sigma) \leq sd(\rho, \rho') + sd(\rho', \sigma) \leq \delta + \delta'.$$

Thus, $d^*(\rho, L(\omega)) \leq \delta + \delta'$ and $(s''_1, s'_3) \in R$. Therefore, R is a $\delta + \delta'$-simulation, and $(s_{1,0}, s_{3,0}) \in R$ because $(s_{1,0}, s_{2,0}) \in R_\delta^{\max}$ and $(s_{2,0}, s_{3,0}) \in R_{\delta'}^{\max}$. Hence, $\mathcal{RT}_1 \preceq_{\delta+\delta'} \mathcal{RT}_3$.

Given an LTS $\mathcal{RT}_1 = (S_1, s_{1,0}, \Sigma, \rightarrow_1)$ and a RE-TS $\mathcal{RT}_2 = (S_2, s_{2,0}, \varpi(\Sigma), \rightarrow_2)$, we say that $\text{Trace}(s_1) \subseteq \text{Trace}(s_2)$ if for each $\sigma_1 = \rho_1 \rho_2 \cdots \in \text{Trace}(s_1)$, there exists $\sigma_2 = \omega_1 \omega_2 \cdots \in \text{Trace}(s_2)$ such that $\rho_i \in L(\omega_i)$ for all $i \in \mathbb{Z}^+$, where $(s_1, s_2) \in S_1 \times S_2$.

Let $\mathcal{RT}_1 = (S_1, s_{1,0}, \Sigma, \rightarrow_1)$ be an LTS, $\mathcal{RT}_2 = (S_2, s_{2,0}, \varpi(\Sigma), \rightarrow_2)$ be a RE-TS, (Σ, d) be a metric space, and $(s_1, s_2) \in S_1 \times S_2$. Given two traces $\sigma_1 = \rho_1 \rho_2 \cdots \in \text{Trace}(s_1)$ and $\sigma_2 = \omega_1 \omega_2 \cdots \in \text{Trace}(s_2)$, the trace distance between σ_1 and σ_2 is defined as

$$td(\sigma_1, \sigma_2) = \sup_{i \in \mathbb{Z}^+} d^*(\rho_i, \omega_i).$$

The trace distance between s_1 and s_2 is defined as

$$\mathcal{T}d(s_1, s_2) = \sup_{\sigma_1 \in \text{Trace}(s_1)} \inf_{\sigma_2 \in \text{Trace}(s_2)} td(\sigma_1, \sigma_2).$$

The language distance between $\mathcal{RT}_1$ and $\mathcal{RT}_2$ is defined as

$$\mathcal{L}d(\mathcal{RT}_1, \mathcal{RT}_2) = \sup_{\sigma_1 \in \text{Trace}(s_{1,0})} \inf_{\sigma_2 \in \text{Trace}(s_{2,0})} td(\sigma_1, \sigma_2).$$

Proposition 4. *Let $\mathcal{RT}_1 = (S_1, s_{1,0}, \Sigma, \rightarrow_1)$ be an LTS, $\mathcal{RT}_2 = (S_2, s_{2,0}, \varpi(\Sigma), \rightarrow_2)$ be a RE-TS, and (Σ, d) be a metric space. Then, the following properties hold:*

(1) $\text{Trace}(s_1) \subseteq \text{Trace}(s_2)$ iff $\mathcal{T}d(s_1, s_2) = 0$, where $s_1 \in S_1$ and $s_2 \in S_2$.
(2) $\mathcal{L}(\mathcal{RT}_1) \subseteq \mathcal{L}(\mathcal{RT}_2)$ iff $\mathcal{L}d(\mathcal{RT}_1, \mathcal{RT}_2) = 0$.

Proof. For property (1): First, for the 'if' part, suppose that $\mathcal{T}d(s_1, s_2) = 0$. It is sufficient to know from the definition of $\mathcal{T}d$ that $\inf_{\sigma_2 \in \text{Trace}(s_2)} td(\sigma_1, \sigma_2) = 0$ for each $\sigma_1 = \rho_1 \rho_2 \cdots \in \text{Trace}(s_1)$. Therefore, there exists $\sigma_2 = \omega_1 \omega_2 \cdots \in \text{Trace}(s_2)$

such that $td(\sigma_1, \sigma_2) = 0$. From the definition of td, we have $d^*(\rho_i, L(\omega_i)) = 0$ for all $i \in \mathbb{Z}^+$. From Proposition 2, there is $\rho_i \in L(\omega_i)$ for all $i \in \mathbb{Z}^+$. Hence, $\text{Trace}(s_1) \subseteq \text{Trace}(s_2)$.

Second, for the 'only if' part, let $\text{Trace}(s_1) \subseteq \text{Trace}(s_2)$. In other words, for each $\sigma_1 = \rho_1\rho_2 \cdots \in \text{Trace}(s_1)$, there exists $\sigma_2 = \omega_1\omega_2 \cdots \in \text{Trace}(s_2)$ such that $\rho_i \in L(\omega_i)$ for all $i \in \mathbb{Z}^+$. And, there is $d^*(\rho_i, L(\omega_i)) = 0$ from Proposition 2 for all $i \in \mathbb{Z}^+$. Moreover, we have $td(\sigma_1, \sigma_2) = 0$ from the definition of td. Hence, $\inf_{\sigma_2 \in \text{Trace}(s_2)} td(\sigma_1, \sigma_2) = 0$ for each $\sigma_1 \in \text{Trace}(s_1)$. Therefore, $\mathcal{T}d(s_1, s_2) = 0$.

For property (2): For the 'if' part, suppose that $\mathcal{L}d(\mathcal{RT}_1, \mathcal{RT}_2) = 0$. It is sufficient to show that $\inf_{\sigma_2 \in \mathcal{L}(\mathcal{RT}_2)} td(\sigma_1, \sigma_2) = 0$ for each $\sigma_1 = \rho_1\rho_2 \cdots \in \mathcal{L}(\mathcal{RT}_1)$. Therefore, there exists $\sigma_2 = \omega_1\omega_2 \cdots \in \mathcal{L}(\mathcal{RT}_2)$ such that $td(\sigma_1, \sigma_2) = 0$. From the definition of td, there is $d^*(\rho_i, L(\omega_i)) = 0$ for all $i \in \mathbb{Z}^+$. Then, we have $\rho_i \in L(\omega_i)$ by Proposition 2. Hence, $\mathcal{L}(\mathcal{RT}_1) \subseteq \mathcal{L}(\mathcal{RT}_2)$.

For the 'only if' part, consider that $\mathcal{L}(\mathcal{RT}_1) \subseteq \mathcal{L}(\mathcal{RT}_2)$. In other words, for each trace $\sigma_1 = \rho_1\rho_2 \cdots$ in $\mathcal{RT}_1$, there exists a trace $\sigma_2 = \omega_1\omega_2 \cdots$ in $\mathcal{RT}_2$ such that $\rho_i \in L(\omega_i)$ for all $i \in \mathbb{Z}^+$. We know from Proposition 2 that $d^*(\rho_i, L(\omega_i)) = 0$. By the definition of td, we have $td(\sigma_1, \sigma_2) = 0$. Hence, $\inf_{\sigma_2 \in \mathcal{L}(\mathcal{RT}_2)} td(\sigma_1, \sigma_2) = 0$ for each $\sigma_1 = \rho_1\rho_2 \ldots \in \mathcal{L}(\mathcal{RT}_1)$. Thus, we have $\mathcal{L}d(\mathcal{RT}_1, \mathcal{RT}_2) = 0$ from the definition of $\mathcal{L}d$.

Let $\mathcal{RT}_1 = (S_1, s_{1,0}, \Sigma, \rightarrow_1)$ be an LTS, $\mathcal{RT}_2 = (S_2, s_{2,0}, \varpi(\Sigma), \rightarrow_2)$ be a RE-TS, and (Σ, d) be a metric space. The simulation distance between $\mathcal{RT}_1$ and $\mathcal{RT}_2$ is defined as

$$\mathcal{S}d(\mathcal{RT}_1, \mathcal{RT}_2) = \inf\{\delta : \mathcal{RT}_1 \preceq_\delta \mathcal{RT}_2\}.$$

Lemma 2. *Let $\mathcal{RT}_1 = (S_1, s_{1,0}, \Sigma, \rightarrow_1)$ be an LTS and $\mathcal{RT}_2 = (S_2, s_{2,0}, \varpi(\Sigma), \rightarrow_2)$ be a RE-TS, and (Σ, d) be a metric space. Then, the following properties hold:*

(1) $\mathcal{S}d(\mathcal{RT}_1, \mathcal{RT}_2) \geq 0$.
(2) $\mathcal{S}d(\mathcal{RT}_1, \mathcal{RT}_2) = 0$ if $\mathcal{RT}_1 \preceq \mathcal{RT}_2$.

Proof. For property (1): It follows immediately from Proposition 2.

For property (2): Consider that $\mathcal{RT}_1 \preceq \mathcal{RT}_2$. By Definition 1, there exists a simulation $R \subset S_1 \times S_2$ such that: For each $(s_1, s_2) \in R$ and $s_1 \xrightarrow{\rho}{}^* s_1'$, there exists $s_2 \xrightarrow{\omega} s_2'$ such that $\rho \in L(\omega)$ and $(s_1', s_2') \in R$. And, we have $d^*(\rho, L(\omega)) = 0$ by Proposition 2. Then, there exists a $\delta = 0$ and R is a δ-simulation. By the definition of $\mathcal{S}d$ and property (1), $\mathcal{S}d(\mathcal{RT}_1, \mathcal{RT}_2) = 0$.

The relationship between the simulation distance and the language distance is captured by the following theorem.

Theorem 1. *Let $\mathcal{RT}_1 = (S_1, s_{1,0}, \Sigma, \rightarrow_1)$ be an LTS, $\mathcal{RT}_2 = (S_2, s_{2,0}, \varpi(\Sigma), \rightarrow_2)$ be a RE-TS, and (Σ, d) be a metric space. Then,*

$$\mathcal{L}d(\mathcal{RT}_1, \mathcal{RT}_2) \leq \mathcal{S}d(\mathcal{RT}_1, \mathcal{RT}_2).$$

Proof. Consider a $\delta \geq Sd(\mathcal{RT}_1, \mathcal{RT}_2)$. By Proposition 3, $\mathcal{RT}_1 \preceq_\delta \mathcal{RT}_2$ holds. Let $\sigma_1 = \rho_1\rho_2\cdots \in \mathcal{L}(\mathcal{RT}_1)$. Then, there exists a path

$$s_1 \xrightarrow{\rho_1}{}^* s_2 \xrightarrow{\rho_2}{}^* s_3 \xrightarrow{\rho_3}{}^* \cdots$$

in $\mathcal{RT}_1$ where $s_1 = s_{0,1}$. By Lemma 1, we know $(s_{0,1}, s_{0,2}) \in R_\delta^{\max}$. Therefore, there exists a path

$$s_1' \xrightarrow{\omega_1} s_2' \xrightarrow{\omega_2} s_3' \xrightarrow{\omega_3} \cdots$$

such that $(s_i, s_i') \in R_\delta^{\max}$ for all $i \in \mathbb{Z}^+$, where $s_1' = s_{0,2}$. Let $\sigma_2 = \omega_1\omega_2\cdots$. By Proposition 2 and the definition of td, it is obvious that $td(\sigma_1, \sigma_2) \leq \delta$. Hence, $\inf\limits_{\sigma_2 \in \mathcal{L}(\mathcal{RT}_2)} td(\sigma_1, \sigma_2) \leq \delta$ for each $\sigma_1 \in \mathcal{L}(\mathcal{RT}_1)$. Thus $\mathcal{L}d(\mathcal{RT}_1, \mathcal{RT}_2) \leq Sd(\mathcal{RT}_1, \mathcal{RT}_2)$.

We next propose two approaches of fixed point characterization of approximate simulation.

We first give a fixed point characterization of maximal δ-simulation for a given δ.

Let $\mathcal{RT}_1 = (S_1, s_{1,0}, \Sigma, \rightarrow_1)$ be an LTS, $\mathcal{RT}_2 = (S_2, s_{2,0}, \varpi(\Sigma), \rightarrow_2)$ be a RE-TS, and (Σ, d) be a metric space. For a given $\delta \geq 0$, we define the following sequence $\{R_\delta^i\}_{i\in\mathbb{N}}$ of subsets of $S_1 \times S_2$:

$$R_\delta^0 = S_1 \times S_2;$$

$$R_\delta^{i+1} = \{(s_1, s_2) \in R_\delta^i : \text{for } s_1 \xrightarrow{\rho}{}^* s_1', \text{there exists } s_2 \xrightarrow{\omega} s_2' \text{ such that } d^*(\rho, L(\omega)) \leq \delta \text{ and } (s_1', s_2') \in R_\delta^i\}.$$

Since S_1 and S_2 are finite, it is clear that $\{R_\delta^i\}_{i\in\mathbb{N}}$ reaches a fixed point in a finite number steps by the definition of $\{R_\delta^i\}_{i\in\mathbb{N}}$.

Lemma 3. *Let $\{R_\delta^i\}_{i\in\mathbb{N}}$ be the sequence of sets defined by definition of $\{R_\delta^i\}_{i\in\mathbb{N}}$. Then, the following properties hold:*

(1) $R_\delta^{i+1} \subseteq R_\delta^i$ for every $i \in \mathbb{N}$.
(2) For each $i \in \mathbb{N}$, $R_\delta^{\max} \subseteq R_\delta^i$.
(3) There exists some $n \in \mathbb{N}$ such that $R_\delta^{\max} = R_\delta^n$.

Proof. For property (1): It follows immediately from the definition of $\{R_\delta^i\}_{i\in\mathbb{N}}$.

For property (2): This will be proved by induction with regard to i.

The initial step is for $i = 1$. It follows from the definition of $\{R_\delta^i\}_{i\in\mathbb{N}}$.

The induction hypothesis is that (2) holds for $i = k$. We now show that (2) holds for $i = k + 1$, i.e., $R_\delta^{\max} \subseteq R_\delta^{k+1}$.

We have to discuss the following two cases.

The first case is that $R_\delta^{\max} = \emptyset$. Then, $R_\delta^{\max} = \emptyset \subseteq R_\delta^{k+1}$ obviously.

The second case is that $R_\delta^{\max} \neq \emptyset$. Then there exists some $(s_1, s_2) \in R_\delta^{\max}$. For each $(s_1, s_2) \in R_\delta^{\max} \subseteq R_\delta^k$, we know rom Lemma 1 that for each $s_1 \xrightarrow{\rho}{}^* s_1'$, there exists $s_2 \xrightarrow{\omega} s_2'$ such that $d^*(\rho, L(\omega)) \leq \delta$, and $(s_1', s_2') \in R_\delta^{\max}$. Then, by the induction hypothesis, $(s_1', s_2') \in R_\delta^{\max} \subseteq R_\delta^k$. Hence, It follows from the definition of $\{R_\delta^i\}_{i\in\mathbb{N}}$ that $(s_1, s_2) \in R_\delta^{k+1}$.

Thus (2) holds for $i = k+1$, which proves the property.

For property (3): By using propery (1) and the definition of $\{R_\delta^i\}_{i\in\mathbb{N}}$, we can obtain that there exists some $k \in \mathbb{N}$ such that $R_\delta^j = R_\delta^k$ for every $j \geq k$.

We have to discuss the following two cases.

The first case is that there exists a k such that $R_\delta^k = \emptyset$. By property (1) and (2), $R_\delta^j \subseteq R_\delta^k$ for every $j \geq k$, and $R_\delta^{\max} \subseteq R_\delta^k$. Hence, $R_\delta^{\max} \subseteq R_\delta^k = \emptyset$.

The second case is that $R_\delta^j \neq \emptyset$ for all $j \in \mathbb{N}$. Let $(s_1, s_2) \in R_\delta^{k+1}$. Then, for each $s_1 \xrightarrow{\rho}{}^* s_1'$, there exists $s_2 \xrightarrow{\omega} s_2'$ such that $d^*(\rho, L(\omega)) \leq \delta$ and $(s_1', s_2') \in R_\delta^k$. Since $R_\delta^{k+1} = R_\delta^k$, it is sufficient to show that R_δ^{k+1} is a δ-simulation and $R_\delta^k = R_\delta^{k+1} \subseteq R_\delta^{\max}$. By property (2), $R_\delta^{\max} \subseteq R_\delta^k$. Hence, $R_\delta^{\max} = R_\delta^k$.

We will introduce another approach, which characterizes the maximal δ-simualtion as the level sets of a function for a given δ.

Let $\mathcal{RT}_1 = (S_1, s_{1,0}, \Sigma, \rightarrow_1)$ be an LTS, $\mathcal{RT}_2 = (S_2, s_{2,0}, \varpi(\Sigma), \rightarrow_2)$ be a RE-TS, and (Σ, d) be a metric space. Define the following sequence $\{f^i\}_{i\in\mathbb{N}}$ of functions from $S_1 \times S_2$ to $\mathbb{R}^+ \cup \{+\infty\}$:

$$f^0(s_1, s_2) = 0;$$
$$f^{i+1}(s_1, s_2) = \sup_{s_1 \xrightarrow{\rho}{}^* s_1'} \inf_{s_2 \xrightarrow{\omega} s_2'} \max(d^*(\rho, L(\omega)), f^i(s_1', s_2')).$$

Lemma 4. *Let $\{f^i\}_{i\in\mathbb{N}}$ be the sequence of functions defined by definition of $\{f^i\}_{i\in\mathbb{N}}$. Then, the sequence $\{f^i(s_1, s_2)\}_{i\in\mathbb{N}}$ is non-decreasing for all $(s_1, s_2) \in S_1 \times S_2$.*

Proof. To prove the lemma, it suffices to verify that

$$f^i(s_1, s_2) \leq f^{i+1}(s_1, s_2)$$

for every $i \geq 0$.

This will be proved by induction on i.

The initial step is for $i = 0$. It follows from the definition of $\{f^i\}_{i\in\mathbb{N}}$ and Proposition 2.

The induction hypothesis is that the lemma holds for $i = k$. We now show that the lemma holds for $i = k+1$, i.e., $f^{k+1}(s_1, s_2) \leq f^{k+2}(s_1, s_2)$ for each $(s_1, s_2) \in S_1 \times S_2$. Then,

$$\begin{aligned} f^{k+2}(s_1, s_2) &= \sup_{s_1 \xrightarrow{\rho}{}^* s_1'} \inf_{s_2 \xrightarrow{\omega} s_2'} \max(d^*(\rho, L(\omega)), f^{k+1}(s_1', s_2')) \\ &\geq \sup_{s_1 \xrightarrow{\rho}{}^* s_1'} \inf_{s_2 \xrightarrow{\omega} s_2'} \max(d^*(\rho, L(\omega)), f^{k}(s_1', s_2')) \\ &= f^{k+1}(s_1, s_2). \end{aligned}$$

This completes the proof of the lemma.

$\{f^i\}_{i\in\mathbb{N}}$ reaches a fixed point in a finite number of steps, which is shown in [8]. Then, for each $(s_1, s_2) \in S_1 \times S_2$, there exists a $k \in \mathbb{N}$ such that; for each $i \leq k$, $f^k(s_1, s_2) \geq f^i(s_1, s_2)$, and for each $j \geq k$, $f^k(s_1, s_2) = f^j(s_1, s_2)$. Let $f^{\min}$ be the branching distance between $\mathcal{RT}_1$ and $\mathcal{RT}_2$. Then, $f^{\min}(s_1, s_2) = f^k(s_1, s_2)$ for all $(s_1, s_2) \in S_1 \times S_2$ [9].

The following theorem will give the relationship between the two approaches.

Theorem 2. *Let $\delta \geq 0$, $\{R^i_\delta\}_{i\in\mathbb{N}}$ be the sequence defined by the definition of $\{R^i_\delta\}_{i\in\mathbb{N}}$, $\{f^i\}_{i\in\mathbb{N}}$ be the sequence defined by the definition of $\{f^i\}_{i\in\mathbb{N}}$, $R^{\max}_\delta$ be the maximal δ-simulation of $\mathcal{RT}_1$ by $\mathcal{RT}_2$, and $f^{\min}$ be the branching distance between $\mathcal{RT}_1$ and $\mathcal{RT}_2$. Then, the following assertions hold:*

(1) $R^i_\delta = \{(s_1, s_2) \in S_1 \times S_2 : f^i(s_1, s_2) \leq \delta\}$ for every $i \in \mathbb{N}$ and $\delta \geq 0$.
(2) $R^{\max}_\delta = \{(s_1, s_2) \in S_1 \times S_2 : f^{\min}(s_1, s_2) \leq \delta\}$ for every $\delta \geq 0$.

Proof. Assertions (1): This will be proved by induction with regard to i.

The initial step is for $i = 0$. Let $\delta \geq 0$, if $i = 0$, then the theorem states that $R^0_\delta = \{(s_1, s_2) \in S_1 \times S_2 : f^0(s_1, s_2) \leq \delta\}$. By definitions of $\{R^i_\delta\}_{i\in\mathbb{N}}$ and $\{f^i\}_{i\in\mathbb{N}}$, $R^0_\delta = \{(s_1, s_2) \in S_1 \times S_2 : f^0(s_1, s_2) \leq \delta\} = S_1 \times S_2$.

The induction hypothesis is that property (1) holds for $i = k$. We now show that property (1) holds for $i = k + 1$, i.e.,

$$R^{k+1}_\delta = \{(s_1, s_2) \in S_1 \times S_2 : f^{k+1}(s_1, s_2) \leq \delta\}.$$

First, assume $(s_1, s_2) \in R^{k+1}_\delta$. Then, it is sufficient to show that there exists $s_2 \xrightarrow{\omega} s'_2$ and $(s'_1, s'_2) \in R$ for each $s_1 \xrightarrow{\rho} {}^* s'_1$. In fact, by induction hypothesis we obtain

$$\sup_{s_1 \xrightarrow{\rho} {}^* s'_1} \inf_{s_2 \xrightarrow{\omega} s'_2} \max(d^*(\rho, L(\omega)), f^k(s'_1, s'_2)) \leq \delta.$$

Since $(s_1, s_2) \in R^{k+1}_\delta \subseteq R^k_\delta$,

$$\sup_{s_1 \xrightarrow{\rho} {}^* s'_1} \inf_{s_2 \xrightarrow{\omega} s'_2} d^*(\rho, L(\omega)) \leq f^k(s_1, s_2) \leq \delta.$$

Therefore, $f^{k+1}(s_1, s_2) \leq \delta$.

Second, let $f^{k+1}(s_1, s_2) \leq \delta$. Then, for each $s_1 \xrightarrow{\rho} {}^* s'_1$, there exists $s_2 \xrightarrow{\omega} s'_2$ such that $f^i(s'_1, s'_2) \leq \delta$. It follows from the definition of $\{R^i_\delta\}_{i\in\mathbb{N}}$, Lemma 4 and induction hypothesis that

$$f^k(s_1, s_2) \leq f^{k+1} \leq \delta.$$

Hence, $(s_1, s_2) \in R^{k+1}_\delta$. This completes the proof of the property (1).

Assertion (2): Consider $(s_1, s_2) \in S_1 \times S_2$. Suppose $f^{\min}(s_1, s_2) \leq \delta$.Then, there exists $f^i(s_1, s_2) \leq \delta$ for each $i \in \mathbb{N}$. We obtain by Lemma 3 and property (1) that $(s_1, s_2) \in R^{\max}_\delta$. On the contrary, let $(s_1, s_2) \in R^{\max}_\delta$. Then, $(s_1, s_2) \in R^i_\delta$ for each $i \in \mathbb{N}$. We obtain by property (1) that $f^i(s_1, s_2) \leq \delta$ for each $i \in \mathbb{N}$. By using Lemma 4, we have $f^{\min}(s_1, s_2) \leq \delta$.

5 Conclusion

In this paper, we have proposed an approximate simulation by using the notion of metrics to measure the behavioral closeness between a RE-TS and an LTS. The approximate simuation has some properties: First, it satisfies reflexivity, and satisfies transitivity with some limitations. Second, like exact simualtion, it has a maximal element. Then we study the relationship between trance disdance and trance inclusion, and the relationship between language distance and language inclusion. Moreover, we give the relationship between language distance and simualtion distance. Finally, we present two approaches to characterize approximate simualtion by using fixed point, and give the relationship between the two approaches.

As a future work, we will use the results of this paper to graph partten matching query, and consider logical characterizations of δ-simulation.

Acknowledgements. The authors would like to thank the anonymous referees for their very helpful suggestions.

References

1. Aceto, L., Ingólfsdóttir, A., Larsen, K.G., Srba, J.: Reactive Systems: Modelling, Specification and Verification. Cambridge University Press, Cambridge (2007)
2. Baier, C., Katoen, J.P.: Principles of Model Checking. MIT Press, Cambridge (2008)
3. Milner, R.: Communication and Concurrency. Prentice Hall, New Jersey (1989)
4. Milner, R.: Communicating and Mobile Systems: The π-Calculus. Cambridge University Press, Cambridge (1999)
5. Pan, H.Y., Cao, Y.Z., Zhang, M., Chen, Y.X.: Simulation for lattice-valued doubly labeled transition systems. Int. J. Approx. Reason. **55**(3), 797–811 (2014). https://doi.org/10.1016/j.ijar.2013.11.009
6. Pan, H., Li, Y., Cao, Y.: Lattice-valued simulations for quantitative transition systems. Int. J. Approx. Reason. **56**, 28–42 (2015). https://doi.org/10.1016/j.ijar.2014.10.001
7. Thrane, C.R., Fahrenberg, U., Larsen, K.G.: Quantitative analysis of weighted transition system. J. Log. Algebraic Program. **79**(7), 689–703 (2010). https://doi.org/10.1016/j.jlap.2010.07.010
8. de Alfaro, L., Faella, M., Stoelinga, M.: Linear and branching system metrics. IEEE Trans. Softw. Eng. **35**(2), 258–273 (2009). https://doi.org/10.1109/TSE.2008.106
9. Girard, A., Pappas, G.J.: Approximation metrics for discrete and continuous systems. IEEE Trans. Autom. Control **52**(5), 782–798 (2007). https://doi.org/10.1109/TAC.2007.895849
10. Ying, M.S.: Bisimulation indexes and their applications. Theor. Comput. Sci. **275**(12), 1–68 (2002). https://doi.org/10.1016/S0304-3975(01)00124-4
11. Desharnais, J., Gupta, V., Jagadeesan, R., Panangaden, P.: Metrics for labelled Markov processes. Theoret. Comput. Sci. **318**(3), 323–354 (2004). https://doi.org/10.1016/j.tcs.2003.09.013

12. Desharnais, J., Laviolette, F., Tracol, M.: Approximate analysis of probabilistic processes: logic, simulation and games. In: 5th International Conference on Quantitative Evaluation of Systems, pp. 264–273. IEEE Computer Society, Saint-Malo (2008). https://doi.org/10.1109/QEST.2008.42
13. Cerný, P., Henzinger, T.A., Radhakrishna, A.: Simulation distances. Theoret. Comput. Sci. **413**(1), 21–35 (2012). https://doi.org/10.1016/j.tcs.2011.08.002
14. Julius, A.A., D'Innocenzo, A., Benedetto, M.D.D., Pappas, G.J.: Approximate equivalence and synchronization of metric transition systems. Syst. Control Lett. **58**(2), 94–101 (2009). https://doi.org/10.1016/j.sysconle.2008.09.001
15. Fan, W.F., Li, J.Z., Ma, S., Tang, N., Wu, Y.H.: Adding regular expressions to graph reachability and pattern queries. Front. Comp. Sci. **6**(3), 313–338 (2012). https://doi.org/10.1007/s11704-012-1312-y
16. Salaün, G.: Quantifying the similarity of non-bisimilar labelled transition systems. Sci. Comput. Program. **202**, 102580 (2021). https://doi.org/10.1016/j.scico.2020.102580
17. Bozzelli, L., Molinari, A., Montanari, A., Peron, A.: Model checking interval temporal logics with regular expressions. Inf. Comput. **272**, 104498 (2020). https://doi.org/10.1016/j.ic.2019.104498
18. Fahrenberg, U., Legay, A., Quaas, K.: Computing branching distances with quantitative games. Theor. Comput. Sci. **847**, 134–146 (2020). https://doi.org/10.1016/j.tcs.2020.10.001
19. Wolper, P.: Temporal logic can be more expressive. Inf. Control **56**(1/2), 72–99 (1983). https://doi.org/10.1016/S0019-9958(83)80051-5
20. Hu, A.J., Vardi, M.Y. (eds.): CAV 1998. LNCS, vol. 1427. Springer, Heidelberg (1998). https://doi.org/10.1007/BFb0028725
21. Juhl, L., Larsen, K.G., Srba, J.: Modal transition systems with weight intervals. J. Log. Algebr. Program. **81**(4), 408–421 (2012). https://doi.org/10.1016/j.jlap.2012.03.008
22. Mateescu, R., Monteriro, P.T., Dumas, E., de Jong, H.: CTRL: extension of CTL with regular expressions and fairness operators to verify genetic regulatory networks. Theoret. Comput. Sci. **412**(26), 2854–2883 (2011). https://doi.org/10.1016/j.tcs.2010.05.009
23. Brázdil, T., Cerná, I.: Model checking of RegCTL. Comput. Artif. Intell. **25**(1), 81–97 (2006)
24. Larsen, K.G.: Modal specifications. In: Automatic Verification Methods for Finite State Systems, pp. 232–246 (1989). https://doi.org/10.1007/3-540-52148-8_19
25. van Breugel, F.: On behavioural pseudometrics and closure ordinals. Inf. Process. Lett. **112**(19), 715–718 (2012). https://doi.org/10.1016/j.ipl.2012.06.019
26. Fahrenberg, U., Legay, A.: The quantitative linear-time-branching-time spectrum. Theor. Comput. Sci. **538**, 54–69 (2014). https://doi.org/10.1016/j.tcs.2013.07.030
27. Zhang, J.J., Zhu, Z.H.: Characterize branching distance in terms of (η,α)-bisimilarity. Inf. Comput. **206**(8), 953–965 (2008). https://doi.org/10.1016/j.ic.2008.06.001
28. Qia, S., Zhu, P.: Limited approximate bisimulations and the corresponding rough approximations. Int. J. Approx. Reason. **130**, 50–82 (2021). https://doi.org/10.1016/j.ijar.2020.12.005
29. Hopcroft, J.E., Motwani, R., Ullman, J.D.: Introduction to Automata Theory, Languages, and Computation. Addison-Wesley Publishing, MA (1979)
30. Sangiorgi, D.: Introduction to Bisimulation and Coinduction. Cambridge University Press, Cambridge (2012)

31. Pan, H.Y., Cao, Y.Z., Chang, L., Qian, J.Y., Lin, Y.M.: Fuzzy alternating refinement relations under the Gödel semantics. IEEE Trans. Fuzzy Syst. **29**(5), 953–964 (2021). https://doi.org/10.1109/TFUZZ.2020.2965860
32. Munkres, J.R.: Typology. Prentice Hall, Englewood Cliffs (1975)
33. Pan, H.Y., Song, F., Cao, Y.Z., Qian, J.Y.: Fuzzy pushdown termination games. IEEE Trans. Fuzzy Syst. **27**(4), 760–774 (2019). https://doi.org/10.1109/TFUZZ.2018.2869127
34. Pan, H., Li, Y., Cao, Y., Li, P.: Nondeterministic fuzzy automata with membership values in complete residuated lattices. Int. J. Approx. Reason. **82**, 22–38 (2017). https://doi.org/10.1016/j.ijar.2016.11.020

On Interval Perturbation of the α-Symmetric Implicational Algorithm

Yiming Tang[1,2(✉)], Jianghui Han[1], Guangqing Bao[1], and Rui Chen[1]

[1] Anhui Province Key Laboratory of Affective Computing and Advanced Intelligent Machine, School of Computer and Information, Hefei University of Technology, Hefei 230601, Anhui, China
tym608@163.com

[2] Engineering Research Center of Safety Critical Industry Measure and Control Technology, Ministry of Education, Hefei University of Technology, Hefei 230601, Anhui, China

Abstract. Perturbation is an important property of fuzzy reasoning algorithms. It is an important task to estimate the upper and lower bounds of the output after a perturbation (such as an interval perturbation) is given to the input. In this study, the interval perturbation problem of a fuzzy reasoning algorithm, i.e., the α-symmetric implicational algorithm, is researched systematically. For FMP (Fuzzy Modus Ponens) and FMT (Fuzzy Modus Tollens) problems of fuzzy reasoning, aiming at the case of a single rule, the estimation for the upper and lower bounds are obtained for the input interval disturbance, in which R-implications and (S, N)-implications are mainly employed. Then the stability is verified on account of the upper and lower bounds. Following that, for the multi-rule case, the upper and lower bounds of the output are also given, and the corresponding stability is validated. In conclusion, it is found that the α-symmetric implicational algorithm has good stability.

Keywords: Fuzzy reasoning · Perturbation · Symmetric implicational algorithm · Fuzzy implications

1 Introduction

Fuzzy set theory was born in 1965 and founded by Professor Zadeh. It is an effective method to deal with problems such as information or data imprecision and uncertainty. Since Zadeh put forward the concept of fuzzy set, the study of fuzzy set has aroused the interest of scholars all over the world. Soon after that, the fuzzy inference theory came into being with the related theory of fuzzy set as a tool to describe uncertain information, and the mathematical logic based on general set theory was expanded. This theory belongs to the category of uncertain inference, which has great significance for the development of artificial intelligence technology.

At present, fuzzy reasoning [1–5] has become an inevitable part of the study of fuzzy set theory, and also an important aspect of its research. It has to be mentioned that the

Y. Chen and S. Zhang (Eds.): AILA 2022, CCIS 1657, pp. 63–76, 2022.
https://doi.org/10.1007/978-981-19-7510-3_5

inference process of fuzzy set is mainly divided into two kinds, Fuzzy Modus Ponens (FMP) problem and Fuzzy Modus Tollens (FMT) problem:

$$\text{FMP model: Give the rule } A \to B \text{ and input } A^*, \text{ calculate the output } B^*, \tag{1}$$

$$\text{FMT model: Give the rule } A \to B \text{ and input } B^*, \text{ calculate the output } A^*, \tag{2}$$

where $A, A^* \in F(X), B, B^* \in F(Y)$ (where $F(X), F(Y)$ respectively denote the set of all fuzzy subsets of universe X and Y).

In dealing with fuzzy reasoning, Zadeh proposed the famous fuzzy reasoning algorithm, i.e., the CRI (compositional rule of inference) algorithm [6] in 1973. The solution of this algorithm was described as follows ($y \in Y$):

$$B^*(y) = \sup_{x \in X}\left\{A^*(x) \wedge (A(x) \to B(y))\right\} \tag{3}$$

Here $\to$ was a fuzzy implication. It is not difficult to find that the CRI algorithm simply combines $A^*(x)$ and $A(x) \to B(y)$ for getting the solution. In 1999, Wang [7] proposed a new algorithm that could effectively improve the CRI algorithm, namely the triple I algorithm.

The basic idea of the triple I algorithm is as follows. Suppose that X, Y are two non-empty sets, and $A \in F(X), B \in F(Y)$, and $A^* \in F(X)$ (or $B^* \in F(Y)$). The algorithm seeks the smallest fuzzy set $B^* \in F(Y)$ (or the largest fuzzy set $A^* \in F(X)$) to make the following formula

$$A(x) \to B(y) \to \left(A^*(x) \to B^*(y)\right) \tag{4}$$

attain its maximum for any $(x, y) \in X \times Y$. Such B^* (A^*) is called the full implication triple I solution for the FMP (FMT) problem.

Many scholars have carried out a series of studies on the triple I algorithm and have achieved rich results, making important contributions to the improvement of the triple I algorithm. Pei [8] systematically discussed the full implication triple I method based on the left-continuous triangular norm. Wang and Fu [9] gave the general form of the triple I algorithm on the basis of regular implication. Pei [10] discussed the same problem based on first-order logic system, and presented a perfect unified logical reasoning system of triple I method. Later, Zheng and Liu [11] proposed the residual intuitionistic implication and intuitionistic triple I fuzzy inference algorithm under multiple rules, which provided a theoretical basis for connecting intuitionistic fuzzy sets with fuzzy inference. Tang et al. [12–14] proposed the differently implicational algorithm based on different fuzzy implications from the perspective of combining CRI and triple I algorithm. In general, triple I algorithm has strong logic foundation, continuity, reducibility, robustness and many other advantages.

From the dual perspectives of logical system and inference model, Tang [15] proposed and studied the symmetric implicational algorithm, which has better results in dealing with (1) and (2). Its algorithm is described as follows:

$$(A(x) \to_1 B(y)) \to_2 \left(A^*(x) \to_1 B^*(y)\right) \tag{5}$$

where $\to_1$ *and* $\to_2$ are different fuzzy implications. It can be seen that triple I algorithm is a special case of the symmetric implicational algorithm. Furthermore, (5) is extended to

$$(A(x) \to_1 B(y)) \to_2 (A^*(x) \to_1 B^*(y) \geq \alpha, \tag{6}$$

and it is called the α-symmetric implicational algorithm. Reference [16] further discussed the symmetric implicational algorithm based on point-by-point support. In Ref. [17], the granular symmetric implicational algorithm was proposed and studied from the perspective of interval values. Dai [18] verified the logical basis of the symmetric implicational algorithm from the perspective of predicate formal representation.

In fuzzy reasoning algorithms, an important property is perturbation research [19–22]. It is an important task to estimate the upper and lower bounds of the output after a perturbation (such as interval perturbation) is given to the input. In [23], for single-rule CRI method, Chen and Fu analyzed the upper and lower bounds of the output error caused by simple perturbation of the input fuzzy set, and determined the oscillation range of the output result when the input interval perturbation occurred. In Ref. [24], we studied the oscillation boundary estimation of BKS (Bandler-Kohout subproduct) perturbation. First, the BKS output variation range for input interval perturbation was estimated. Secondly, aiming at the complex problem of fuzzy inference chain, the oscillation bounds of BKS output caused by perturbation of input interval were given. Thirdly, for the simple perturbation of the input fuzzy set, the upper and lower bounds of BKS output deviation were constructed. Finally, the stability of all BKS strategies was verified.

Similarly, for the α-symmetric implicational algorithm, its perturbation is obviously also a concern of us.

In this study, we discuss the upper and lower bounds estimation of the output of the α-symmetric implicational algorithm for input interval perturbation, and analyze its stability.

The structure of this study is arranged as follows. Section 2 provides the necessary preliminaries. In Sect. 3, the corresponding upper and lower bounds estimates for R-implication and (S,N)-implication are given for the case of input interval perturbation with a single rule, and the corresponding stability is verified. Section 4 continues to discuss the case of multiple rules. Section 5 gives a summary.

2 Preliminaries

Definition 2.1. ([25]) Let $a, b, c \in L$, where $L = [0, 1]$. If the binary operation $\otimes$ satisfies the following four properties,

$$a \otimes b = b \otimes a, \tag{T1}$$

$$a \otimes (b \otimes c) = (a \otimes b) \otimes c, \tag{T2}$$

$$a \otimes b \leq a \otimes c \text{ if } b \leq c \tag{T3}$$

$$a \otimes 1 = a. \tag{T4}$$

then ⊗ is called a triangular norm (abbreviated by t-norm) on L.

Definition 2.2. ([25]) Let a, b, $c \in L$. If the binary operation ⊕ satisfies (T1), (T2), (T3) and

$$a \oplus 0 = a, \tag{S1}$$

Then ⊕ is called a triangular conorm (abbreviated by t-conorm) on L.

Let ⊗ be a t-norm on L. If the binary operation ⊕ satisfies

$$a \oplus b = 1 - (1 - a) \otimes (1 - b) \tag{7}$$

for any $a,\ b \in L$, then ⊕ is called the dual t-conorm of ⊗.

Definition 2.3. ([25]) A t-norm ⊗ is said to be left continuous, if ⊗ satisfies (for any $a_i, b \in L$):

$$(\vee_{i \in I} a_i) \otimes b = \vee_{i \in I} (a_i \otimes b). \tag{8}$$

Proposition 2.1. ([25]) The t-norm is left continuous, if and only if its dual t-conorm is right-continuous.

Definition 2.4. ([26]) Suppose a mapping $\rightarrow: [0, 1]^2 \rightarrow [0, 1]$ which satisfies

$$0 \rightarrow 0 = 0 \rightarrow 1 = 1 \rightarrow 1 = 1, 1 \rightarrow 0 = 0, \tag{9}$$

then → is called fuzzy implication on [0, 1].

Definition 2.5. ([25]) Let ⊗ and → be two $[0, 1]^2 \rightarrow [0, 1]$ mappings, then (⊗, →) is called a residual pair, or ⊗ *and* → are residual to each other, if the following residuation condition holds (for any $a, b, c \in [0, 1]$),

$$\mathrm{a \otimes b \leq c} \quad \text{if and only if} \quad a \leq b \rightarrow c \tag{10}$$

For fuzzy set A, we denote $A'(x) = (A(x))' = 1 - A(x)$.

Lemma 2.1. ([27]) Suppose ⊗ is a left continuous t-norm on [0, 1], → is obtained from (where $p, q \in [0, 1]$)

$$\mathrm{p \rightarrow q = sup\{y \in [0, 1] | p \otimes y \leq q\}}, \tag{11}$$

then (⊗, →) is a residual pair, and → satisfies $(a, b, c \in [0, 1], P \neq \phi)$,

$$a \rightarrow b \text{ is non - decreasing w.r.t. } b; \tag{C1}$$

$$a \rightarrow b \text{ is right - continuous w.r.t. } b; \tag{C2}$$

$$a \rightarrow b \text{ is non - increasing w.r.t. } a; \tag{C3}$$

$$a \leq b \text{ if and only if } a \rightarrow b = 1 \tag{C4}$$

$$1 \rightarrow a = 1 \tag{C5}$$

$$a \leq b \rightarrow c \text{ if and only if } b \leq a \rightarrow c \tag{C6}$$

$$a \rightarrow (b \rightarrow c) = b \rightarrow (a \rightarrow c) \tag{C7}$$

$$\inf\{a \rightarrow x_i | i \in P\} = a \rightarrow \inf\{x_i | i \in P\} \tag{C8}$$

$$\inf\{x_i \rightarrow b | i \in P\} = \sup\{x_i | i \in P\} \rightarrow b \tag{C9}$$

Definition 2.6. ([25]) A negation on [0, 1] is a decreasing mapping N: [0, 1] → [0, 1] which satisfies N(0) = 1 and N(1) = 0. If

$$\mathrm{N(N(a)) = a}$$

holds for any a ∈ [0, 1], then N is said to be involutive.

Definition 2.7. ([28]) A mapping → is a (S, N)-implication, when there is a fuzzy negation N making

$$a \rightarrow b = N(a) \oplus b$$

hold ($a \in [0, 1], b \in [0, 1]$).

Definition 2.8. ([23]) (i) Suppose that $\beta^-, \beta^+ \in F(Z)$ and that

$$\beta^-(z) \leq \beta^+(z)(z \in Z)$$

then $[\beta^-, \beta^+]$ is known as a fuzzy interval on Z.

(ii) Suppose that $C \in F(Z)$ and that $[\beta^-, \beta^+]$ be a fuzzy interval on Z. If

$$\beta^-(z) \leq C(z) \leq \beta^+(z),$$

then $[\beta^-, \beta^+]$ is referred to as an interval perturbation of C, where

$$C \in [\beta^-, \beta^+].$$

Definition 2.9. ([23]) Suppose that $A, B \in F(Z)$. If there is a mapping $\beta : Z \rightarrow [-1, 1]$ making

$$B(z) = A(z) + \beta$$

hold, then B is called a simple perturbation, and β is said to be a perturbation factor of A.

Definition 2.10. ([23]) Suppose that $C, E \in F(Z), D \in F(W)$, and that

$$C \in [\alpha^-, \alpha^+], D \in [\eta^-, \eta^+], E \in [\gamma^-, \gamma^+]$$

A fuzzy inference method f is said to be a stable function, if for any $\varepsilon > 0$, there is a fuzzy interval $[\lambda^-, \lambda^+]$ on W and $\delta > 0$ making

$$\lambda^+ - \lambda^- < \varepsilon$$

be effective for any $w \in W$ and

$$f(C, D, E) \in [\lambda^-, \lambda^+]$$

If

$$\alpha^+(z) - \alpha^-(z) < \delta, \eta^+(w) - \eta^-(w) < \delta, \gamma^+(z) - \gamma^-(z) < \delta \quad (z \in Z, w \in W)$$

Lemma 2.2. ([23]) Let a_1 and a_2 be real-valued, bounded mappings on Z, and C, $D \in F(Z)$. Thus, one has:

(i) $a_1 \vee a_2 = max\{a_1, a_2\} = (a_1 + a_2)/2 + |a_1 - a_2|/2$,
(ii) $a_1 \wedge a_2 = min\{a_1, a_2\} = (a_1 + a_2)/2 - |a_1 - a_2|/2$,
(iii) $-|a_1 - a_2| \leq |a_1| - |a_2| \leq |a_1 - a_2|$,
(iv) $inf_{z \in Z} C(z) + inf_{z \in Z} D(z) \leq inf_{z \in Z}(C(z) + D(z))$,
(v) $sup_{z \in Z}(C(z) + D(z)) \leq sup_{z \in Z} C(z) + sup_{z \in Z} D(z)$,
(vi) $sup_{z \in Z} C(z) = -inf_{z \in Z}(-C(z))$,
(vii) $inf_{z \in Z}(C(z) \wedge D(z)) = inf_{z \in Z} C(z) \wedge inf_{z \in Z} D(z)$,
(viii) $inf_{z \in Z}(C(z) \vee D(z)) \geq inf_{z \in Z} C(z) \vee inf_{z \in Z} D(z)$,
(ix) $inf_{z \in Z}(C(z) \vee D(z)) \geq inf_{z \in Z} C(z) \vee inf_{z \in Z} D(z)$,
(x) $sup_{z \in Z}(C(z) \vee D(z)) = sup_{z \in Z} C(z) \vee sup_{z \in Z} D(z)$.

Lemma 2.3. ([24]) Let a_1 and a_2 be real-valued, bounded mappings on Z, then:

(i) $sup_{z \in Z} a_1(z) - sup_{z \in Z} a_2(z) \leq sup_{z \in Z}(a_1(z) - a_2(z))$,
(ii) $inf_{z \in Z} a_1(z) - inf_{z \in Z} a_2(z) \leq sup_{z \in Z}(a_1(z) - a_2(z))$,
(iii) $inf_{z \in Z} a_1(z) - inf_{z \in Z} a_2(z) \geq inf_{z \in Z}(a_1(z) - a_2(z))$,
(iv) $sup_{z \in Z} a_1(z) - sup_{z \in Z} a_2(z) \geq inf_{z \in Z}(a_1(z) - a_2(z))$.

Proposition 2.2. ([15]) Suppose that $A, A^* \in F(X)$ and that $B \in F(Y)$. If $\rightarrow_1, \rightarrow_2$ are two fuzzy implications satisfying (C1), (C2), (C5), and $\otimes_1$, $\otimes_2$ are respectively the operations residual to $\rightarrow_1, \rightarrow_2$, then the optimal solution of the α-symmetric implicational algorithm for FMP is as follows ($y \in Y$).

$$SIP(y) = sup_{x \in X}\{A^*(x) \otimes_1 ((A(x) \rightarrow_1 B(y)) \otimes_2 \alpha)\}. \quad (12)$$

Proposition 2.3. ([15]) Suppose that $A \in F(X)$ and that $B, B^* \in F(Y)$. If $\rightarrow_1$ is a fuzzy implication satisfying (C6), and $\rightarrow_2$ is a fuzzy implication satisfying (C1), (C2), (C5), and $\otimes_2$ is the operation residual to $\rightarrow_2$, then the optimal solution of the α-symmetric implicational algorithm for FMT is as follows ($x \in X$).

$$SIT(x) = inf_{y \in Y}\{(A(x) \rightarrow_1 B(y)) \otimes_2 \alpha) \rightarrow_2 B^*(y)\}. \quad (13)$$

3 Interval Perturbation for the α-Symmetric Implicational Algorithms with One Rule

3.1 The Interval Perturbation with One Rule for the FMP Problem

Note

$$\tilde{i}_X(f) = inf_{x\in X}\{f(x)\}, \tilde{s}_X(f) = sup_{x\in X}\{f(x)\}. \tag{14}$$

Theorem 3.1. Let $A, A^* \in F(X)$ and $B \in F(Y)$. Suppose that

$$A \in [\beta^-, \beta^+], A^* \in [\gamma^-, \gamma^+], B \in [\eta^-, \eta^+]$$

If $\rightarrow_1$ is a $R-$ implication, and $\rightarrow_2$ is a $R-$ implication or an $(S, N)-$ implication satisfying (C2). $\otimes_1, \otimes_2$ are respectively the operation residual to $\rightarrow_1, \rightarrow_2$. Then for the α-symmetric implicational algorithm for FMP, we have

$$\tilde{i}_X(\gamma^-) \otimes_1 ((\tilde{s}_X(\beta^+) \rightarrow_1 \eta^-) \otimes_2 \alpha) \le SIP(y) \le \tilde{s}_X(\gamma^+) \otimes_1 \left(\left(\tilde{i}_X(\beta^-) \rightarrow_1 \eta^+\right) \otimes_2 \alpha\right). \tag{15}$$

Proof: In accordance to the characteristics of $\rightarrow_1$ and $\rightarrow_2$, obviously $\rightarrow_1$ and $\rightarrow_2$ are decreasing with respect to the first variable and increasing with respect to the second variable, then

$$sup(\beta^+) \rightarrow_1 \eta^- \le A(x) \rightarrow_1 B(y) \le inf(\beta^-) \rightarrow_1 \eta^+.$$

Based on the fact that $\otimes_1$ and $\otimes_2$ are increasing, it can be obtained:

$$inf(\gamma^-) \otimes_1 ((sup(\beta^+) \rightarrow_1 \eta^-) \otimes_2 \alpha)$$

$$\le A^*(x) \otimes_1 ((A(x) \rightarrow_1 B(y)) \otimes_2 \alpha)$$

$$\le sup(\gamma^+) \otimes_1 ((inf(\beta^-) \rightarrow_1 \eta^+) \otimes_2 \alpha),$$

By combining these two formulas and considering Proposition 2.2, it can be obtained

$$\tilde{i}_X(\gamma^-) \otimes_1 ((\tilde{s}_X(\beta^+) \rightarrow_1 \eta^-) \otimes_2 \alpha)$$

$$\le sup_{x\in X}\left\{A^*(x) \otimes_1 ((A(x) \rightarrow_1 B(y)) \otimes_2 \alpha)\right\} = SIP(y)$$

$$\le \tilde{s}_X(\gamma^+) \otimes_1 \left(\left(\tilde{i}_X(\beta^-) \rightarrow_1 \eta^+\right) \otimes_2 \alpha\right).$$

The proof is done.

Similar to Theorem 3.1, Theorem 3.2 can be proved.

Theorem 3.2. Let $A, A^* \in F(X)$ and $B \in F(Y)$. Suppose that

$$A \in [\beta^-, \beta^+], A^* \in [\gamma^-, \gamma^+], B \in [\eta^-, \eta^+]$$

$\to_1$ is an $(S, N)-$ implication satisfying (C2) (associated with N and $\oplus_1$), and $\to_2$ is a $R-$ implication or an $(S, N)-$ implication satisfying (C2). $\otimes_1, \otimes_2$ are respectively the operation residual to $\to_1, \to_2$. Then for the α-symmetric implicational algorithm for FMP, we have

$$\begin{aligned} \tilde{i}_x(\gamma^-) \otimes_1 ((N(\tilde{s}_x(\beta^+)) \oplus_1 \eta^-) \otimes_2 \alpha) \le SIP(y) \le \\ \tilde{s}_x(\gamma^+) \otimes_1 ((N(\tilde{i}_x(\beta^-)) \oplus_1 \eta^+) \otimes_2 \alpha). \end{aligned} \tag{16}$$

3.2 The Interval Perturbation with One Rule for the FMT Problem

Theorem 3.3. Let $B, B^* \in F(Y)$ and $A \in F(X)$. Suppose that

$$B \in [\beta^-, \beta^+], B^* \in [\gamma^-, \gamma^+], A \in [\eta^-, \eta^+].$$

$\to_1$ is a $R-$ implication, and $\to_2$ is a $R-$ implication or an $(S, N)-$ implication satisfying (C2). $\otimes_1, \otimes_2$ are respectively the operation residual to $\to_1, \to_2$. Then for the α-symmetric implicational algorithm for FMT, one has

$$((\eta^- \to_1 \tilde{s}_y(\beta^+)) \otimes_2 \alpha) \to_1 \tilde{i}_y(\gamma^-) \le SIT(x) \le \left(\left(\eta^+ \to_1 \tilde{i}_y(\beta^-)\right) \otimes_2 \alpha\right) \to_1 \tilde{s}_y(\gamma^+). \tag{17}$$

Proof: In accordance to the characteristics of $\to_1$ and $\to_2$, obviously $\to_1$ and $\to_2$ are decreasing with respect to the first variable and increasing with respect to the second variable, and $\otimes_1$ and $\otimes_2$ are increasing, then

$$\eta^+ \to_1 inf(\beta^-) \le A(x) \to_1 B(y) \le \eta^- \to_1 sup(\beta^+).$$

Furthermore, we can get

$$((A(x) \to_1 B(y)) \otimes_2 \alpha) \to_1 inf(\gamma^-)$$

$$\le ((A(x) \to_1 B(y)) \otimes_2 \alpha) \to_1 B^*(y)$$

$$\le ((A(x) \to_1 B(y)) \otimes_2 \alpha) \to_1 sup(\gamma^+),$$

Combining with Proposition 2.3, it can be obtained

$$((\eta^- \to_1 \tilde{s}_y(\beta^+)) \otimes_2 \alpha) \to_1 \tilde{i}_y(\gamma^-)$$

$\le$ $inf_{y \in Y}\{((A(x) \to_1 B(y)) \otimes_2 \alpha) \to_1 B^*\}$ $=$ $SIT(x) \le$ $\left(\left(\eta^+ \to_1 \tilde{i}_y(\beta^-)\right) \otimes_2 \alpha\right) \to_1 \tilde{s}_y(\gamma^+)$.

The proof is completed.

Similar to Theorem 3.3, Theorem 3.4 can be proved.

Theorem 3.4. Let $B, B^* \in F(Y)$, and $A \in F(X)$. Suppose that

$$B \in [\beta^-, \beta^+], B^* \in [\gamma^-, \gamma^+], A \in [\eta^-, \eta^+]$$

$\rightarrow_1$ is an $(S, N)-$ implication satisfying (C2) and (C6) (associated with $N and \oplus_1$), and $\rightarrow_2$ is a $R-$ implication or an $(S, N)-$ implication satisfying (C2). $\otimes_1, \otimes_2$ are respectively the operation residual to $\rightarrow_1, \rightarrow_2$. Then for the α-symmetric implicational algorithm for FMT, one has

$$\begin{aligned} &N\left(\left(N(\eta^-) \oplus_1 \tilde{s}_y(\beta^+)\right) \otimes_2 \alpha\right) \oplus_1 \tilde{i}_y(\gamma^-) \leq SIT(x) \leq \\ &N\left(\left(N(\eta^+) \oplus_1 \tilde{i}_y(\beta^-)\right) \otimes_2 \alpha\right) \oplus_1 \tilde{s}_y(\gamma^+). \end{aligned} \tag{18}$$

If the general operators are continuous (including t-norm, t-conorm and fuzzy negation), then the α-symmetric implicational algorithm in which $\rightarrow_1$ $isan(S, N)-$ implication is stable. As for the situation of $R-$ implication, if the $R-$ implication and t-norm are continuous, then the α-symmetric implicational algorithm in which $\rightarrow_1$ employs an $R-$ implication is also stable. Specifically, suppose that there is a perturbation sequence

$$\left([\beta_m^-, \beta_m^+], [\eta_m^-, \eta_m^+], [\gamma_m^-, \gamma_m^+]\right)$$

of the input (A, B, A^*) with respect to the α-symmetric implicational algorithm for FMP, which means that

$$\begin{aligned} &lim_{m\rightarrow\infty} sup_{x\in X}\left(\beta_m^+(x) - \beta_m^-(x)\right) = lim_{m\rightarrow\infty} sup_{y\in Y}\left(\eta_m^+(y) - \eta_m^-(y)\right) \\ &= lim_{m\rightarrow\infty} sup_{x\in X}\left(\gamma_m^+(x) - \gamma_m^-(x)\right) = 0 \end{aligned} \tag{19}$$

is effective, where $m = 1, 2, ...$ In Theorem 3.1 and 3.2, we adopt λ_m^+, λ_m^- to indicate the corresponding lower and upper bounds of the output of α-symmetric implicational algorithm, viz.,

$$SIP(A, B, A^*) \in [\lambda_m^-, \lambda_m^+]$$

For $\left([\beta_m^-, \beta_m^+], [\eta_m^-, \eta_m^+], [\gamma_m^-, \gamma_m^+]\right)$, when the continuous condition is satisfied, one has from Theorem 3.1 and 3.2 that

$$lim_{m\rightarrow\infty} sup_{y\in Y}\left(\lambda_m^+(y) - \lambda_m^-(y)\right) = 0. \tag{20}$$

That is, the output of the α-symmetric implicational algorithm converges steadily to a value when the continuous conditions mentioned above is effective. Similarly, for FMT input (A, B, B^*), Theorem 3.3 and Theorem 3.4 have similar results. In short, the α-symmetric implicational algorithm for situation of one rule is stable from the perspective of interval perturbation.

4 Interval Perturbation for the α-Symmetric Implicational Algorithms with Multiple Rules

4.1 The Interval Perturbation with Multiple Rules for the FMP Problem

For the situation of multiple rules, the solutions of FMP and FMT are denoted as

$$SIP_n(A_1, ..., A_n, B_1, ..., B_n, A^*)(y)$$

and

$$SIT_n\big(A_1, ..., A_n, B_1, ..., B_n, B^*\big)(x),$$

respectively abbreviated by $SIP_n(y)$ and $SIT_n(x)$. For n rules, in Proposition 2.2 and 2.3, we replace $(A(x) \rightarrow_1 B(y))$ with

$$\vee_{i=1}^{n}(A_i(x) \rightarrow_1 B_i(y)).$$

Theorem 4.1. Let $A_i, A^* \in F(X), B_i \in F(Y)$. Suppose that

$$A_i \in \big[\beta_i^-, \beta_i^+\big], A^* \in \big[\gamma^-, \gamma^+\big], B_i \in \big[\eta_i^-, \eta_i^+\big](i = 1, 2, \ldots, n).$$

If $\rightarrow_1$ is a $R-$ implication, and $\rightarrow_2$ is a $R-$ implication or an $(S, N)-$ implication satisfying (C2). $\otimes_1, \otimes_2$ are respectively the operation residual to $\rightarrow_1, \rightarrow_2$. Then for the α-symmetric implicational algorithm for FMP, we have

$$\begin{gathered}\tilde{i}_x\big(\gamma^-\big) \otimes_1 \big(V_{i=1}^{n}\big(\tilde{s}_x\big(\beta_i^+\big) \rightarrow_1 \eta_i^-\big) \otimes_2 \alpha\big) \leq SIP_n(y) \leq \\ \tilde{s}_x\big(\gamma^+\big) \otimes_1 \Big(V_{i=1}^{n}\Big(\tilde{i}_x\big(\beta_i^-\big) \rightarrow_1 \eta_i^+\Big) \otimes_2 \alpha\Big).\end{gathered} \tag{21}$$

Proof: Obviously $\rightarrow_1$ and $\rightarrow_2$ are decreasing with respect to the first variable and increasing with respect to the second variable, then

$$\vee_{i=1}^{n}\big(sup\big(\beta_i^+\big) \rightarrow_1 \eta_i^-\big) \leq \vee_{i=1}^{n}(A_i(x) \rightarrow_1 B_i(y)) \leq \vee_{i=1}^{n}(inf\big(\beta_i^-\big) \rightarrow_1 \eta_i^+).$$

Based on the fact that $\otimes_1$ and $\otimes_2$ are increasing, it can be obtained:

$$inf\big(\gamma^-\big) \otimes_1 \big(\big(sup\big(\beta_i^+\big) \rightarrow_1 \eta_i^-\big) \otimes_2 \alpha\big)$$

$$\leq A^*(x) \otimes_1 \big(\vee_{i=1}^{n}(A_i(x) \rightarrow_1 B_i(y)) \otimes_2 \alpha\big)$$

$$\leq sup\big(\gamma^+\big) \otimes_1 \big(\big(inf\big(\beta_i^-\big) \rightarrow_1 \eta_i^+\big) \otimes_2 \alpha\big),$$

By combining these two formulas and considering Proposition 2.2, it can be obtained

$$\tilde{i}_x\big(\gamma^-\big) \otimes_1 \big(\vee_{i=1}^{n}\big(\tilde{s}_x\big(\beta_i^+\big) \rightarrow_1 \eta_i^-\big) \otimes_2 \alpha\big)$$

$$\leq sup_{x \in X}\big\{A^*(x) \otimes_1 \big(\vee_{i=1}^{n}(A_i(x) \rightarrow_1 B_i(y)) \otimes_2 \alpha\big)\big\} = SIP_n(y).$$

$$\leq \tilde{s}_x\big(\gamma^+\big) \otimes_1 \Big(\vee_{i=1}^{n}\Big(\tilde{i}_x\big(\beta_i^-\big) \rightarrow_1 \eta_i^+\Big) \otimes_2 \alpha\Big).$$

The proof is completed.

Similar to Theorem 4.1, Theorem 4.2 can be proved.

Theorem 4.2. Let $A_i, A^* \in F(X), B_i \in F(Y)$. Suppose that

$$A_i \in [\beta_i^-, \beta_i^+], A^* \in [\gamma^-, \gamma^+], B_i \in [\eta_i^-, \eta_i^+](i = 1, 2, \ldots, n).$$

$\rightarrow_1$ is an $(S, N)-$ implication satisfying (C2) (associated with $N$and $\oplus_1$), and $\rightarrow_2$ is a $R-$ implication or an $(S, N)-$ implication satisfying (C2). $\otimes_1, \otimes_2$ are respectively the operation residual to $\rightarrow_1, \rightarrow_2$. Then for the α-symmetric implicational algorithm for FMP, we obtain

$$\begin{aligned} &\tilde{i}_x(\gamma^-) \otimes_1 (\vee_{i=1}^n (N(\tilde{s}_x(\beta_i^+)) \otimes_1 \eta_i^-) \otimes_2 \alpha) \leq SIP_n(y) \\ &\leq \tilde{s}_x(\gamma^+) \otimes_1 \left(V_{i=1}^n \left(N\left(\tilde{i}_x(\beta_i^-)\right) \otimes_1 \eta_i^+\right) \otimes_2 \alpha\right). \end{aligned} \tag{22}$$

4.2 The Interval Perturbation with Multiple Rules for the FMT Problem

Theorem 4.3. Let $B_i, B^* \in F(Y), A_i \in F(X)$. Suppose that

$$B_i \in [\beta_i^-, \beta_i^+], B^* \in [\gamma^-, \gamma^+], A_i \in [\eta_i^-, \eta_i^+](i = 1, 2, \ldots, n).$$

$\rightarrow_1$ is a $R-$ implication, and $\rightarrow_2$ is a $R-$ implication or an $(S, N)-$ implication satisfying (C2). $\otimes_1, \otimes_2$ are respectively the operation residual to $\rightarrow_1, \rightarrow_2$. Then for the α-symmetric implicational algorithm for FMT, we achieve

$$\begin{aligned} &(\vee_{i=1}^n (\eta_i^- \rightarrow_1 \tilde{s}_y(\beta_i^+)) \otimes_2 \alpha) \rightarrow_1 \tilde{i}_y(\gamma^-) \leq SIT_n(x) \\ &\leq \left(V_{i=1}^n \left(\eta_i^+ \rightarrow_1 \tilde{i}_y(\beta_i^-)\right) \otimes_2 \alpha\right) \rightarrow_1 \tilde{s}_y(\gamma^+). \end{aligned} \tag{23}$$

Proof: Evidently $\rightarrow_1$ and $\rightarrow_2$ are decreasing with respect to the first variable and increasing with respect to the second variable. Meanwhile $\otimes_1$ and $\otimes_2$ are increasing, then one has

$$\vee_{i=1}^n (\eta_i^+ \rightarrow_1 inf(\beta_i^-)) \leq \vee_{i=1}^n (A_i(x) \rightarrow_1 B_i(y)) \leq \vee_{i=1}^n (\eta_i^- \rightarrow_1 sup(\beta_i^+)).$$

Furthermore, we can get

$$\begin{aligned} &(\vee_{i=1}^n (A_i(x) \rightarrow_1 B_i(y)) \otimes_2 \alpha) \rightarrow_1 inf(\gamma^-) \\ &\leq (\vee_{i=1}^n (A_i(x) \rightarrow_1 B_i(y)) \otimes_2 \alpha) \rightarrow_1 B^*(y) \\ &\leq (\vee_{i=1}^n (A_i(x) \rightarrow_1 B_i(y)) \otimes_2 \alpha) \rightarrow_1 sup(\gamma^+), \end{aligned}$$

Combining with Proposition 2.3, it can be obtained

$$\begin{aligned} &(\vee_{i=1}^n (\eta_i^- \rightarrow_1 \tilde{s}_y(\beta_i^+)) \otimes_2 \alpha) \rightarrow_1 \tilde{i}_y(\gamma^-) \\ &\leq inf_{y \in Y}\{(\vee_{i=1}^n (A_i(x) \rightarrow_1 B_i(y)) \otimes_2 \alpha) \rightarrow_1 B^*\} = SIT_n(x) \\ &\leq \left(\vee_{i=1}^n \left(\eta_i^+ \rightarrow_1 \tilde{i}_y(\beta_i^-)\right) \otimes_2 \alpha\right) \rightarrow_1 \tilde{s}_y(\gamma^+) \end{aligned}$$

The proof is completed.

Similar to Theorem 4.3, Theorem 4.4 can be proved.

Theorem 4.4. Let B_i,$B^* \in F(Y)$, $A_i \in F(X)$. Suppose that

$$B_i \in [\alpha_i^-, \alpha_i^+], B^* \in [\gamma^-, \gamma^+], A_i \in [\eta_i^-, \eta_i^+] (i = 1, 2, \ldots, n)$$

$\to_1$ is an $(S, N)-$ implication satisfying (C2) and (C6) (associated with N *and* $\oplus_1$), and $\to_2$ is a $R-$ implication or an $(S, N)-$ implication satisfying (C2). $\otimes_1, \otimes_2$ are respectively the operation residual to $\to_1, \to_2$. Then for the α-symmetric implicational algorithm for FMT, we achieve

$$\begin{aligned} &N\left(\vee_{i=1}^n \left(N(\eta_i^-) \oplus_1 \tilde{s}_y(\beta_i^+)\right) \otimes_2 \alpha\right) \oplus_1 \tilde{i}_y(\gamma^-) \leq SIT_n(x) \leq \\ &N\left(\vee_{i=1}^n \left(N(\eta_i^+) \oplus_1 \tilde{i}_y(\beta_i^-)\right) \otimes_2 \alpha\right) \oplus_1 \tilde{s}_y(\gamma^+) \end{aligned} \tag{24}$$

If the general operators are continuous (including t-norm, t-conorm and fuzzy negation), then α-symmetric implicational algorithm in which $\to_1$ is an $(S, N)-$ implication with multiple rules is stable. If the $R-$ implication and t-norm are continuous, then the α-symmetric implicational algorithm in which $\to_1$ is an $R-$ implication with multiple rules is also stable. Specifically, suppose that there is a perturbation sequence

$$\left([\beta_{im}^-, \beta_{im}^+], [\eta_{im}^-, \eta_{im}^+], [\gamma_m^-, \gamma_m^+]\right)$$

of the input (A_i, B_i, A^*) with respect to the α-symmetric implicational algorithm for FMP, which means that $(m = 1, 2, \ldots)$

$$\begin{aligned} &lim_{m\to\infty} sup_{x\in X}\left(\beta_{im}^+(x) - \beta_{im}^-(x)\right) = lim_{m\to\infty} sup_{y\in Y}\left(\eta_{im}^+(y) - \eta_{im}^-(y)\right) \\ &= lim_{m\to\infty} sup_{x\in X}\left(\lambda_m^+(x) - \gamma_m^-(\gamma)\right) = 0 \end{aligned} \tag{25}$$

is effective. In Theorem 4.1 and 4.2, we adopt λ_m^+, λ_m^- to indicate the corresponding lower and upper bounds of output of the α-symmetric implicational algorithm, viz., $SIP_n(y) \in [\lambda_m^-, \lambda_m^+]$. For

$$\left([\beta_{im}^-, \beta_{im}^+], [\eta_{im}^-, \eta_{im}^+], [\gamma_m^-, \gamma_m^+]\right),$$

when the continuous condition is satisfied, one has from Theorem 4.1 and 4.2 that

$$lim_{m\to\infty} sup_{y\in Y}\left(\lambda_m^+(y) - \lambda_m^-(y)\right) = 0. \tag{26}$$

That is, the output of the α-symmetric implicational algorithm converges steadily to a value when the continuous conditions mentioned above is effective. Similarly, for the FMT input (A_i, B_i, B^*), Theorem 4.3 and Theorem 4.4 obtain similar results. In short, the α-symmetric implicational algorithm for situation of multiple rules is stable from the perspective of interval perturbation.

5 Summary and Prospect

In this study, the interval perturbation problem of the α-symmetric implicational algorithm is studied. Aiming at the α-symmetric implicational algorithm, for the FMP and

FMT problems with fuzzy reasoning, the corresponding upper and lower bound estimators for $R-$ implications and $(S, N)-$ implications are given for the input interval perturbation with one rule, and the corresponding stability is verified. Then, for the situation of multiple rules, the upper and lower bound estimators for the output are also given, and the corresponding stability is validated. To sum up, it is found that the α-symmetric implicational algorithm has good stability.

Furthermore, the interval perturbation problem will be studied for other kinds of symmetric implicational algorithms in the future.

In future research, we try to combine fuzzy reasoning and fuzzy clustering [29–31]. In detail, we can integrate the idea of fuzzy reasoning into the design of the objective function of the fuzzy clustering algorithm, so as to establish a new fuzzy clustering strategy.

Acknowledgment. This work has been supported by the National Natural Science Foundation of China (Nos. 62176083, 61673156, 62176084, 61877016, 61976078), the National Key Research and Development Program of China under Grant 2020YFC1523100, and the Fundamental Research Funds for the Central Universities of China (No. PA2021GDSK0092).

References

1. Li, P., Wang, X., Liang, H., et al.: A fuzzy semantic representation and reasoning model for multiple associative predicates in knowledge graph. Inf. Sci. **599**, 208–230 (2022)
2. Yang, X.Y., Yu, F.S., Pedrycz, W.: Long-term forecasting of time series based on linear fuzzy information granules and fuzzy inference system. Int. J. Approximate Reasoning **81**, 1–27 (2017)
3. Chen, S.M., Hsin, W.C.: Weighted fuzzy interpolative reasoning based on the slopes of fuzzy sets and particle swarm optimization techniques. IEEE Trans. Cybern. **45**(7), 1250–1261 (2015)
4. Tang, Y.M., Li, L., Liu, X.P.: State-of-the-art development of complex systems and their simulation methods. Complex Syst. Model. Simul. **1**(4), 271–290 (2021)
5. Chrysafiadi, K., Papadimitriou, S., Virvou, M.: Cognitive-based adaptive scenarios in educational games using fuzzy reasoning. Knowl.-Based Syst. **250**, 109111 (2022)
6. Zadeh, L.A.: Outline of a new approach to the analysis of complex systems and decision processes. IEEE Trans. Syst. Man Cybern. **3**(1), 28–44 (1973)
7. Wang, G.J.: On the logic foundation of fuzzy reasoning. Inf. Sci. **117**(1), 47–88 (1999)
8. Pei, D.W.: Unified full implication algorithms of fuzzy reasoning. Inf. Sci. **178**(2), 520–530 (2008)
9. Wang, G.J., Fu, L.: Unified forms of triple I method. Comput. Math. Appl. **49**(5–6), 923–932 (2005)
10. Pei, D.W.: Formalization of implication based fuzzy reasoning method. Int. J. Approximate Reasoning **53**(5), 837–846 (2012)
11. Zheng, M.C., Liu, Y.: Multiple-rules reasoning based on triple I method on Atanassov's intuitionistic fuzzy sets. Int. J. Approximate Reasoning **113**, 196–206 (2019)
12. Tang, Y.M., Ren, F.J.: Universal triple I method for fuzzy reasoning and fuzzy controller. Iran. J. Fuzzy Syst. **10**(5), 1–24 (2013)
13. Tang, Y.M., Ren, F.J.: Variable differently implicational algorithm of fuzzy inference. J. Intell. Fuzzy Syst. **28**(4), 1885–1897 (2015)

14. Tang, Y.M., Ren, F.J.: Fuzzy systems based on universal triple I method and their response functions. Int. J. Inf. Technol. Decis. Mak. **16**(2), 443–471 (2017)
15. Tang, Y.M., Yang, X.Z.: Symmetric implicational method of fuzzy reasoning. Int. J. Approximate Reasoning **54**(8), 1034–1048 (2013)
16. Tang, Y.M., Pedrycz, W.: On the α(u, v)-symmetric implicational method for R- and (S, N)-implications. Int. J. Approximate Reasoning **92**, 212–231 (2018)
17. Tang, Y.M., Pedrycz, W., Ren, F.J.: Granular symmetric implicational method. IEEE Trans. Emerging Top. Comput. Intell. **6**(3), 710–723 (2022)
18. Dai, S.: Logical foundation of symmetric implicational methods for fuzzy reasoning. J. Intell. Fuzzy Syst. **39**(1), 1089–1095 (2020)
19. Pappis, C.P.: Value approximation of fuzzy systems variables. Fuzzy Sets Syst. **39**(1), 111–115 (1991)
20. Hong, D.H., Hwang, S.Y.: A note on the value similarity of fuzzy systems variables. Fuzzy Sets Syst. **66**(3), 383–386 (1994)
21. Ying, M.S.: Perturbation of fuzzy reasoning. IEEE Trans. Fuzzy Syst. **7**(5), 625–629 (1999)
22. Cai, K.Y.: δ-equalities of fuzzy sets. Fuzzy Sets Syst. **76**(1), 97–112 (1995)
23. Cheng, G.S., Yu, Y.X.: Error estimation of perturbations under CRI. IEEE Trans. Fuzzy Syst. **14**(6), 709–715 (2006)
24. Tang, Y.M., Pedrycz, W.: Oscillation bound estimation of perturbations under Bandler-Kohout subproduct. IEEE Trans. Cybern. **52**(7), 6269–6282 (2022)
25. Klement, E.P., Mesiar, R., Pap, E.: Triangular Norms. Kluwer Academic Publishers, Dordrecht (2000)
26. Mas, M., Monserrat, M., Torrens, J., Trillas, E.: A survey on fuzzy implication functions. IEEE Trans. Fuzzy Syst. **15**(6), 1107–1121 (2007)
27. Fodor, J., Roubens, M.: Fuzzy Preference Modeling and Multicriteria Decision Support. Kluwer Academic Publishers, Boston (1994)
28. Baczynski, M., Jayaram, B.: On the characterizations of (S, N)-implications. Fuzzy Sets Syst. **158**(15), 1713–1727 (2007)
29. Tang, Y.M., Hu, X.H., Pedrycz, W., Song, X.C.: Possibilistic fuzzy clustering with high-density viewpoint. Neurocomputing **329**, 407–423 (2019)
30. Tang, Y.M., Ren, F.J., Pedrycz, W.: Fuzzy c-means clustering through SSIM and patch for image segmentation. Appl. Soft Comput. **87**, 1–16 (2020)
31. Dang, B.Z., Wang, Y.X., Zhou, J., et al.: Transfer collaborative fuzzy clustering in distributed peer-to-peer networks. IEEE Trans. Fuzzy Syst. **30**(2), 500–514 (2022)

Fuzzy Logic

(α, β) - Colored Resolution Method of Linguistic Truth-Valued Intuitionistic Fuzzy Logic

Tie Hou[1], Nan Li[2], Yanfang Wang[3], Kuo Pang[4], and Li Zou[1](✉)

[1] School of Computer Science and Technology, Shandong Jianzhu University, Jinan 250102, China
zoulicn@163.com
[2] School of Computing and Mathematics, Ulster University, Northern Ireland BT456DB, UK
[3] School of Information Engineering, Liaoyang Vocational College of Technology, Liaoyang 111000, China
[4] Information Science and Technology College, Dalian Maritime University, Dalian 116026, China

Abstract. To improving the (α, β)-resolution efficiency for linguistic truth-valued intuitionistic fuzzy propositional logic (LTV-IFPL), colored resolution is applied to (α, β)-resolution. By setting a threshold (α, β), the set of clauses is divided into two types, the (α, β)-unsatisfiable clauses are colored as clash electrons, uncolored clauses as clash core, which requires resolution between the two sets, and provides the ordering of the generalized literals. The (α, β)-colored resolution method is introduced into the linguistic truth-valued intuitionistic fuzzy logic system, and its completeness and reliability are proved. Finally, we provide the steps of the automatic reasoning algorithm based on (α, β)-colored resolution of the linguistic truth-valued intuitionistic fuzzy logic system, and illustrate the effectiveness of the algorithm by examples.

Keywords: Automatic reasoning · LTV-IFPL · (α, β)-Colored resolution

1 Introduction

Automatic reasoning based on resolution serves as a pivotal vehicle to achieve computational reasoning for intelligent or complex systems. Due to the large amount of fuzzy and uncertain information in real life, it is a significant challenge to extend resolution principles to effectively mimic human reasoning in imprecise and uncertain environments. Fuzzy set theory proposed by Zadeh [1] is an alternative to traditional two-valued logic for describing situations containing imprecise and vague data. However, people tend to have positive and negative attitudes when evaluating. For example, the evaluation of a student by his/her lecturer could be "good but not exceptional", in which case fuzzy set theory cannot capture the complete information well, resulting in information loss. In order to overcome this drawback and preserve as much full information as possible, Atanassov [2] proposes intuitionistic fuzzy set theory, which can express people's thoughts more accurately. Liu et al. [3] propose intuitionistic fuzzy logic on the basis of

Y. Chen and S. Zhang (Eds.): AILA 2022, CCIS 1657, pp. 79–90, 2022.
https://doi.org/10.1007/978-981-19-7510-3_6

the attribution principle of fuzzy logic, using two real numbers from 0 to 1 to represent the "degree of truth" and "degree of falsity" of a proposition. On this basis, Zou et al. [4] introduce the concepts of clause (α, β)-satisfiability and (α, β)-resolution formula, argue the satisfiability of generalized clauses and their resolution formula, propose a method for (α, β)-generalized lock resolution based on intuitionistic fuzzy logic and give the proof of its soundness and completeness. Xu et al. conquer the challenge of generalized literal α-resolution in LF(X) by transforming to LP(X), in order to support α-resolution-based automated reasoning algorithms [5]. Recently, linguistic approaches are widely studied and applied to address linguistic assessment issues, such as personnel management [6], web information processing [7], multi-criteria decision making [8], fuzzy risk analysis [9, 10] and so on. He et al. introduce a method of α-lock resolution for lattice-valued propositional logic LP(X) and prove its weak completeness [11]. To deal with linguistic truth-valued uncertainty reasoning, a linguistic truth-valued propositional logic framework is proposed by Zou et al. and a reasoning method for linguistic truth-valued logic system is developed [12, 13].

If people want to make machines more intelligent, computers must necessarily simulate the ability of people to process linguistic information, and people usually use natural language to reason and make decisions, such as, "somewhat good", "fair to poor", "very good", etc. We usually consider "very good" to be better than "very good" and "very good" to be better than "average". In many cases, linguistic values are not comparable. For example, "somewhat good" and "generally bad" are not comparable and cannot be defined simply as good or bad. In the past, linguistic values have been usually converted into numerical values for calculation, but linguistic values are fuzzy while numerical values are precise, and there is information missing in the conversion process. Therefore, we apply a linguistic truth-valued intuitionistic fuzzy lattice [14], which enables computers to reason directly about linguistic values, effectively avoiding the missing information caused when converting linguistic values into numerical values for calculation, and at the same time solving the problem of incomparable information about linguistic values due to the characteristics of the lattice. The outputs of this paper would provide a valuable flexibility for uncertainty reasoning in linguistic-valued intuitionistic fuzzy logic, which benefits the development of artificial intelligence.

The approximate structure of this paper is as follows. Section 2 briefly reviews the (α, β)-reduction principle of LTV-IFPL. Section 3 presents the (α, β)-colored resolution method for LTV-IFPL and proves the completeness and reliability of the proposed method. Section 4 provides the steps of the automatic inference algorithm based on (α, β)-colored resolution for linguistic truth-valued intuitionistic fuzzy logic systems and illustrates the effectiveness of the algorithm with examples. Conclusions and future work are described in Sect. 5.

2 Preliminaries

Definition 2.1. [15] Let $L = (L, \wedge, \vee, O, I)$ be a bounded lattice with inverse order pairwise union "$'$" and I and O be the largest and smallest elements of L. If

(1) $x \to (y \to z) = y \to (x \to z)$;

(2) $x \to x = I$;
(3) $x \to y = y' \to x'$;
(4) if $x \to y = y \to x$, then $x = y$;
(5) $(x \to y) \to y = (y \to x) \to x$;
(6) $(x \vee y) \to z = (x \to z) \vee (y \to z)$;
(7) $(x \wedge y) \to z = (x \to z) \wedge (y \to z)$;

Then $L = (L, \wedge, \vee, O, I)$ is the lattice implication algebra.

Definition 2.2. [15] $AD_n = \{h_1, h_2, \cdots, h_n\}$ is the set of n tone operators, $h_1 < h_2 < \cdots < h_n$, $M_T = \{t, f\}$ the set representing the meta-linguistic truth values and $c_1 < c_2$, defining $L_{v(n\times 2)} = AD_n \times M_T$ and defining the image g as follows:

$$g((h_i, M_T)) = \begin{cases} (d_i', b_1), M_T = c_1 \\ (d_i', b_2), M_T = c_2 \end{cases}$$

Then g is a bijection whose inverse map is denoted g^{-1}, and for any $x, y \in L_{v(n\times 2)}$, define:

$$(x \vee y) = g^{-1}(g(x) \vee g(y));$$

$$(x \wedge y) = g^{-1}(g(x) \wedge g(y));$$

$$x' = g^{-1}(g(x))';$$

$$(x \to y) = g^{-1}(g(x) \to g(y));$$

Then $L_{v(n\times 2)} = \left(L_{v(n\times 2)}, \wedge, \vee, ', (h_n, f), (h_n, t)\right)$ is called the linguistic truth-valued lattice implication algebra generated with AD_n and M_T. Its Hasse diagram is shown in Fig. 1.

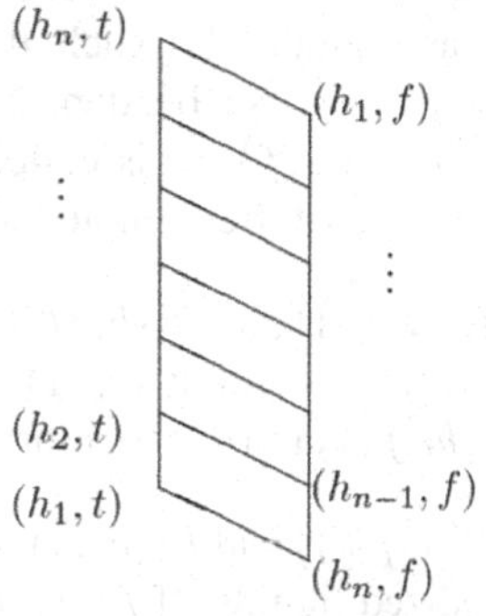

Fig. 1. Hasse Diagrams of $L_{v(n\times 2)}$

Definition 2.3. [15] In $2n$-element linguistic truth-valued implication algebra $L_{V(n\times 2)} = \{(h_i, t), (h_j, f) | h_i \in L_n, t, f \in L_2\}$, for any $(h_i, t), (h_j, f) \in L_{V(n\times 2)}$, call $((h_i, t), (h_j, f))$ a linguistic truth-valued intuitionistic fuzzy pair and $S=\{(h_i, t), (h_j, f) | i, j \in \{1, 2, \ldots, n\}\}$ a set of $2n$-element linguistic truth-valued intuitionistic fuzzy pairs. If $((h_i, t), (h_j, f))$ satisfies $(h_i, t)'(h_j, f)$, where the "$'$" operation is an inverse-order pair ensemble in $L_{V(n\times 2)}$.

Let the set of tone words $L_5 = \{h_i | i = 1, 2, 3, 4, 5\}$, where the tone words h_1 denote "slightly", h_2 denote "somewhat", h_3 denote "generally", h_4 denote "very", h_5 denotes "very", $h_1 < h_2 < h_3 < h_4 < h_5$, and the elementary linguistic truth value $\{t, f\}$, wo can obtain a ten-element linguistic truth value intuitionistic fuzzy lattice LI_{10}, its Hasse diagram is shown in Fig. 2.

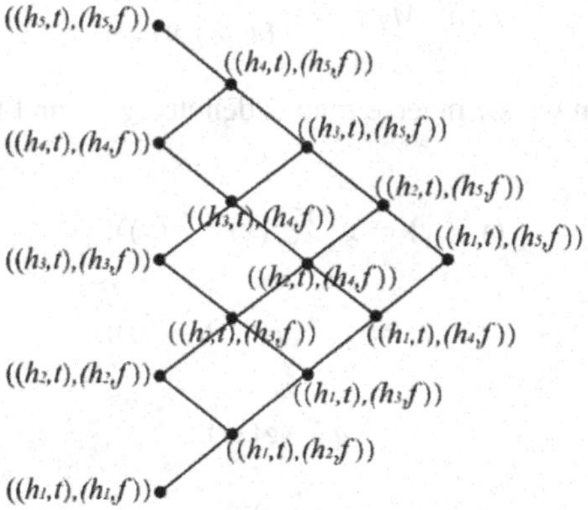

Fig. 2. Structure diagrams of 10-element linguistic truth-valued intuitionistic fuzzy lattice LI_{2n}

We consider linguistic truth-valued intuitionistic fuzzy inference with intermediate elements $((h_{(n+1)/2}, t), (h_{(n+1)/2}, f))$ (n of which are odd).

Definition 2.4. [15] Let G be a linguistic truth-valued intuitionistic fuzzy generalized clause. If there exists an assignment V, such that $V(G) = ((h_i, t), (h_j, f))$ satisfies $(h_i, t) \geq \alpha$, $(h_j, f) \leq \beta$, and can be denoted as $V(G) \geq (\alpha, \beta)$, where $\alpha \geq (h_{(n+1)/2}, \cdots t)$ and $\beta \leq (h_{(n+1)/2}, f)$, G is called (α, β)-satisfiable, otherwise G is called (α, β)-unsatisfiable or (α, β)-false, denoted as $V(G) < (\alpha, \beta)$.

Definition 2.5. [15] Let $P((h_i, t), (h_j, f))$ and $P((h_k, t), (h_l, f))$ be two LTV-IFPL generalized literals. If $(h_i, t) \geq \alpha$, $(h_j, f) \leq \beta$,but $(h_k, t) \leq \alpha$,$(h_l, f) \geq \beta$,then we call $P((h_i, t), (h_j, f))$ and $P((h_k, t), (h_l, f))$ are (α, β)-complementary literals.

Definition 2.6. [15] Let $P((h_i, t), (h_j, f))$ and $P((h_k, t), (h_l, f))$ be two linguistic truth-valued intuitionistic fuzzy generalized literals. If $P((h_i, t), (h_j, f))$ is (α, β)-satisfiable and $P((h_k, t), (h_l, f))$ is (α, β)-satisfiable, or $P((h_i, t), (h_j, f))$ is (α, β)-unsatisfiable and $P((h_k, t), (h_l, f))$ is (α, β)-unsatisfiable, then we say that $P((h_i, t), (h_j, f))$ and $P((h_k, t), (h_l, f))$ are (α, β)-similar literals.

Definition 2.7. [15] Let A and B be generalized clauses as follows:

$$A = A_1 \vee \cdots \vee A_i \vee \cdots \vee A_n$$

$$B = B_1 \vee \cdots \vee B_j \vee \cdots \vee B_n$$

where A_i and B_j are generalized clauses. If A_i and B_j are (α, β)-complementary literals, then $R = A_1 \vee \cdots \vee A_{i-1} \vee A_{i+1} \cdots \vee A_n \vee B_1 \vee \cdots \vee B_{j-1} \vee B_{j+1} \vee \cdots \vee B_n$ is called the (α, β)- resolution form of A and B, denoted as $R = R(A, B)$.

Theorem 2.1. Let a, b be linguistic truth-valued intuitionistic fuzzy logic propositions, then the following definitions and operations are commonly used in linguistic truth-valued intuitionistic fuzzy logic.

(1) Negative operation

$$V(\sim P) = \left(\left(h_{n-j-1}, t\right), \left(h_{n-i-1}, f\right)\right)$$

(2) Combining operation

$$V(P \wedge Q) = \left(\left(h_{\min(i,k)}, t\right), \left(h_{\min(j,l)}, f\right)\right)$$

(3) Analytic operation

$$V(P \vee Q) = \left(\left(h_{\max(i,k)}, t\right), \left(h_{\max(j,l)}, f\right)\right)$$

(4) Implication operation

$$V(P \to Q) = \left(\left(h_{\min(n, n-i+k, n-j+l)}, t\right), \left(h_{\min(n, n-i+l)}, f\right)\right)$$

For the three LTV-IFPL formulas, $F\left((h_i, t), \left(h_j, f\right)\right), G((h_k, t), (h_l, f))$, $H((h_m, t), (h_n, f)))$, the following equivalence properties hold:

(1) Idempotent law

$$F\left((h_i, t), \left(h_j, f\right)\right) \vee F\left((h_i, t), \left(h_j, f\right)\right) = F\left((h_i, t), \left(h_j, f\right)\right)$$

(2) Commutative law

$$F\left((h_i, t), \left(h_j, f\right)\right) \vee G((h_k, t), (h_l, f)) = G((h_k, t), (h_l, f)) \vee F\left((h_i, t), \left(h_j, f\right)\right)$$
$$F\left((h_i, t), \left(h_j, f\right)\right) \wedge G((h_k, t), (h_l, f)) = G((h_k, t), (h_l, f)) \wedge F\left((h_i, t), \left(h_j, f\right)\right)$$

(3) Associative law

$$F\left((h_i, t), \left(h_j, f\right)\right) \vee (G((h_k, t), (h_l, f)) \vee H((h_m, t), (h_n, f)))$$
$$= \left(F\left((h_i, t), \left(h_j, f\right)\right) \vee G((h_k, t), (h_l, f))\right) \vee H((h_m, t), (h_n, f))$$
$$F\left((h_i, t), \left(h_j, f\right)\right) \wedge (G((h_k, t), (h_l, f)) \wedge H((h_m, t), (h_n, f)))$$
$$= \left(F\left((h_i, t), \left(h_j, f\right)\right) \wedge G((h_k, t), (h_l, f))\right) \wedge H((h_m, t), (h_n, f))$$

(4) Absorption law

$$F\big((h_i,t),(h_j,f)\big)\vee\big(F\big((h_i,t),(h_j,f)\big)\wedge G((h_k,t),(h_l,f))\big)=F\big((h_i,t),(h_j,f)\big)$$
$$F\big((h_i,t),(h_j,f)\big)\wedge\big(F\big((h_i,t),(h_j,f)\big)\vee G((h_k,t),(h_l,f))\big)=F\big((h_i,t),(h_j,f)\big)$$

(5) Distribution law

$$F\big((h_i,t),(h_j,f)\big)\vee(G((h_k,t),(h_l,f))\wedge H((h_m,t),(h_n,f)))$$
$$=\big(F\big((h_i,t),(h_j,f)\big)\vee G((h_k,t),(h_l,f))\big)\wedge\big(F\big((h_i,t),(h_j,f)\big)\vee H((h_m,t),(h_n,f))\big)$$
$$F\big((h_i,t),(h_j,f)\big)\wedge(G((h_k,t),(h_l,f))\vee H((h_m,t),(h_n,f)))$$
$$=\big(F\big((h_i,t),(h_j,f)\big)\wedge G((h_k,t),(h_l,f))\big)\vee\big(F\big((h_i,t),(h_j,f)\big)\wedge H((h_m,t),(h_n,f))\big)$$

(6) De Morgan's law

$$\sim\big(F\big((h_i,t),(h_j,f)\big)\wedge G((h_k,t),(h_l,f))\big)$$
$$=\sim F\big((h_i,t),(h_j,f)\big)\vee\sim G((h_k,t),(h_l,f))$$
$$\sim\big(F\big((h_i,t),(h_j,f)\big)\vee G((h_k,t),(h_l,f))\big)$$
$$=\sim F\big((h_i,t),(h_j,f)\big)\vee\sim G((h_k,t),(h_l,f))$$

3 (α,β) -Colored Resolution of Linguistic Truth-Valued Intuitionistic Fuzzy Propositional Logic

Definition 3.1. We add a special symbol (put a horizontal bar on the literals) to the literals in the LTV-IFPL generalization clause, and these literals are called colored literals, and the LTV-IFPL generalized clause with colored literals is called a LTV-IFPL colored clause, and the set of colored text is denoted as $col(C)$, and the set of uncolored text is denoted as $uncol(C)$.

Definition 3.2. If C is a LTV-IFPL colored clause in which at least two literals have $mgu\sigma$, and it is satisfied iff $L((h_m,t),(h_n,f))\in col(C)$, $L^{\sigma}((h_m,t),(h_n,f))\in col(C^{\sigma})$ established, then C^{σ} is called the colored genetic factor of the LTV-IFPL colored clause.

Definition 3.3. If C is a l LTV-IFPL colored clause in which at least two literals have $mgu\sigma$, and it is satisfied iff $L((h_m,t),(h_n,f))\in uncol(C)$, $L^{\sigma}((h_m,t),(h_n,f))\in uncol(C^{\sigma})$ established, then C^{σ} is called the uncolored genetic factor of the LTV-IFPL colored clause.

Definition 3.4. Let C_1 and C_2 be LTV-IFPL colored clauses without common variables, which $L_1\big((h_i,t),(h_j,f)\big)$ and $L_2((h_k,t),(h_l,f))$ are the literals in C_1 and C_2, respectively, they are (α,β)-complementary literals. If there exists a $mgu\sigma$ between $L_1\big((h_i,t),(h_j,f)\big)$ and $\sim L_2((h_k,t),(h_l,f))$, the formula of (α,β)-resolution $Res_J(C_1,C_2)$ is called the formula of (α,β)-colored resolution. The literals colored in the formula of (α,β)-colored resolution is: if there is $L((h_m,t),(h_n,f))\big(uncol\big(C_1^{\sigma}\big)-L_1\big((h_i,t),(h_j,f)\big)\big)$ or $L((h_m,t),(h_n,f))\in\big(uncol\big(C_2^{\sigma}\big)-L_2((h_k,t),(h_l,f))\in\big(uncol\big(C_2^{\sigma}\big)-L_2((h_k,t),$

$(h_l,f)) \wedge L((h_m,t),(h_n,f)) \notin \left(col\left(C_1^{\sigma}\right) - L_1^{\sigma}((h_i,t),(h_j,f)\right) \wedge L((h_m,t),(h_n,f)) \notin \wedge L((h_m,t),(h_n,f)) \notin L_2^{\sigma}((h_k,t),(h_l,f))$, then $L((h_m,t),(h_n,f))$ is a LTV-IFPL colored literal and the other literals are called l LTV-IFPL uncolored literals.

Definition 3.5. LTV-IFPL colored clause C_1 for C_2 the formula of (α,β)-colored resolution are C_1 or C_1 genetic factor for C_2 or C_2 genetic factor's the formula of (α,β)-colored resolution.

Specifies that the designation of a coloring literal is (α,β)-unsatisfiable.

Definition 3.6. Let V be an assignment of the set of clauses S in LTV-IFPL, a finite sequence of clauses $(E_1,\cdots E_q,N), q \geq 1$ and P is an ordering of all generalized literals, called (α,β)-colored clash of V iff $E_1,\cdots E_q,N$ satisfied the following requirements:

(1) $V(E_i) < (\alpha,\beta)$, where $\alpha \geq (h_{(n+1)/2},t)$ and $\beta \leq ((h_{(n+1)/2},f), i = 1,\cdots q$, that is, all the characters in E_i are colored literals.
(2) Let $R_1 = N$, for every $i = 1,\cdots q$, there exists the formula of (α,β)-colored resolution R_{i+1} of R_i for E_i.
(3) The (α,β)- resolution literal of E_i is the largest generalized literal in E_i.
(4) $V(R_{q+1}) < (\alpha,\beta)$, then R_{q+1} is (α,β)-unsatisfiable.

R_{q+1} is called the formula of (α,β)-colored resolution of (α,β)-colored clash, $E_1,\cdots E_q$ is called electrons of (α,β)-colored clash and N is called cores of (α,β)-colored clash.

Definition 3.7. Let C be a colored clause of LTV-IFPL and C' a colored example of C. If $C' = C^{\theta}$, and the colored literal in is specified as $L'((h_i,t),(h_j,t)) \in uncol(C')$, iff there exists $L((h_i,t),(h_j,t)) \in uncol(C)$ and $L'((h_i,t),(h_j,t)) = L^{\theta}((h_i,t),(h_j,t))$.

Definition 3.8. Let V be an assignment of the clause set of LTV-IFPL, a finite sequence of clauses $(E_1,\cdots E_q,N), q \geq 1$, and P be an ordering of all generalized literals. The resolution deduction from S is called (α,β)-colored resolution deduction iff any clauses in the deduction is the formula of (α,β)-colored resolution or a clause in S.

Theorem 3.1. (Reliability). Let V be an assignment of the clause set S of LTV-IFPL, a finite sequence of clauses $(E_1,\cdots E_q,N), q \geq 1$, and P be an ordering of all generalized literals, the clause set S of LTV-IFPL is (α,β)-unsatisfiable if there exists an (α,β)-colored resolution deduction from S to $(\alpha,\beta)-\square$.

Proof. We use the converse method and assume that there exists an (α,β)-colored resolution deduction from S to $(\alpha,\beta)-\square$, and the clause set S of LTV-IFPL is (α,β)-satisfiable. Then, we can obtain that there exist two resolution unit clauses $L_m((h_{i_m},t),(h_{j_m},f))$ and $L_n((h_{i_n},t),((h_{j_n},f))$, such that $V(L_m \wedge L_n) = ((h_{\min(i_m,i_n)},t),(h_{\min(j_m,j_n)},f)) < (\alpha,\beta)$, and that $L_m((h_{i_m},t),(h_{j_m},f))$ and $L_n((h_{i_n},t),(h_{j_n},f))$ are (α,β)-satisfiable, and we know that the formula of (α,β)-resolution is also (α,β)-satisfiable, contradicting the assumption, and the theorem is proved.

Theorem 3.2. (Completeness). Let the clause set S of LTV-IFPL be the set of (α, β)-unsatisfiable clauses, V be an assignment of the clause set S of LTV-IFPL, a finite sequence of clauses $\left(E_1, \cdots E_q, N\right)$,$q \geq 1$, and P be an ordering of all generalized literals, then there exists an (α, β)-colored resolution deduction from S to $(\alpha, \beta)-\square$.

Proof. We use induction method and set the number of distinct literals appearing in the LTV-IFPL clause set to $|O|$.

When $|O|$=1, since the LTV-IFPL clause set S is an unsatisfiable clauses set, there is $P \in S, \sim P \in S$, then there exists an (α, β)-colored resolution deduction from S to $(\alpha, \beta)-\square$.

When $|O| \leq n$, the theorem holds.

When $|O|$=n+1, discussed in two cases:

(1) There is a unit clause $K\left((h_i, t), \left(h_j, t\right)\right)$ in S, assign the $K\left((h_i, t), \left(h_j, t\right)\right)$ value to (α, β)-unsatisfiable, then removing all clauses containing $K\left((h_i, t), \left(h_j, t\right)\right)$ from S, and removing $\sim K\left((h_i, t), \left(h_j, t\right)\right)$ from the remaining clauses, we obtain the set of clauses S'. It is obvious that S' is (α, β)-unsatisfiable and the number of atoms is less than n in S', by the induction hypothesis there exists an (α, β)-colored resolution deduction D' that introduces $(\alpha, \beta)-\square$ from S'. Transform D' a bit: for every (α, β)-colored clash $\left\{E_1', \cdots E_q', N'\right\}$ in D', and $E_1', \cdots E_q', N'$ is the initial clause. If N' is obtained by removing $\sim K\left((h_i, t), \left(h_j, t\right)\right)$ from N, then we can replace $\left\{E_1', \cdots E_q', N\right\}$ by constituting a new (α, β)-colored clash; If E_i' is obtained by removing $\sim K\left((h_i, t), \left(h_j, t\right)\right)$ from E_i, then we can replace E_i' by a (α, β)-colored clash $\{K, E_i\}$. Others remain the same, and the modified deduction D can also introduce $(\alpha, \beta)-\square$.

(2) There is no unit clause that be assigned to (α, β)-unsatisfiable in S. Taking the smallest literal $J\left((h_i, t), \left(h_j, t\right)\right)$, set $L((h_m, t), (h_n, t))$ to an (α, β)-unsatisfiable clause in both $J\left((h_i, t), \left(h_j, t\right)\right)$ and $\sim J\left((h_i, t), \left(h_j, t\right)\right)$. By removing $\sim L((h_m, t), (h_n, t))$ from S and removing $L((h_m, t), (h_n, t))$ from the remaining literals, we can obtain the set of clauses S'. It is obvious that S' is (α, β)-unsatisfiable and the number of all atoms in S' is less than n. By induction it is assumed that there exists an (α, β)- colored resolution deduction from S' to D'. By restoring the clauses of S in D' to the clause of S, we can get an (α, β)-colored resolution deduction S from D_1 and can launch $(\alpha, \beta)-\square$ or launch $L((h_m, t), (h_n, t))$. If we can launch $(\alpha, \beta)-\square$, the theorem is proved; if we launch $L((h_m, t), (h_n, f))$, the set of clauses is updated to $S \cup \{L((h_m, t), (h_n, t))\}$, because $L((h_m, t), (h_n, t))$ is an (α, β)-unsatisfiable unit clause, so there exists an (α, β)-colored resolution deduction D_2 from $S \cup \{L((h_m, t), (h_n, t))\}$. Connecting D_1 and D_2, we can get an (α, β)-colored resolution deduction from S to $(\alpha, \beta)-\square$.

The induction method is completed and the theorem is proved. There exists an (α, β)-colored resolution deduction from S to $(\alpha, \beta)-\square$.

4 (α, β)-Colored Resolution Algorithms of Linguistic Truth-Valued Intuitionistic Fuzzy Propositional Logic

We present the (α, β)-colored resolution algorithm for LTV-IFPL as follows:

(1) Input known conditions;
(2) We convert the known conditions into generalized clauses of the LTV-IFPL to obtain the set of generalized clauses S of the LTV-IFPL;
(3) We sort all generalized literals in the set of generalized clauses S of LTV-IFPL;
(4) If there exists $(\alpha, \beta)-\square$, then the theorem is proved, end. Otherwise go to (5);
(5) We determine whether the generalized clause of LTV-IFPL is (α, β)-unsatisfiable, and if so, store the clause in S_1, add a horizontal bar to the literals in the clauses, i.e., colored, and treat the clause as (α, β)-colored clash electrons; otherwise store the clause in S_2, as (α, β)-colored clash core;
(6) We perform (α, β)- colored resolution using Definition 3.6 to obtain the formula R_q of (α, β)-colored resolution of LTV-IFPL, go to (5).

The flow chart is shown in Fig. 3.

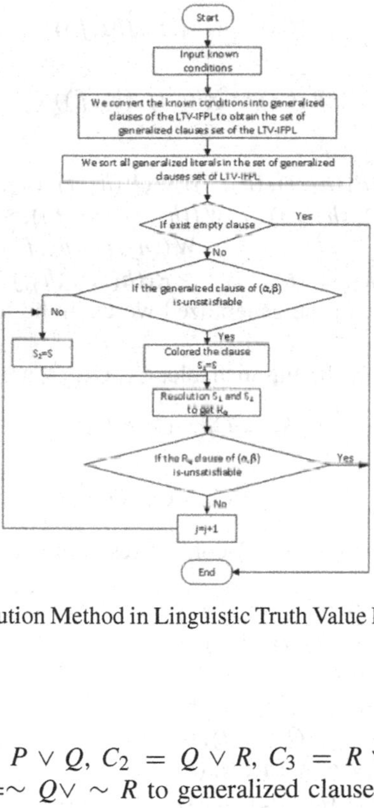

Fig. 3 (α, β)-Colored Resolution Method in Linguistic Truth Value Intuitive Fuzzy Propositional Logic

Example 4.1. Set $C_1 = P \vee Q$, $C_2 = Q \vee R$, $C_3 = R \vee W$, $C_4 =\sim R\vee \sim P$, $C_5 =\sim W\vee \sim Q$, $C_6 =\sim Q\vee \sim R$ to generalized clauses of LTV-IFPL, and $S = C_1 \wedge C_2 \wedge C_3 \wedge C_4 \wedge C_5 \wedge C_6$, V to an assignment of the clause set S of LTV-IFPL such that:

$$V(P) = ((h_1, t), (h_2, f)),$$

$$V(Q) = ((h_2, t), (h_2, f)),$$

$$V(R) = ((h_1, t), (h_3, f)),$$

$$V(W) = ((h_2, t), (h_3, f))$$

Taking (α, β)=$((h_3, t), (h_4, t))$, then we have,

$$V(C_1) = ((h_2, t), (h_2, f))$$

$$V(C_2) = ((h_2, t), (h_3, f))$$

$$V(C_3) = ((h_2, t), (h_3, f))$$

$$V(C_4) = ((h_4, t), (h_5, f))$$

$$V(C_5) = ((h_4, t), (h_4, f))$$

$$V(C_6) = ((h_4, t), (h_5, f))$$

The clause set of LTV-IFPL $S = \{P((h_1, t), (h_2, f)) \vee Q((h_2, t), (h_2, f)), Q((h_2, t), (h_2, f)) \vee R((h_1, t), (h_3, f)), R((h_1, t), (h_3, f)) \vee W((h_2, t), (h_3, f)), \sim R((h_1, t), (h_3, f)) \vee \sim P((h_1, t), (h_2, f)), \sim W((h_2, t), (h_3, f)) \vee \sim Q((h_2, t), (h_2, f)), \sim Q((h_2, t), (h_2, f)) \vee \sim R((h_1, t), (h_3, f))\}$.

We specify an ordering of the generalized words in the set of clauses as $P > Q > W > R$.

Where C_1, C_2, C_3 are (α, β)-unsatisfiable, C_4, C_5, C_6 are (α, β)-satisfiable, then

$$S_1 = C_1 \wedge C_2 \wedge C_3$$

$$S_2 = C_4 \wedge C_5 \wedge C_6$$

We finish coloring S_1 and its (α, β)-colored resolution as follow:

(1) $P((h_1, t), (h_2, f)) \vee Q((h_2, t), (h_2, f))$
(2) $Q((h_2, t), (h_2, f)) \vee R((h_1, t), (h_3, f))$
(3) $R((h_1, t), (h_3, f)) \vee W((h_2, t), (h_3, f))$
(4) $\sim R((h_1, t), (h_3, f)) \vee \sim P((h_1, t), (h_2, f))$
(5) $\sim W((h_2, t), (h_3, f)) \vee \sim Q((h_2, t), (h_2, f))$
(6) $\sim Q((h_2, t), (h_2, f)) \vee \sim R((h_1, t), (h_3, f))$
(7) $P((h_1, t), (h_2, f)) \vee Q((h_2, t), (h_2, f))$ (1) (2) (6) resolution
(8) $Q((h_2, t), (h_2, f))$ (2)(7)(4) resolution
(9) $R((h_1, t), (h_3, f))$ (3)(8)(5) resolution
(10) $(\alpha, \beta)-\square$ (8)(9)(6) resolution

There exists an (α, β)-colored resolution deduction from S to $(\alpha, \beta)-\square$ with assignment V, and the conclusion holds.

5 Conclusion

We have proposed the (α, β)-colored resolution method of LTV-IFPL system, which divides the LTV-IFPL generalized literals into two types, namely, colored literals and uncolored literals, and provided the ordering between the literals, requiring the resolution literals to be the maximum literal symbol. The more restrictions on the resolution, the higher the efficiency of the resolution. Our proposed (α, β)-colored resolution method restricts not only the clauses, but also the resolution literals, which avoids the generation of some redundant resolution clauses, thus significantly improving the resolution efficiency. As future work we intend to introduce this resolution method to linguistic truth-valued intuitionistic fuzzy first-order logic.

Acknowledgements. This work is partially supported by the National Natural Science Foundation of P.R. China (Nos. 61772250, 62176142), Foundation of Liaoning Educational Committee (No. LJ2020007) and Special Foundation for Distinguished Professors of Shandong Jianzhu University.

References

1. Zadeh, L.A.: Fuzzy sets. Information and Control **8**(3), 338–353 (1965)
2. Atanassov, K.: Intuitionistic fuzzy sets. Fuzzy Sets and Systems **20**(1), 87–96 (1986)
3. Liu, X., Sun, F., Zou, L.: Soft-Resolution Method of Six-Element Linguistic Truth-Valued Intuitionistic Fuzzy Propositional Logic. Journal of Donghua University (English Edition) **27**(02), 135–138 (2010)
4. Zou, L., Liu, D., Zheng, H.: (α, β)-Generalized Lock Resolution of Intuitionistic Fuzzy Logic. Journal of Frontiers of Computer Science and Technology (2015)
5. Xu, Y., Li, X., Liu, J., Ruan, D.: Determination of alpha-resolution for lattice-valued first-order logic based on lattice implication algebra. In: Proc. 2007 International Conference on Intelligent Systems and Knowledge Engineering, pp. 1638–1645 (2007)
6. Pei, Z., Xu, Y., Ruan, D., Qin, K.: Extracting complex linguistic data summaries from personnel database via simple linguistic aggregations. Information Sciences **179**, 2325–2332 (2009)
7. Pei, Z., Ruan, D., Xu, Y., Liu, J.: Handling linguistic web information based on a Multi-agent system. Int. J. Intelligent Systems **22**, 435–453 (2007)
8. Chen, S.W., Liu, J., Wang, H., Xu, Y., Augusto, J.C.: A linguistic multi-criteria decision making approach based on logical reasoning. Information Sciences **258**, 266–276 (2013)
9. Pei, Z.: Fuzzy risk analysis based on linguistic information fusion. ICIC Express Letters **3**(3), 325–330 (2009)
10. Pei, Z., Shi, P.: Fuzzy risk anal sis based on linguistic aggregation operators. Int. J. Innovative Computing, Information and Control **7**(12), 7105–7118 (2011)
11. He, X., Xu, Y., Li, Y., Liu, J., Martinez, L., Ruan, D.: α-Satisfiability and α-Lock Resolution for a Lattice-Valued Logic LP(X). In: The 5th International Conference on Hybrid Artificial Intelligence Systems, pp. 320–327. San Sebastian, Spain (2010)
12. Zou, L., Ruan, D., Pei, Z., Xu, Y.: A linguistic truth-valued reasoning approach in decision making with incomparable information. Journal of Intelligent and Fuzzy Systems **19**(4–5), 335–343 (2008)
13. Zou, L., Liu, X., Wu, Z., Xu, Y.: A uniform approach of linguistic truth values in sensor evaluation. Int. J. Fuzzy Optimiz. Decision Making **7**(4), 387–397 (2008)

14. Zou, L., et al.: Linguistic-valued layered concept lattice and its rule extraction. International Journal of Machine Learning and Cybernetics, 1–16 (2021)
15. Zou, L.: Studies on Lattice-Valued Propositional Logic and its Resolution-Based Automatic Reasoning Based on Linguistic Truth - valued Lattice implication Algebra. Southwest Jiaotong University (2010)

Viewpoint-Driven Subspace Fuzzy C-Means Algorithm

Yiming Tang(✉), Rui Chen, and Bowen Xia

Affective Computing and Advanced Intelligent Machine, School of Computer and Information, Hefei University of Technology, Hefei 230601, Anhui, China
tym608@163.com

Abstract. Most of the current fuzzy clustering algorithms are sensitive to cluster initialization and do not cope well with high dimensionality. To alleviate these problems, we come up with a viewpoint-driven subspace fuzzy c-means (VSFCM) algorithm. First of all, based on the DPC (clustering by fast search and find of density peaks) algorithm, a new cut-off distance is proposed, and the cut-off distance-induced cluster initialization (CDCI) method is established as a new strategy for initialization of cluster centers and viewpoint selection. Moreover, by taking the viewpoint obtained by CDCI as the entry point of knowledge, a new fuzzy clustering strategy driven by knowledge and data is formed. We introduce the subspace clustering mode, fuzzy feature weight processing mechanism, and derive the separation formula between the clusters of viewpoint optimization. Based upon these points, we put forward the VSFCM algorithm. Finally, by comparing experiments and using multiple advanced clustering algorithms and experimenting with artificial and UCI data sets, it is demonstrated that the VSFCM algorithm has the best performance expressed in terms of five indexes.

Keywords: Fuzzy clustering · Fuzzy c-means · Separation between clusters · Viewpoint · Cluster center initialization

1 Introduction

Clustering algorithms have been widely studied in various fields such as pattern recognition, biology, engineering system and so forth [1–3]. In the early stage, hard clustering was mainly studied where every object strictly belonged to a single cluster. For example, the DPC algorithm [4] is a recent excellent representative.

In fuzzy clustering, the fuzzy c-means (FCM) algorithm was one of the most commonly used method [5, 6]. Then, the use of weighted processing had become an important direction, which included the weighted FCM (WFCM) algorithm [7], the feature weighted fuzzy k-means (FWFKM) algorithm [8], the attribute weight algorithm (AWA) [9], the fuzzy weighted k-means (FWKM) algorithm [10], and the fuzzy subspace clustering (FSC) algorithm [11, 12]

However, when clustering was completed in a high-dimensional space, the traditional clustering algorithms had shortcomings [13]. A key challenge was that in many real-world problems, data points in different clusters were often related to different feature

Y. Chen and S. Zhang (Eds.): AILA 2022, CCIS 1657, pp. 91–105, 2022.
https://doi.org/10.1007/978-981-19-7510-3_7

subsets, that was, clusters could exist in different subspaces. These subspaces were composed of different feature subsets [14]. The typical algorithms incorporated the simultaneous clustering and attribute discrimination (SCAD) algorithm [15], the enhanced soft subspace clustering (ESSC) [16], and the feature-reduction FCM (FRFCM) [17].

Clustering was mainly a data-driven optimization process. In fact, domain knowledge could be used to assist the development of clustering, e.g., the viewpoint-based FCM (V-FCM) algorithm [18] and the density viewpoint induced possibilistic FCM (DVPFCM) algorithm.

The current problems of fuzzy clustering algorithms are mainly as follows:

- Issues sensitive to cluster initialization

Most fuzzy clustering algorithms are sensitive to the initial results of clustering. For example, FCM, V-FCM, SCAD, ESSC, FRFCM are all sensitive to the initialization of the method. The better processing mechanism for this is the HDCCI initialization method given by the DVPFCM algorithm, and its idea is based on the DPC algorithm. We know that it is extremely important for the selection of the cut-off distance in the DPC algorithm, because it directly affects the accuracy of clustering. The cut-off distance calculation method in the DVPFCM algorithm uses a fixed and rigid formula, which lacks solid basis and cannot adapt to various data sets.

- Problems with weak adaptability to high dimensionality

Nowadays the amount of data information is getting larger and the dimensions are getting higher and higher, which require clustering algorithms to have certain requirements for the ability to process high-dimensional data. Most algorithms are still weak for it. FCM, V-FCM, and DVPFCM do not have any special measures to deal with high-dimensional data, and they appear to be unable to do so. SCAD, ESSC, and FRFCM all use subspace processing methods with different weights, which are relatively better, but these are purely data-driven, and the efficiency and accuracy of high-dimensional data clustering cannot reach the ideal level.

So far, in this study we have put forward the VSFCM algorithm. On the one hand, using the dual-driven pattern of data and knowledge, the cut-off distance of DPC is improved in terms of viewpoint selection, and the point with the highest density is selected as the viewpoint more accurately, and a cut-off distance-induced clustering initialization method CDCI is proposed. This provides a new initialization strategy and viewpoint. On the other hand, we introduce viewpoints into subspace clustering to improve the convergence speed of each subspace. And adding the separation part between clusters can minimize the compactness of the subspace clusters and maximize the projection subspace where each cluster is located. The fuzzy feature weight processing mode is introduced, and the VSFCM algorithm is established on the basis of CDCI and viewpoint.

2 Related Works

Assuming that the data set $X = \{x_j\}_{j=1}^{n}$ is a set of n samples, we cluster the data into c $(2 \leq c \leq n)$ classes, and produce a cluster center set $V = \{v_i\}_{i=1}^{c}$. Each sample x_j and the cluster center v_i are l dimensional data.

Frigui and Nasraoui [15] proposed the SCAD algorithm that simultaneously performed clustering and feature weighting. It used continuous feature weighting, so it provided a richer feature correlation representation than feature selection. Moreover, SCAD independently learned the feature association representation of each cluster in an unsupervised way. Later, Deng [16] studied the use of intra-class information and inter-class information, and proposed the ESSC algorithm, in which the intra-class compactness in the subspace was combined with separation between clusters. Yang and Nataliani [17] proposed the FRFCM algorithm, which automatically calculated the weight of a single feature while it reduced these unrelated feature components.

Pedrycz et al. [18] introduced knowledge into a data-driven process, where knowledge was embodied through viewpoints, thus giving the V-FCM algorithm. Tang et al. [19] proposed a new knowledge and data-driven fuzzy clustering algorithm, which is called the DVPFCM algorithm. Thereinto, a new calculation method of density radius is proposed, and a hypersphere density-based cluster center initialization (HDCCI) algorithm was established. This method could obtain the initial cluster centers in densely sampled areas. Then, the high-density points obtained by the HDCCI method were used as new viewpoints, and it was proposed to integrate them into the DVPFCM algorithm.

The comparison of related algorithms is summarized in Table 1. In a word, the existing algorithms still have big deficiencies in the initialization of cluster centers and the processing of spatial dimensions. For this reason, we focus on solving these two types of problems.

Table 1. Comparison of advantages and disadvantages of each algorithm

Algorithms	Advantages	Disadvantages
FCM	As the classic algorithm, it can automatically find cluster centers	Sensitive to cluster center initialization, poor noise immunity
V-FCM	Simplify the clustering process, fast convergence	Poor noise immunity, sensitive to cluster center initialization
DPC	Can quickly determine cluster centers	The density radius is difficult to determine, and there are often human errors
SCAD	Fuzzy weighted index is introduced to obtain better weight value	Sensitive to cluster center initialization

(continued)

Table 1. (*continued*)

Algorithms	Advantages	Disadvantages
ESSC	Taking into account the distance between classes and classes, using entropy information	Sensitive to cluster center initialization, more parameters need to be set manually
FRFCM	Select important features by weighting, and reduce feature dimension by discarding unimportant features	Sensitive to cluster center initialization, without considering the spatial characteristics of the data
DVPFCM	By new viewpoints and typical values, there is relatively stronger robustness; there is a better initialization strategy	Processing high-dimensional data appears weak, and the cut-off distance is not perfect

3 The Proposed VSFCM Algorithm

3.1 Cluster Initialization Method Induced by Cut-off Distance

Here is a new cluster center initialization method. The DPC algorithm is still used as the starting point, where the local density ρ_j of the sample and its minimum distance δ_j to other points with higher local density still use the previous calculation formulas, namely the following (1) and (3).

$$\rho_j = \sum f(d_{jk} - r), \tag{1}$$

$$f(x) = \begin{cases} 1, & x = d_{jk} - r < 0 \\ 0, & \text{other} \end{cases}, \tag{2}$$

$$\delta_j = \min\{d_{jk} | \rho_k > \rho_j, \quad k \in \{1, 2, \cdots, n\}\}. \tag{3}$$

Here d_{jk} is the distance between two data points and r is the density radius. The local density ρ reflects the number of data points within the radius r. It has been found that the cut-off distance is critical to this algorithm.

In this study, we put forward a new cut-off distance as follows:

$$cd = \min((cd_1 + cd_2)/2, cd_1). \tag{4}$$

Among them,

$$cd_1 = D_{position}, cd_2 = \frac{d_{\max}}{2c}. \tag{5}$$

The cut-off distance recommended by the DPC algorithm should let that the average number of neighbors of each data point is about 1%–2% of the total number of data. n is the number of data points and c is the number of clusters. d_{kj} is the distance from the data point x_k to x_j. There are $M = \frac{1}{2}n(n-1)$ distances. We sort d_{kj} from small to large, and we might as well write the resulting ordered sequence as D $(d_1 \leq d_2 \leq \ldots \leq d_{\max})$. Then we cut off according to the upper limit ratio of 2%, which can be taken as $position = round(2\% \times M)$. Hence we get $D_{position}$, namely cd_1.

Furthermore, we use the maximum distance $d_{\max}$ to divide the c categories, and take their radius, then the reference number is $\frac{d_{\max}}{2c}$, which is recorded as cd_2. Then the two are combined and $(cd_1 + cd_2)/2$ is used as another factor.

Taking into account the limited range of $1\% - 2\%$ of the total number of data, taking the smaller of the two factors, (4) is obtained. Then, naturally we use.

$$\rho_j = \sum f(d_{kj} - cd). \tag{6}$$

Next, we introduce parameters τ_j $(j = 1, \cdots, n)$ to calculate the initial cluster centers directly. The formula is as below:

$$\tau_j = \rho_j \times \delta_j \tag{7}$$

The proposed initialization method can automatically select a more appropriate cut-off radius, so that the selected initial cluster center is closer to the true value. The traditional DPC algorithm uses the $\rho - \delta$ distribution map to subjectively select the cluster centers, which is easy to cause human error. We calculate the parameters τ_j $(j = 1, \cdots, n)$ and sort them, and then select the largest number of data points τ_j (that are not within the same cut-off radius) as the initial cluster centers. This ensures that we can select the initial cluster centers and viewpoints conveniently, efficiently and relatively accurately.

The obtained method is called the cut-off distance-induced clustering initialization (CDCI) method.

3.2 The Mechanism of VSFCM Algorithm

Here we show the main idea of the viewpoint-driven subspace fuzzy C-means (VSFCM) algorithm.

The first cluster center selected from the CDCI method is recorded as x_e (i.e., the largest point τ_j), which is taken as the viewpoint. The position of our viewpoint is constantly changing with iteration. The row number of the viewpoint in the cluster center matrix is $q = arc(\min(d_{qd}))$. We replace the cluster center closest to the viewpoint as the viewpoint.

We use 3 parts to complete the construction of the objective function. The first part is to introduce fuzzy feature weights based on the conventional objective function of FCM, and integrate the original cluster centers into the viewpoint. The second part is an adaptive fuzzy weight penalty term, which uses a pilot parameter δ_i that can be automatically calculated. The third part is the separation item between clusters. Thereinto, fuzzy feature weights are used, and the cluster centers of the viewpoints are merged, and the centroid of the initialized cluster centers is used as the reference point for separation between clusters. The objective function is as follows:

$$J_{VSFCM} = \sum_{i=1}^{c}\sum_{j=1}^{n} u_{ij}^{m} \sum_{k=1}^{l} w_{ik}^{t} ||x_{jk} - h_{ik}||^2 + \\ \sum_{i=1}^{c} \delta_i \sum_{k=1}^{l} w_{ik}^{t} - \eta \sum_{i=1}^{c} (\sum_{j=1}^{n} u_{ij}^{m}) \sum_{k=1}^{l} w_{ik}^{\tau} ||v_{0k} - h_{ik}||^2. \tag{8}$$

Among them ($i = 1, ..., c$).

$$h_i = \begin{cases} v_i, \ i \neq q \\ x_e, \ i = q \end{cases}, v_0 = \sum_{i=1}^{c} v_i / c. \tag{9}$$

The following constraints are imposed ($i = 1, ..., c, j = 1, ..., n$).

$$\sum_{i=1}^{c} u_{ij} = 1, \quad \sum_{k=1}^{l} w_{ik} = 1. \tag{10}$$

Here h_i is the cluster center related to the viewpoint. When $i = q$, h_i is replaced with the point of maximum density or viewpoint. The reference point of our separation between clusters is the above (9). The fuzzy weight w_{ik}^{τ} is the weight for the k-th feature of the i-th category, in which the fuzzy coefficient τ is used, general settings $\tau > 1$. δ_i is the leading parameter used to implement fuzzy weight penalty. Parameter η is used to adaptively adjust the separation terms between clusters.

Note that (8) can be transformed into:

$$J = \sum_{i=1}^{c}\sum_{j=1}^{n} u_{ij}^{m} \sum_{k=1}^{l} w_{ik}^{\tau}[(x_{jk} - h_{ik})^2 - \eta(v_{0k} - h_{ik})^2] + \sum_{i=1}^{c} \delta_i \sum_{k=1}^{l} w_{ik}^{\tau}. \tag{11}$$

The solution process is given below. Using the Lagrangian multiplier method for (11), we get

$$u_{ij} = \frac{D_{ij}^{-\frac{1}{m-1}}}{\sum_{l=1}^{c} D_{lj}^{-\frac{1}{m-1}}}. \tag{12}$$

This results in the iterative formula of membership degree, which involves the value of η. Note that when η is very large, D_{ij} may become negative, which is obviously not what we want. For this reason, we can naturally give the following qualifications ($i = 1, ..., c, j = 1, ..., n, k = 1, ..., l$):

$$(x_{jk} - h_{ik})^2 - \eta(v_{0k} - h_{ik})^2 \geq 0. \quad (13)$$

Therefore, it is natural to make.

$$\eta = \alpha_0 \min_{i,j,k} \frac{(x_{jk} - h_{ik})^2}{(v_{0k} - h_{ik})^2}. \quad (14)$$

Here α_0 is a constant, and $\alpha_0 \in [0, 1]$ [0, 1].

Secondly, we show the solution process of h_{ik}. Starting from $\frac{\partial J'}{\partial h_{ik}} = 0$, one has ($i = 1, ..., c, k = 1, ..., l$)

$$h_{ik} = \begin{cases} x_{ek}, & i = q, \\ \frac{\sum_{j=1}^{n}(x_{jk} - \eta v_{0k})u_{ij}^m}{\sum_{j=1}^{n}(1-\eta)u_{ij}^m}, & i \neq q \end{cases}. \quad (15)$$

Finally, we supply the solution process of w_{ik}. Starting from $\frac{\partial J'}{\partial w_{ik}}$=0 ($i = 1, ..., c, k = 1, ..., l$), we get:

$$w_{ik} = \frac{T_{ik}^{-\frac{1}{\tau-1}}}{\sum_{p=1}^{l} T_{ip}^{-\frac{1}{\tau-1}}}. \quad (16)$$

Among them, $\tau \in (1, +\infty)$, and δ_i is a penalty item. If δ_i is too large, then in (16), each feature in the cluster is assigned a weight close to $\frac{1}{l}$. If δ_l is too small, when u_{ij} is 0, the weight of one feature in the cluster will be assigned as 1, and the weight of other features will be assigned as 0. δ_i reflects the contribution of each attribute to the cluster center. In actual processing, we can define δ_i as the ratio of the sum of the previous part of (11) and the fuzzy feature weight:

$$\delta_i = K\frac{\sum_{j=1}^{n} u_{ij}^m D_{ij}}{\sum_{k=1}^{l} w_{ik}^{\tau}}. \quad (17)$$

Here K is a positive constant.

So far, the derivation process of cluster centers, membership degree matrix and weight matrix of the VSFCM algorithm has been explained.

3.3 Framework of the VSFCM Algorithm

The execution process of the CDCI method and the VSFCM algorithm is shown in Table 2 and Table 3.

Table 2. The execution process of the CDCI method

Algorithm 1 Cut-off Distance-induced Clustering Initialization (CDCI)

Input: Data set $X=\{x_k\}_{k=1}^{N}$, number of clusters C.

Output: cluster center matrix $H=\{h_i\}_{i=1}^{C}$.

CDCI (Data X, Number C){

According to (4), the cut-off radius cd is obtained;

According to (6), the local density ρ_j of each point is achieved;

Compute the distance δ_j of each point according to (3);

Obtain τ_j according to (7);

Rearrange $\tau=\{\tau_j\}_{j=1}^{n}$ from largest to smallest, and get the corresponding X' after the original data set X is re-sorted by τ;

Select τ_1 corresponding to x_1' as the first cluster center x_e, and let $H=H\cup x_e$;

Let tt=1, k=2;

repeat

while $\|x_k' - H\| < dc \quad k = k+1;$

$H=H\cup x_k'$;

tt = tt+1;

until $tt = c$

return H;}

4 Experimental Studies

In terms of comparison algorithms, it is compared with V-FCM, SCAD, ESSC, FRFCM, DVPFCM algorithms. Among the two algorithms of SCAD, the structure of the first one is more complex and closer to the one in this study, so we compare it with the first one. In terms of initialization methods, we compare our method with the DPC algorithm alone. In the experiment, the default parameter settings are adopted.

The testing data sets include 2 artificial data sets and 8 UCI machine learning data sets. The artificial data sets DATA1 and DATA2 are composed of Gaussian distribution points obtained by our own generation tools. The tested UCI data set [20] includes Iris, Wireless Indoor, Wine, Breast Cancer Wisconsin, Seeds, Letter(A, B), Ionosphere, SPECT heart data. These UCI data sets are relatively common and representative data sets in the field of machine learning.

Table 4 counts the basic information of two artificial data sets and 8 UCI machine learning data sets. For all experiments, the parameters are selected by default, and the specific settings are as follows: $m = 2$, $\tau = 2$, $\varepsilon = 10^{-5}$. For convenience, the selection of *cd* in the DPC algorithm is calculated based on (6).

Table 3. The VSFCM algorithm

Algorithm 2 The VSFCM algorithm

Input: $X=\{x_j\}_{j=1}^{n}$, number of clusters c.

Output: Membership $U=\{u_{ij}\}_{i,j=1}^{c,n}$, cluster center $H=\{h_i\}_{i=1}^{c}$, weight $W=\{w_{ik}\}_{i,k=1}^{c,n}$.

```
VSFCM (Data X, Number C) {
    Set threshold ε and maximum number of iterations iM;
    Run Algorithm 1, and get H^(0) and the point x_e with the highest density;
    do
      iter = iter + 1;
      Calculate u_ik by (12) to obtain the updated U^(iter) = [u_ij];
      Compute h_i by (15) to obtain the updated H^(iter) = [h_i];
      Calculate w_ik by (16) to obtain the updated W^(iter) = [w_ik];
    while ||H^(iter) − H^(iter−1)|| ≥ ε and iter ≤ iM;
    return U^(iter), H^(iter), W^(iter);}
```

Table 4. Testing data sets

ID	data set	# of instances	# of features	# of clusters
D1	Iris	150	4	3
D2	Wireless Indoor	2000	7	4
D3	Wine	178	13	3
D4	Breast cancer	569	30	2
D5	Seeds	210	7	3
D6	Letter(A, B)	1155	16	2
D7	Ionosphere	351	33	2
D8	SPECT heart data	267	22	2
D9	DATA1	300	2	3
D10	DATA2	180	3	3

We use hard and soft clustering validity indexes. Next, we use the superscript (+) to indicate that the larger the index value, the better. The reverse superscript (−) indicates that the smaller the index value, the better. The hard clustering validity index adopts the following three kinds. Classification rate (ACC) reflects the proportion of samples that are correctly classified. Normalized mutual information (NMI) [21] reflects the statistical information shared between two categories. The Calinski-Harabasz (CH) index [22] is a measure from the perspective of distance within a class and dispersion between classes. The validity index of soft clustering adopts the following two kinds. The EARI index is a fuzzy extension of the Adjusted Rand index (ARI) [23], and its idea is to describe the similarity of two clustering results. The Xie-Beni (XB) index [24] is a highly recognized index of the validity of fuzzy clustering.

Figure 1 is a data distribution diagram of DATA1 and DATA2.

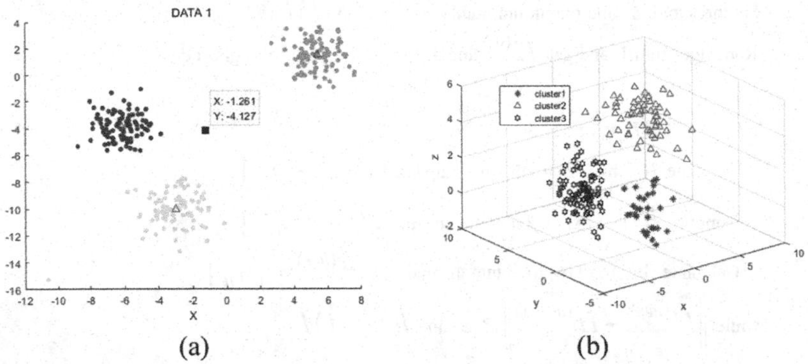

Fig. 1. (a) DATA1 data distribution map. (b) DATA2 data distribution map

As shown in Fig. 1(a), "Δ" represents the clustering center of each cluster, and "■" represents the separation reference point between clusters. The cluster centers of the three clusters are $V_1 = [-6.1545, -3.8668]$, $V_2 = [5.3800, 1.4988]$, $V_3 = [-3.0078, -10.0121]$. By computing, we get $X_0 = [-1.2809, -4.1416]$, $V_0 = [-1.2607, -4.1267]$. We express the weight of the full space as $w_1' = w_2' = w_3' = [0.5, 0.5]$. Three subspace weights are expressed as $w_1 = [0.9, 0.1]$, $w_2 = [0.4, 0.6]$, $w_3 = [0.6, 0.4]$. The corresponding separation between clusters can be expressed by $J_{full\_space} = \sum_{i=1}^{3} ((w_{i1}')^2 (V_{i1} - V_{01})^2 + (w_{i2}')^2 (V_{i2} - V_{02})^2) = 34.36$, and then $J_{ESSC\_subspace} = 44.38$, $J_{VSFCM\_subspace} = 44.49$. Among them, $v_0 = \sum_{i=1}^{c} v_i / c$, $x_0 = \sum_{j=1}^{n} x_j / n$, $v_0 = \sum_{i=1}^{c} v_i^{hddci} / c$. Here v_i is obtained using our CDCI.

In this example, the separation between clusters in the subspace is significantly greater than the separation between clusters in the full space, which means that the subspace is easier to cluster than the full space. In particular, the separation between full-space clusters and the separation between subspace clusters can also be used as separations under different distance metrics. Our algorithm replaces x_0 in the ESSC algorithm with v_0. The effect of subspace separation is slightly better than the previous one, but it is more stable. When faced with some complex, more scattered data sets, and irregular data sets, we can still maintain a good effect. Compared with ESSC, it can get better separation between clusters, which can get better results.

Table 5 is the ACC values of several algorithms for DATA2. Obviously, the performance of our algorithm is better than other algorithms. Moreover, the clustering accuracy of the subspace clustering algorithm is obviously higher than that of other algorithms, and the weight distribution may be an important influencing factor.

Breast Cancer is a common medical data set in machine learning. This data set can be divided into two categories. Figure 2 shows the $\rho - \delta$ distribution diagram of Breast Cancer for two cluster center initialization methods. In Fig. 2, we can see that the first cluster center is well determined, that is, the data point in the upper right corner. In the second cluster center, the boundaries between the selectable points of the CDCI algorithm and other data points are very clear, while the HDCCI algorithm is more scattered. Then when we use the HDCCI algorithm to select the initial cluster center, it is easy to choose the wrong one. The initial cluster centers selected by our cluster center initialization method CDCI are more in line with the characteristics of ideal cluster centers.

Table 5. ACC values of clustering results for DATA2

V-FCM	SCAD1	ESSC	DVPFCM	FRFCM	VSFCM
0.7000	0.9833	0.9833	0.8333	0.9899	0.9986

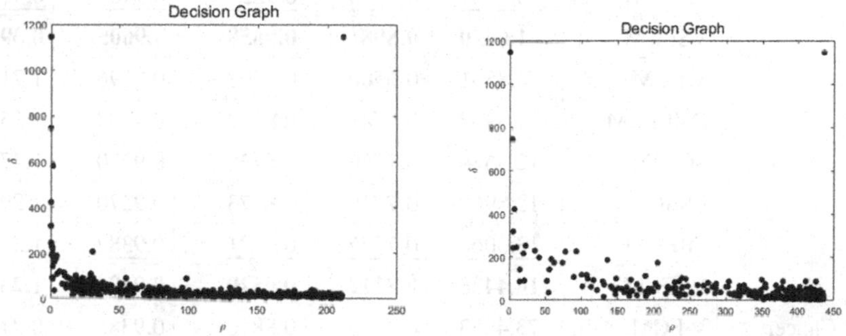

(a) (b)

Fig. 2. (a) CDCI $\rho - \delta$ distribution map. (b) HDCCI $\rho - \delta$ distribution map

Table 6 shows the running results of each clustering algorithm (including V-FCM, SCAD, ESSC, FRFCM, DVPFCM and our algorithm) on the UCI data sets. The adopted evaluation indexes are the above-mentioned five indexes. The performance of the VSFCM algorithm is evaluated and compared with the existing three subspace clustering algorithms and two viewpoint fuzzy clustering algorithms. Each algorithm was run 20 times, and the EARI, ACC, XB, CH, and NMI values of the clustering results are saved in Table 6. In Table 6, we bold the best result and underlined the second best result.

It can be seen from Table 6 that in the six algorithms, the overall can be divided into 3 grades. The worst is V-FCM and DVPFCM, and then FRFCM, SCAD1, ESSC are relatively better. The best is the VSFCM algorithm.

VSFCM has the best clustering performance on 7 UCI data sets. The performance of FRFCM and SCAD1 is equal to or better than ESSC. However, although DVPFCM and V-FCM are generally inferior to the other three algorithms, they can achieve better clustering performance on the Iris and Letter(A, B) measured by NMI and ACC, respectively.

Table 6. Clustering results of UCI data sets

	Algorithms	CH(+)	NMI(+)	EARI(+)	ACC(+)	XB(−)
Iris	V-FCM	11.8364	0.8498	0.9312	0.9533	0.2786
	DVPFCM	11.8423	0.8705	0.9400	0.9600	0.2731
	SCAD1	11.5894	0.7277	0.8712	0.8800	0.3205
	ESSC	11.6442	0.7578	0.8861	0.9000	0.3117
	FRFCM	11.6853	0.7665	0.8895	0.9067	0.2880
	VSFCM	**12.1360**	**0.8801**	**0.9568**	**0.9667**	**0.2628**
Wireless Indoor	V-FCM	66.8676	0.8113	0.8730	0.9325	0.4845
	DVPFCM	67.0487	0.8117	0.8733	0.9330	0.4764
	SCAD1	79.9004	0.8508	0.9478	0.9445	0.4262
	ESSC	80.1303	0.8549	0.9494	0.9460	0.4240
	FRFCM	80.3587	0.8620	0.9529	0.9465	0.4113
	VSFCM	**81.6420**	**0.8981**	**0.9658**	**0.9605**	**0.3988**
Wine	V-FCM	8.8531	0.6500	0.8598	0.8596	1.7169
	DVPFCM	12.3937	0.6748	0.8712	0.8708	1.5531
	SCAD1	12.5666	0.7710	0.8728	0.9270	1.4794
	ESSC	12.8969	0.7710	0.8773	0.9270	1.2942
	FRFCM	13.3068	0.7955	0.9031	0.9382	1.2786
	VSFCM	**14.4425**	**0.8212**	**0.9520**	**0.9494**	**1.2302**
Breast Cancer	V-FCM	73.4053	0.6555	0.8850	0.9385	0.3178
	DVPFCM	74.0104	0.6555	0.8850	0.9385	0.3146
	SCAD1	108.1275	0.6320	0.9250	0.9297	0.2913
	ESSC	108.7064	0.6507	0.9400	0.9332	0.2903
	FRFCM	107.3007	0.6106	0.9244	0.9262	0.2940
	VSFCM	**110.5897**	**0.7084**	**0.9825**	**0.9420**	**0.2779**
Seeds	V-FCM	16.0498	0.6423	0.8450	0.8762	0.6296
	DVPFCM	16.0668	0.6545	0.8478	0.8810	0.5999
	SCAD1	16.1163	0.6654	0.8487	0.8857	0.2508
	ESSC	16.1378	0.6795	0.8580	0.8905	0.2493
	FRFCM	16.2337	0.7026	0.8601	0.8952	0.2280
	VSFCM	**16.7541**	**0.7173**	**0.8745**	**0.9095**	**0.2269**
Ionosphere	V-FCM	19.9278	0.1036	0.4159	0.6809	0.7634
	DVPFCM	21.5964	0.1036	0.4180	0.6809	0.6566

(continued)

Table 6. (*continued*)

	Algorithms	CH(+)	NMI(+)	EARI(+)	ACC(+)	XB(−)
	SCAD1	32.2422	**0.1320**	0.4762	**0.7094**	0.4649
	ESSC	31.6597	0.1271	0.4732	0.7066	0.5057
	FRFCM	32.1711	0.1292	0.4738	0.7066	0.4804
	VSFCM	**32.3488**	**0.1320**	**0.4764**	**0.7094**	**0.4635**
Letter(A,B)	V-FCM	155.3391	0.7197	0.8948	0.9408	0.6633
	DVPFCM	156.9256	0.7249	0.8965	0.9415	0.6561
	SCAD1	195.6678	0.6655	0.9181	0.9293	0.6486
	ESSC	198.4907	0.7189	0.9353	0.9395	0.6352
	FRFCM	196.1744	0.6917	0.9288	0.9344	0.6390
	VSFCM	**199.5879**	**0.7249**	**0.9385**	**0.9415**	**0.6333**

From Table 6, V-FCM and DVPFCM (belonging to viewpoint-oriented fuzzy clustering algorithms) have better results on some relatively small data sets. It may because these data sets have low dimensions and weights have little effect on the results. Moreover, the NMI and ACC indexes of V-FCM and DVPFCM on some data sets perform better. This is because the existence of viewpoints can guide the clustering algorithm to run in a more correct direction.

Moreover, the weighted fuzzy clustering algorithms of SCAD, ESSC, and FRFCM are more complex and have great advantages in data sets with more dimensions. Weight distribution enhances the efficiency and effect of clustering.

Finally, our proposed algorithm is obviously superior over the other algorithms mentioned above. Our algorithm integrates the advantages of the two types of algorithms well and improves the clustering effect.

In general, the performance of the VSFCM algorithm is more ideal in terms of obtaining the initial cluster centers and various evaluation indexes.

5 Conclusions

In this study, we develop the VSFCM algorithm, and has achieved good clustering results. First of all, we propose a new cut-off distance under the system of the DPC algorithm, and further provide a cut-off distance-induced cluster initialization method CDCI. Secondly, by taking the viewpoint obtained by CDCI as being reflective of domain knowledge, the fuzzy clustering idea driven by knowledge and data is presented. We comprehensively establish the VSFCM algorithm. Finally, by comparing experimental results produced by V-FCM, SCAD, ESSC, FRFCM, DVPFCM on artificial and UCI data sets, it is concluded that the VSFCM algorithm performs best in terms of the five indexes.

In our future work, we can develop our fuzzy clustering algorithm to the field of fuzzy reasoning [25–29] and carry on clustering for the fuzzy rules.

Acknowledgements. This work was supported by the National Natural Science Foundation of China (Nos. 62176083, 61673156, 62176084, 61877016, 61976078), and the Fundamental Research Funds for the Central Universities of China (No. PA2021GDSK0092).

References

1. Chakraborty, S., Das, S.: Detecting meaningful clusters from high-dimensional data: a strongly consistent sparse center-based clustering approach. IEEE Transactions on Pattern Analysis and Machine Intelligence **44**(6), 2894–2908 (2022)
2. Tang, Y.M., Li, L., Liu, X.P.: State-of-the-art development of complex systems and their simulation methods. Complex System Modeling and Simulation **1**(4), 271–290 (2021)
3. Averbuch-Elor, H., Bar, N., Cohen-Or, D.: Border-peeling clustering. IEEE Trans. Pattern Analysis and Machine Intelli. **42**(7), 1791–1797 (2020)
4. Rodriguez, A., Laio, A.: Clustering by fast search and find of density peaks. Science **344**(6191), 1492–1496 (2014)
5. Bezdek, J.C.: Pattern Recognition With Fuzzy Objective Function Algorithms. Academic, New York (1981)
6. Tang, Y.M., Ren, F.J., Pedrycz, W.: Fuzzy c-means clustering through SSIM and patch for image segmentation. Applied Soft Computing **87**, 1–16 (2020)
7. Wang, X., Wang, Y., Wang, L.: Improving fuzzy c-means clustering based on feature-weight learning. Pattern Recognition Letters **25**(10), 1123–1132 (2004)
8. Li, J., Gao, X., Jiao, L.: A new feature weighted fuzzy clustering algorithm. In: Rough Sets, Fuzzy Sets, Data Mining, and Granular Computing, pp. 412–420. Springer, Berlin, Heidelberg (2005)
9. Chan, E.Y., et al.: An optimization algorithm for clustering using weighted dissimilarity measures. Pattern Recognition **37**(5), 943–952 (2004)
10. Jing, L., et al.: Subspace clustering of text documents with feature weighting k-means algorithm. In: Advances in Knowledge Discovery and Data Mining, pp. 802–812. Springer, Berlin, Heidelberg (2005)
11. Gan, G., Wu, J., Yang, Z.: A fuzzy subspace algorithm for clustering high dimensional data. In: Advanced Data Mining and Applications, pp. 271–278. Springer, Berlin, Heidelberg (2006)
12. Gan, G., Wu, J.: A convergence theorem for the fuzzy subspace clustering (FSC) algorithm. Pattern Recognition **41**(6), 1939–1947 (2008)
13. Cao, Y., Wu, J.: Projective ART for clustering data sets in high dimensional spaces. Neural Networks **15**, 105–120 (2002)
14. Lu, C., et al.: Subspace clustering by block diagonal representation. IEEE Trans. Pattern Analysis and Machine Intelligence **41**(2), 487–501 (2019)
15. Frigui, H., Nasraoui, O.: Unsupervised learning of prototypes and attribute weights. Pattern Recognition **37**(3), 567–581 (2004)
16. Deng, Z., et al.: Enhanced soft subspace clustering integrating within-cluster and between-cluster information. Pattern Recognition **43**(3), 767–781 (2010)
17. Yang, M.S., Nataliani, Y.: A feature-reduction fuzzy clustering algorithm based on feature-weighted entropy. IEEE Transactions on Fuzzy Systems **26**(2), 817–835 (2018)
18. Pedrycz, W., Loia, V., Senatore, S.: Fuzzy clustering with viewpoints. IEEE Trans. Fuzzy Sys. **18**(2), 274–284 (2010)
19. Tang, Y.M., Hu, X.H., Pedrycz, W., Song, X.C.: Possibilistic fuzzy clustering with high-density viewpoint. Neurocomputing **329**, 407–423 (2019)

20. Asuncion, A., Newman, D.J.: UCI Machine Learning Repository, School of Information and Computer Science. University of California, Irvine, CA, USA (2007). [online], Available: http://archive.ics.usi.edu/ml/datasets.html
21. Strehl, A., Ghosh, J.: Cluster ensembles-A knowledge reuse framework for combining multiple partitions. J. Machine Learn. Res. **3**, 583–617 (2002)
22. Calinski, T., Harabasz, J.: A dendrite method for cluster analysis. Communications in Statistics **3**(1), 1–27 (1972)
23. Huang, H.C., Chuang, Y.Y., Chen, C.S.: Multiple kernel fuzzy clustering. IEEE Trans. Fuzzy Sys. **20**(1), 120–134 (2012)
24. Xie, X.L., Beni, G.: A validity measure for fuzzy clustering. IEEE Trans. Pattern Analy. Machine Intelli. **13**(8), 841–847 (1991)
25. Tang, Y.M., Ren, F.J.: Universal triple I method for fuzzy reasoning and fuzzy controller. Iranian Journal of Fuzzy Systems **10**(5), 1–24 (2013)
26. Tang, Y.M., Ren, F.J.: Variable differently implicational algorithm of fuzzy inference. J. Intelli. Fuzzy Sys. **28**(4), 1885–1897 (2015)
27. Tang, Y.M., Ren, F.J.: Fuzzy systems based on universal triple I method and their response functions. Int. J. Info. Technol. Decision Making **16**(2), 443–471 (2017)
28. Tang, Y.M., Pedrycz, W., Ren, F.J.: Granular symmetric implicational method. IEEE Trans. Emerg. Topics in Computational Intelligence **6**(3), 710–723 (2022)
29. Tang, Y.M., Pedrycz, W.: Oscillation bound estimation of perturbations under Bandler-Kohout subproduct. IEEE Transactions on Cybernetics **52**(7), 6269–6282 (2022)

TOPSIS Method Based on Intuitionistic Fuzzy Linguistic Concept

Yue Zhou[1(✉)] and Li Zou[2]

[1] School of Mathematics, Liaoning Normal University, Dalian 116029, China
lnsfdxzy@163.com
[2] School of Computer Science and Technology, Shandong Jianzhu University, Jinan 250101, China

Abstract. MADM always been the focus of academic research. In order to solve MADM problem, it is very important to introduce intuitionistic fuzzy theory. This paper integrates linguistic concepts into traditional intuitionistic fuzzy theory and proposes intuitionistic fuzzy decision matrix with linguistic concepts. Based on the positive and negative forms of attributes, we propose the definition of upper and lower bounds of linguistic terms, and take linguistic value as the classification criteria. Specifically, we define the interval-valued linguistic terms by combining upper and lower bounds of linguistic terms, and obtain the solving models of membership degree and non-membership degree of IFN. Then, considering the steps of TOPSIS method, we propose (a) the weighting method of multi-attribute decision matrix of intuitionistic fuzzy linguistic concepts; (b) positive and negative ideal solution; (c) pseudo distance; and (d) relative progress. Finally, the reliability and effectiveness of the method are verified by analyzing the possibility of flight delay at 11 airports in China.

Keywords: Intuitionistic fuzzy linguistic concept · Positive and negative attributes · TOPSIS

1 Introduction

In the real world, people often have more than one factor to consider when making a decision. For example, when people decide to travel, they need to take into account "weather, distance, cost performance" and other factors. Therefore, the problem of multi-attribute decision making (MADM) has been widely studied by many experts and scholars. Fuzzy sets introduced by Zadeh [1] in 1965 are used to describe uncertain environments. In recent years, many experts and scholars put forward a series of extensions of fuzzy sets, such as IFSs [2–4] and HFSs [5, 6], to deal with decision problems because there are many uncertain factors in decision problems. Xu et al. [7] proposed a new method to solve the MADM problem, using the maximum deviation method, in which HFS is used to describe the evaluation of decision makers. Li [8] provided an effective approach to

This work was supported by the Chinese Natural Science Foundation (Nos. 6217023422).

Y. Chen and S. Zhang (Eds.): AILA 2022, CCIS 1657, pp. 106–116, 2022.
https://doi.org/10.1007/978-981-19-7510-3_8

using the IFS expressed with membership and non-membership functions, and applied it to game theory, operations research, industrial engineering and other fields. Ashraf et al. [9] developed a MADM method based on weighted averaging and weighted geometric aggregation operators of spherical fuzzy numbers. Siregar et al. [10] uses VIKOR method to get the optimal solution computationally and helps parties develop a model any solutions. In order to use the data of membership degree and non-membership degree, this paper uses IFN to describe the relationship between objects and linguistic concepts, which greatly avoids the loss of information.

The MADM problem in this paper is realized by TOPSIS method. In fact, TOPSIS method, also known as ideal solution method [11], is often used to solve decision-making problems. If the result is close to the positive ideal solution and far from the negative ideal solution, then it is said to be optimal. Vafaei et al. [12] proposed a new evaluation method, which provided a new idea for the evaluation technology in TOPSIS and made the results more reliable. Zulqarnain et al. [13] proposed a new method to solve the order preference problem by using the TOPSIS method, and applied the method to a concrete example to verify its effectiveness. Afsordegan et al. [14] solved the problem that the TOPSIS method could not deal with decisions with qualitative linguistic labels, and provided a new solution for decision making in fuzzy environment.

In the real world, MADM problem can be seen everywhere, so it has great research significance. This paper focuses on improving multi-attribute decision matrix. Traditionally, multi-attribute decision matrix [15–17] only describes the relationship between the attribute and the object, but the linguistic concept [25] is proposed to realize the use of linguistic values to describe attributes, which is more consistent with human thinking. Therefore, this paper proposes the multi-attribute decision matrix of intuitionistic fuzzy linguistic concepts, which is used to describe the relationship between the object and the linguistic concept, so as to realize the unification and datalization of people's language thinking. In previous decision-making problems [18–21], data comparison was generally used to select the optimal scheme, which was not consistent with human language thinking mode. Therefore, this paper uses linguistic terms as classification criteria, which makes the classification results more real and effective. The properties of positive and negative attributes are embodied in the upper and lower bounds of linguistic terms, and the optimal solution is obtained by using TOPSIS method.

The rest of the paper is as follows. Section 2 reviews the definition and operational properties of IFSs and the basic knowledge of linguistic term set and linguistic concept set. Section 3 introduces the definition of upper and lower bounds of linguistic terms. Taking the interval-valued linguistic terms as the classification criteria, the evaluation formulas of membership degree, non-membership degree and hesitancy degree are proposed, and the steps of TOPSIS method are given based on the multi-attribute decision matrix of intuitionistic fuzzy linguistic concepts.

2 Preliminaries

Definition 2.1. [22] Let a finite set X be fixed, the *IFS* A in X is defined as:

$$A = \{< x, \mu_A(x), \nu_A(x) > |x \in X\},$$

where the functions $\mu_A(x) : X \to [0, 1], x \in X \to \mu_A(x) \in [0, 1], \nu_A(x) : X \to [0, 1], x \in X \to \nu_A(x) \in [0, 1]$ $[0, 1]$ satisfy the condition: $0 \le \mu_A(x) + \nu_A(x) \le 1$. x denotes the principal component, and $\mu_A(x)$, $\nu_A(x)$ denote the degree of membership and the degree of non-membership of the element $x \in X$ to the set A, respectively. In addition, for each *IFS* $A \subseteq X$, $\pi_A(x) = 1 - \mu_A(x) - \nu_A(x)$ is called the degree of indeterminacy of x to A, or is called the degree of hesitancy of x to A.

For convenience, the intuitionistic fuzzy number is called *IFN*, the intuitionistic fuzzy set is called *IFS*, and the set of all *IFSs* on X is denoted as *IFSs*(X). Then for $\forall A, B \in IFSs(X)$, its operational properties [23] are as follows:

(1) $A \le B$ if and only if $\mu_A(x) \le \mu_B(x)$ and $\nu_A(x) \ge \nu_B(x)$ for all x in X;
(2) $A = B$ if and only if $A \le B$ and $B \le A$;
(3) $A \cap B = \{< x, \min(\mu_A(x), \mu_B(x)), \max(\nu_A(x), \nu_B(x)) > | x \in X\}$;
(4) $A \cup B = \{< x, \max(\mu_A(x), \mu_B(x)), \min(\nu_A(x), \nu_B(x)) > | x \in X\}$;
(5) The complementary of an *IFS* A is $A_c = \{< x, \nu_A(x), \mu_A(x) > | x \in X\}$.

In fact, the MADM problem is the decision makers on the basis of n attributes from m alternatives to choose or sorting. If the decision maker's evaluation of alternatives based on attributes is represented by $f =< \mu_{ij}, \nu_{ij} >$, where μ_{ij} and ν_{ij} represent the membership degree and non-membership degree of decision makers' evaluation of alternatives based on attributes respectively. Then the intuitionistic fuzzy multi-attribute decision making matrix can be defined as $F = [f_{ij}]_{m \times n}$.

Definition 2.2. [24] Let $S = \{s_i | i = 0, 1, 2, \cdots, g\}$ be a linguistic term set with odd cardinality. Any label, s_i represents a possible value for a linguistic variable, and it should satisfy the following characteristics:

(1) Order relation. $s_i > s_j$, if $i > j$;
(2) Negation operator. $Neg(s_i) = s_j$, where $j = g - i$;
(3) Maximization operator. $\max\{s_i, s_j\} = s_i$, if $i \ge j$;
(4) Minimization operator. $\min\{s_i, s_j\} = s_j$, if $i \ge j$.

Definition 2.3. [25] Let $S = \{s_i | i = 0, 1, 2, \cdots, g\}$ be a linguistic term set, $L = \{l^1, l^2, \cdots, l^n\}$ be a set of attributes, then L_{s_α} is called a linguistic concept.

3 Multi-attribute Decision Making Problem of Intuitionistic Fuzzy Linguistic Concepts

3.1 Classification Criteria Based on Interval-Valued Linguistic Terms

In general, the decision-making problem is to choose an optimal one from some given alternatives, or to obtain the order of all alternatives. The attributes considered based on decision-making are often not all positive. For example, in order to select the appropriate factory to purchase, the purchasing group needs to consider three attributes: C_1: product quality, C_2: delivery performance, and C_3: product price. Then factories with better quality, higher performance and lower price will be preferentially selected. Therefore, C_1 and C_2 are positive attributes, while C_3 is negative attributes.

In daily decision-making, linguistic terms are often used as evaluation information for alternatives. However, because of experts' experience, ability, resources and social status are different, their subjective understanding of linguistic terms also be biased. For example, expert A considers a student with a score of 100 to be a good student, while expert B considers a student with a score of 95 or more to be a good student. In order to coordinate the opinions of all experts and consider the positive and negative attributes, this paper proposes to divide classification levels by using the interval value of linguistic terms. And by defining the upper and lower bounds of linguistic terms, the computational complexity caused by positive and negative attributes is cleverly solved, so it is defined as follows:

Definition 3.1.1. Let $S = \{s_\alpha | \alpha = 1, 2, \cdots, l\}$ be a linguistic term set, $C = \{c_1, c_2, \cdots, c_n\}$ is an attribute set. If for $\forall c_j \in C$, the linguistic term $s_i \in S$ has upper and lower bounds, $s_i(c_j) = [s_i^-(c_j), s_i^+(c_j)]$ is called interval-valued linguistic terms based attribute, where $s_i^+(c_j)$ and $s_i^-(c_j)$ are the upper and lower bounds of s_i based on c_j, respectively.

Different from the traditional upper and lower bounds in mathematics, the upper and lower bounds of linguistic terms are only related to the positive and negative of attributes, and have nothing to do with the value they represent.

Theorem 3.1.1. Let $C = \{c_1, c_2, \cdots, c_n\}$ be a set of attributes containing positive and negative attributes, then the interval-valued linguistic terms satisfy the following properties:

(1) $s_i^-(c_j) \leq s_i^+(c_j)$ if and only if c_j is a positive attribute;
(2) $s_i^-(c_j) \geq s_i^+(c_j)$ if and only if c_j is a negative attribute.

Proof. Clearly, it can be proved by Definition 3.1.1

Obviously, the interval-valued linguistic terms is not regular.

Example 3.1.1. Take choosing the appropriate factory for example. The linguistic term set $S = \{s_1(very\, bad), s_2(bad), s_3(average), s_4(good), s_5(very\, good)\}$ is the evaluation value of the purchasing group on the alternative factories based on the attribute set $C = \{c_1, c_2, c_3\}$. Where c_1: product quality, c_2: delivery performance, and c_3: product price. The classification criteria of linguistic value obtained based on the opinions of all experts is shown in Table 1, in which attributes c_1 and c_2 are evaluated by experts' scores. For $s_4 \in S$, its interval-valued linguistic terms based on three attributes are $s_4(c_1) = [80, 90]$, $s_4(c_2) = [85, 95]$, $s_4(c_3) = [3, 1.5]$. According to Theorem 3.1.1, c_1 and c_2 are positive attributes, while c_3 is a negative attribute.

Displayed equations are centered and set on a separate line.

Table 1. Classification criteria of linguistic value.

Classification level	c_1(score)	c_2(score)	c_3(million)
s_1	[0,60]	[0,65]	[4.5,4]
s_2	[60,70]	[65,75]	[4,3.5]
s_3	[70,80]	[75,85]	[3.5,3]
s_4	[80,90]	[85,95]	[3,1.5]
s_5	[90,100]	[95,100]	[1.5,1]

3.2 Multi-attribute Decision Matrix of Intuitionistic Fuzzy Linguistic Concepts

In the intuitionistic fuzzy multi-attribute decision matrix, the DM evaluates the object according to each attribute, but not for each classification level. Therefore, it may cause a certain degree of information loss in the application process. In this section, linguistic terms are used as the classification levels and the decision matrix is redefined for each linguistic concept.

Definition 3.2.1. For the object set $X = \{x_1, x_2, \cdots, x_m\}$, attribute set $C = \{c_1, c_2, \cdots, c_n\}$, linguistic term set $S = \{s_\alpha | \alpha = 1, 2, \cdots, l\}$ and linguistic concept set $C_{s_\alpha} = \{c^1_{s_1}, \cdots, c^1_{s_l}, c^2_{s_1}, \cdots, c^2_{s_l}, \cdots, c^n_{s_1}, \cdots, c^n_{s_l}\}$, $\tilde{f} =< \mu^j_{i\alpha}, v^j_{i\alpha} >$ is an IFN, representing the evaluation value of DMs on the relationship between object and linguistic concept, where $\mu^j_{i\alpha}$ and $v^j_{i\alpha}$ represent the membership degree and non-membership degree of the *ith* object to the *jth* attribute based on linguistic term s_α respectively. Then the multi-attribute decision matrix of intuitionistic fuzzy linguistic concepts is $\tilde{F} = [\tilde{f}_{st}]_{m \times nl}$.

Intuitionistic fuzzy numbers don't just have two parameters. When solving decision problems, the DM can often increase his evaluation by increasing the degree of hesitation. Therefore, we should not only consider the corresponding relationship between membership and non-membership degrees and classification criteria, but also consider the influence of DMs' hesitation on membership and non-membership degrees, when calculating IFNs based on the given classification criteria of linguistic value. The formula is as follows:

$$\mu^n_{mi} = \left(\frac{1-\beta}{2}\right)^{\left(\frac{x_m - R_i}{s_i^+ - R_i}\right)^2} \tag{1}$$

$$v^n_{mi} = 1 - \left(1 - \frac{1-\beta}{2}\right)^{\left(\frac{x_m - R_i}{s_i^+ - R_i}\right)^2} \tag{2}$$

$$\pi^n_{mi} = \left(\frac{1+\beta}{2}\right)^{\left(\frac{x_m - R_i}{s_i^+ - R_i}\right)^2} - \left(\frac{1-\beta}{2}\right)^{\left(\frac{x_m - R_i}{s_i^+ - R_i}\right)^2} \tag{3}$$

$$R_i = \frac{s_i^- + s_i^+}{2} \tag{4}$$

where, R_i represents the median of the interval-valued linguistic terms, $\beta \in [0, 1]$ represents the decision maker's hesitancy degree. The greater β is, the more hesitant the decision maker is. Then, the smaller the membership degree of the object x_m belonging to the classification criteria s_i is, and the smaller the non-membership degree is, the greater the hesitancy degree is.

3.3 TOPSIS Method for Multi-attribute Decision Matrix of Intuitionistic Fuzzy Linguistic Concepts

For MADM problems, if the importance of each attribute is different, the flexibility will be greater. In this paper, intuitionistic fuzzy number is used to represent the weight, i.e.

$$W = \{w_1, w_2, \cdots, w_n\} = \{< s_1, t_1 >, < s_2, t_2 >, \cdots, < s_n, t_n >\}$$

For $w_k =< s_k, t_k >, k = 1, 2, \cdots, n$, satisfying $\sum_{k=1}^{n} s_k \leq 1$. Let $\tau_k = 1 - s_k - t_k$ be the hesitancy degree over the importance of the attribute, where s_k and t_k are respectively the membership degree and non-membership degree of the importance of attribute to fuzzy problems. In practical problems, the choice of the appropriate objects may not need to consider all the attributes, so the important degree of each attribute change with the changes in the environment with the problem.

The main steps of TOPSIS decision-making are as follows:

Step 1. Construct a weighted multi-attribute decision matrix of intuitionistic fuzzy linguistic concepts:

$$F^* =< s_j \mu_{i\alpha}^j, t_j + \nu_{i\alpha}^j - t_j \nu_{i\alpha}^j >=< \mu_{i\alpha}^{j*}, \nu_{i\alpha}^{j*} > \tag{5}$$

Step 2. Define positive ideal solutions of intuitionistic fuzzy linguistic concepts:

$$\begin{array}{lccccccccccc} & a_{s_1}^1 & a_{s_2}^1 \cdots & a_{s_l}^1 & a_{s_1}^2 & a_{s_2}^2 \cdots & a_{s_l}^2 & \cdots & a_{s_1}^n & a_{s_2}^n & \cdots & a_{s_l}^n \\ R^+ = \{ & < s_1, t_1 >, & < 0, 1 >, \cdots, & < 0, 1 > & < s_2, t_2 >, & < 0, 1 >, \cdots, & < 0, 1 > & \cdots & < s_n, t_n >, & < 0, 1 >, & \cdots, & < 0, 1 > \} \\ R^- = \{ & < 0, 1 >, & < 0, 1 >, \cdots, & < s_1, t_1 > & < 0, 1 >, & < 0, 1 >, \cdots, & < s_2, t_2 > & \cdots & < 0, 1 >, & < 0, 1 >, & \cdots, & < s_n, t_n > \} \end{array} \tag{6}$$

Step 3. Calculate the pseudo-distances of each alternative $x_i (i = 1, 2, \cdots, n)$ to positive ideal solution R^+ and negative ideal solution R^-, respectively:

$$d(x_i, R^+) = \sqrt{\frac{1}{2}\sum_{j=1}^{n}[(\mu_{i1}^{j*} - s_j)^2 + (\nu_{i1}^{j*} - t_j)^2 + (\pi_{i1}^{j*} - \tau_j)^2] + \frac{1}{2}\sum_{j=1}^{n}\sum_{\alpha=2}^{l}[(\mu_{i\alpha}^{j*})^2 + (\nu_{i\alpha}^{j*} - 1)^2 + (\pi_{i\alpha}^{j*})^2]} \tag{7}$$

$$d(x_i, R^-) = \sqrt{\frac{1}{2}\sum_{j=1}^{n}[(\mu_{il}^{j*} - s_j)^2 + (\nu_{il}^{j*} - t_j)^2 + (\pi_{il}^{j*} - \tau_j)^2] + \frac{1}{2}\sum_{j=1}^{n}\sum_{\alpha=1}^{l-1}[(\mu_{i\alpha}^{j*})^2 + (\nu_{i\alpha}^{j*} - 1)^2 + (\pi_{i\alpha}^{j*})^2]} \tag{8}$$

Step 4. Calculate the relative progress of each alternative $x_i (i = 1, 2, \cdots, n)$ and positive ideal solution R^+:

$$\rho_i = \frac{d(x_i, R^-)}{d(x_i, R^+) + d(x_i, R^-)} \tag{9}$$

where, the larger $d(x_i, R^-)$ is, the closer it is to the positive ideal solution, and the alternative x_i is optimal.

4 A Case Study

Based on the above steps, this section takes 11 domestic airports as the main research samples and selects three attributes as the evaluation basis to conduct TOPSIS decision analysis.

4.1 Data Gathering

This article uses data from Flightstats.com. Taking Chongqing Jiangbei International Airport, Xi'an Xianyang International Airport, Beijing Capital International Airport and other 11 domestic airports as alternatives, this paper classifies the flight delay probability of each major airport by linguistic value with annual punctuality rate, annual average departure delay and annual average delay time as attributes. The data of 11 airports are shown in Table 2.

Table 2. Flight delay data of 11 airports in China in 2021.

Domestic airport	Annual punctuality rate (%)	Annual average departure delay	Annual average delay time (min)
Chongqing Jiangbei International Airport	88.49	52	27.43
Xi'an Xianyang International Airport	88.73	64	20.57
Beijing Capital International Airport	89.15	49	22.44
Shenzhen Bao'an International Airport	88.94	53	33.02
Shanghai Pudong International Airport	91.04	52	34.23
Shanghai Hongqiao International Airport	92.78	55	36.86
Hangzhou International Airport	88.24	56	38.45
Guangzhou Baiyun International Airport	92.52	47	29.65
Chengdu Shuangliu International Airport	87.27	69	22.61
Zhengzhou Xinzheng International Airport	93.81	51	36.86
Hong Kong International Airport	82.90	58	39.32

Let $X = \{x_1, x_2, x_3, x_4, x_5, x_6, x_7, x_8, x_9, x_{10}, x_{11}\}$ be the object set and $A = \{a_1, a_2, a_3\}$ the attribute set, corresponding to the 11 airports and 3 attributes in Table 2 respectively. According to the data survey, the expert group gives the linguistic term set $S = \{s_1(verylow), s_2(low), s_3(fair), s_4(high), s_5(veryhigh)\}$, and the classification criteria of linguistic value of flight delay possibility is shown in Table 3.

Table 3. Classification criteria of linguistic value.

Classification level	$a_1(\%)$	a_2	$a_3(\min)$
s_1	[96,100]	[40,46]	[20,24]
s_2	[92,96]	[46,52]	[24,28]
s_3	[88,92]	[52, 58]	[28,32]
s_4	[84,88]	[58,64]	[32,36]
s_5	[80,84]	[64,70]	[36,40]

Obviously, c_1 is a negative attribute, c_2 and c_3 are positive attributes.

4.2 TOPSIS Decision Analysis

Taking x_1 as an example, the first row of multi-attribute decision matrix of intuitionistic fuzzy linguistic concepts can be obtained according to Eq. (1)–(4) and Table 2 and Table 3:

$$\begin{aligned}&\{< 0.001, 0.996 >, < 0.001, 0.979 >, < 0.593, 0.253 >, < 0.242, 0.547 >,\\&< 0.001, 0.995 >, < 0.001, 0.99 >, < 0.4, 0.4 >, < 0.4, 0.4 >, < 0.001, 0.99 >,\\&< 0.001, 0.998 >, < 0.001, 0.977 >, < 0.626, 0.23 >, < 0.22, 0.57 >,\\&< 0.001, 0.996 >, < 0.001, 0.998 >\}\end{aligned} \tag{10}$$

Given the weight set $W = \{< 0.35, 0.6 >, < 0.4, 0.5 >, < 0.15, 0.75 >\}$ of three attributes, the first row of the weighting matrix F^* as follows:

$$\begin{aligned}&\{< 0.0004, 0.9984 >, < 0.0004, 0.9916 >, < 0.2076, 0.7012 >, < 0.0847, 0.8188 >,\\&< 0.0004, 0.998 >, < 0.0004, 0.995 >, < 0.16, 0.7 >, < 0.16, 0.7 >, < 0.0004, 0.995 >,\\&< 0.0004, 0.999 >, < 0.0002, 0.9943 >, < 0.0939, 0.8075 >, < 0.033, 0.8925 >,\\&< 0.0002, 0.999 >, < 0.0002, 0.9995 >\}\end{aligned} \tag{11}$$

According to Steps 2–4, the pseudo distance and relative progress of 11 domestic airports and positive and negative ideal solutions were calculated, the linguistic value classification of alternatives were carried out according to the classification criteria of linguistic value range. The results are shown in Table 4.

Where, the interval-valued linguistic terms classification of airport flight delay possibility is shown in Table 5.

Table 4. Classification results of 11 airports.

Domestic airport	Positive distance	Negative distance	Relative progress	Linguistic value classification
x_1	0.8132	0.8157	0.5008	s_2
x_2	0.7699	0.6621	0.4624	s_4
x_3	0.8062	0.8869	0.5238	s_2
x_4	0.8399	0.8396	0.4999	s_3
x_5	0.8317	0.8262	0.4983	s_4
x_6	0.8699	0.8261	0.4871	s_4
x_7	0.8537	0.7973	0.4829	s_4
x_8	0.7417	0.8356	0.5297	s_2
x_9	0.7632	0.5909	0.4364	s_4
x_{10}	0.8426	0.8129	0.4910	s_4
x_{11}	0.8277	0.5984	0.4196	s_4

Table 5. The interval-valued linguistic terms.

Linguistic terms	Interval value
s_1	[0.6317,1]
s_2	[0.5001,0.6317]
s_3	[0.4989,0.5001]
s_4	[0.3683,0.4989]
s_5	[0,0.3683]

Therefore, according to Table 5, we can classify the delay possibility of 11 airports by linguistic values, as shown in Table 4. For example, the relative progress of x_1 is 0.5008, which belongs to [0.5001,0.6317], so the delay possibility of Chongqing Jiangbei International Airport is $s_2(low)$.

5 Conclusions

In this paper, based on the MADM problem, the linguistic concept is integrated into the intuitionistic fuzzy multi-attribute decision matrix, which makes the matrix not only reflect the relationship between the object and the attribute, but also reflect the relationship between the object and the classification level, and the decision result is more accurate. In order to distinguish the definition of upper bound and maximum value, this paper proposes the upper and lower bound of linguistic terms to define the interval-valued linguistic terms, and gives the formulas of membership degree, non-membership degree and hesitancy degree based on this. Finally, the linguistic value classification of flight delay possibility of 11 domestic airports is realized by using TOPSIS method.

References

1. Zadeh, L.A.: A fuzzy-set-theoretic interpretation of linguistic hedges. J. Cybernet. **2**(3), 4–34 (1972)
2. Atanassov, K.T.: Intuitionistic fuzzy sets. Int. J. Bioautom. **1**(20), 1–6 (2016)
3. Xu, Z.S., Zhao, N.: Information fusion for intuitionistic fuzzy decision making: an overview. Inf. Fusion **28**, 10–23 (2016)
4. Sahoo, S., Pal, M.: Intuitionistic fuzzy competition graphs. J. Appl. Math. Comput. **52**(1–2), 37–57 (2015). https://doi.org/10.1007/s12190-015-0928-0
5. Torra, V.: Hesitant fuzzy sets. Int. J. Intell. Syst. **25**(6), 529–539 (2010)
6. Xu, Z.S., Xia, M.M.: Distance and similarity measures for hesitant fuzzy sets. Inf. Sci. **181**(11), 2128–2138 (2011)
7. Xu, Z.S., Zhang, X.L.: Hesitant fuzzy multi-attribute decision making based on TOPSIS with incomplete weight information. Knowl.-Based Syst. **52**, 53–64 (2013)
8. Li, D.-F.: Matrix games with goals of intuitionistic fuzzy sets and linear programming method. In: Decision and Game Theory in Management with Intuitionistic Fuzzy Sets. SFSC, vol. 308, pp. 399–420. Springer, Heidelberg (2014). https://doi.org/10.1007/978-3-642-40712-3_10
9. Ashraf, S., Abdullah, S., Mahmood, T., Ghani, F., Mahmood, T.: Spherical fuzzy sets and their applications in multi-attribute decision making problems. J. Intell. Fuzzy Syst. **36**(3), 2829–2844 (2019)
10. Siregar, D., et al.: Multi-attribute decision making with VIKOR method for any purpose decision. J. Phys. Conf. Ser. **1019**(1), 012034 (2018)
11. Li, M.J., Pan, Y.X., Xu, L.M., Lu, J.H.: Improved dynamic TOPSIS evaluation method of interval numbers. J. Syst. Sci. Math. Sci. **41**(7), 1891–1904 (2021)
12. Vafaei, N., Ribeiro, R.A., Camarinha-Matos, L.M.: Data normalisation techniques in decision making: case study with TOPSIS method. Int. J. Inf. Decision Sci. **10**(1), 19–38 (2018)
13. Zulqarnain, R.M., Saeed, M., Ahmad, N., Dayan, F., Ahmad, B.: Application of TOPSIS method for decision making. Int. J. Scientific Math. Stat. Sci. **7**(2), 76–81 (2020)
14. Afsordegan, A., Sánchez, M., Agell, N., Zahedi, S., Cremades, L.V.: Decision making under uncertainty using a qualitative TOPSIS method for selecting sustainable energy alternatives. Int. J. Environ. Sci. Technol. **13**(6), 1419–1432 (2016). https://doi.org/10.1007/s13762-016-0982-7
15. Zhao, H., Xu, Z.S.: Intuitionistic fuzzy multi-attribute decision making with ideal-point-based method and correlation measure. J. Intell. Fuzzy Syst. **30**(2), 747–757 (2016)
16. Ouyang, Y., Pedrycz, W.: A new model for intuitionistic fuzzy multi-attributes decision making. Eur. J. Oper. Res. **249**(2), 677–682 (2016)
17. Xu, Z.: Multi-person multi-attribute decision making models under intuitionistic fuzzy environment. Fuzzy Optim. Decis. Making **6**(3), 221–236 (2007)
18. Garg, H., Kumar, K.: Improved possibility degree method for ranking intuitionistic fuzzy numbers and their application in multiattribute decision-making. Granular Comput. **4**(2), 237–247 (2019)
19. Gupta, P., et al.: A new method for intuitionistic fuzzy multiattribute decision making. IEEE Trans. Syst. Man Cybernet. Syst. **46**(9), 1167–1179 (2015)
20. Wei, G.: Some induced geometric aggregation operators with intuitionistic fuzzy information and their application to group decision making. Appl. Soft Comput. **10**(2), 423–431 (2010)

21. Li, D.F.: TOPSIS-based nonlinear-programming methodology for multiattribute decision making with interval-valued intuitionistic fuzzy sets. IEEE Trans. Fuzzy Syst. **18**(2), 299–311 (2010)
22. Xian, S., et al.: A novel approach for linguistic group decision making based on generalized interval-valued intuitionistic fuzzy linguistic induced hybrid operator and TOPSIS. Int. J. Intell. Syst. **33**(2), 288–314 (2018)
23. Mishra, A.R., Rani, P.: Information measures based TOPSIS method for multicriteria decision making problem in intuitionistic fuzzy environment. Iranian J. Fuzzy Syst. **14**(6), 41–63 (2017)
24. Liu, P., Cui, H., Cao, Y., Hou, X., Zou, L.: A method of multimedia teaching evaluation based on fuzzy linguistic concept lattice. Multimedia Tools Appl. **78**(21), 30975–31001 (2019). https://doi.org/10.1007/s11042-019-7669-2
25. Zou, L., et al.: A knowledge reduction approach for linguistic concept formal context. Inf. Sci. **524**, 165–183 (2020)

Properties of Fuzzy λ-Approximate Context-Free Languages

Ping Li, Huanhuan Sun(✉), Yongxia He, and Yanping Yang

School of Mathematics and Statistics, Shaanxi Normal University, Xi'an 710119, China
{liping,he-yx1203}@snnu.edu.cn
2506978810@qq.com

Abstract. The approximation of fuzzy languages is one of the important problems, it is more practical to consider its approximate implementation if the fuzzy languages cannot be realized by a automaton. In this paper, for a real number $\lambda \in [0,1]$, we give the definition of fuzzy λ-approximate context-free languages and their Pumping lemma. Then we study the algebraic properties of fuzzy λ-approximate context-free languages. Firstly, we give a hierarchical characterization of fuzzy λ-approximate context-free languages different from the previous fuzzy languages. Furthermore, we show that fuzzy λ-approximate context-free languages are closed under the operations union, concatenation and Kleene closure, but not closed under the operations intersection, complement, Łukasiewicz addition, Łukasiewicz product and Łukasiewicz implication. Finally, we discuss the relationships between fuzzy λ-approximate context-free languages and fuzzy λ-approximate regular languages, and prove that the intersection of a fuzzy λ-approximate context-free languages and a fuzzy λ-approximate regular languages is a fuzzy λ-approximate context-free languages.

Keywords: Fuzzy λ-approximate regular languages · Fuzzy λ-approximate context-free languages · Pumping lemma · Closure of operations

1 Introduction

Automata and formal languages give a model of abstract and formal representation of problems, which play important roles in the applications of computer science. In 1965, Professor L.A. Zadeh [26] creatively put forward fuzzy sets theory, which provides a powerful mathematical tool for the development of artificial intelligence field. With the introduction of fuzzy sets, the ability of

Supported by the National Natural Science Foundation of China [grant numbers 61673250]; the Fundamental Research Funds for the Central Universities [grant number Gk201803008]; and the China CEE 410 University Joint Education Project [grant number 202008].

Y. Chen and S. Zhang (Eds.): AILA 2022, CCIS 1657, pp. 117–130, 2022.
https://doi.org/10.1007/978-981-19-7510-3_9

automata to recognize languages and the ability of grammars to generate languages have been studied by employing fuzzy sets theory, then fuzzy automata have been produced. Since then, fuzzy languages and fuzzy automata have been widely studied as methods to bridge the gap between the accuracy and fuzziness of computer languages [8–10,14,15]. In the 1970s, Gaines and Kohout discussed the logical basis of fuzzy automata [6]. Santos further studied the fuzzy languages of fuzzy automata recognition and its closure under various operations, proposed the context-free fuzzy grammars, and discussed the properties of the language generated by it [18,19]. Bucurescu, Pascu [5] and Xing [23] generalized pushdown automata to fuzzy pushdown automata. Asveld studied basic properties of fuzzy pushdown automata [2,3]. Since then, many scholars have done a lot of work and achieved fruitful results [1,4,12,16,17,20,21,24].

The approximation of fuzzy languages is one of the important problems. In the case that fuzzy languages cannot be realized by a automaton, it is more practical to consider its approximate implementation. Some achievements have been made in the approximation of fuzzy languages. For example, Li Yongming [11] shows that a nondeterministic fuzzy automaton under *max*-$*$ compositional inference for some t-norm $*$ can be approximated by some deterministic fuzzy automata with arbitrary given accuracy if the t-norm $*$ satisfies the weakly finite generated condition. That is, for a t-norm $*$ which satisfies the weakly finite generated condition, nondeterministic fuzzy automata under *max*-$*$ compositional inference are equivalent to nondeterministic fuzzy automata under *max-min* compositional inference in the approximate sense. Finally, a sufficient condition for fuzzy languages to be approximated by deterministic fuzzy automata is given. In [22], the approximation of fuzzy languages is considered by fuzzy context-free grammars (shortly, $FCFG$). Firstly, this paper discusses the $FCFG$ under different kind of t-norms, and gives the necessary and sufficient conditions for the equivalence of fuzzy context-free *max-min* grammars and fuzzy context-free *max*-$*$ grammars. Secondly, the condition that fuzzy context-free *max*-$*$ grammars can be approximated by fuzzy context-free *max-min* grammars with arbitrary accuracy is given. Finally, a sufficient condition for fuzzy languages to be approximated by fuzzy context-free *max-min* grammars is given. In [25], the definition of fuzzy λ-approximate regular languages (shortly, $F\lambda$-ARL) and their Pumping lemma are given. Then the sufficient and necessary conditions for fuzzy languages to be ϵ-approximated by deterministic fuzzy automata are given. Finally, according to an ϵ-approximate equivalence relation, a polynomial-time algorithm to construct a minimal deterministic fuzzy automaton ϵ-accepting a fuzzy regular language is given. The study about the approximation of fuzzy regular languages is well established, but the study about the approximation of fuzzy context-free languages is relatively rare. In this paper, for a real number $\lambda \in [0,1]$, we define the concept of fuzzy λ-approximate context-free languages (shortly, $F\lambda$-$ACFL$) and discuss their algebraic properties.

The main work of this paper is arranged as follows. In Sect. 2, we review some relevant concepts of fuzzy languages and fuzzy grammars and give the Pumping lemma in fuzzy context-free languages (shortly, $FCFL$). In Sect. 3, for a real number $\lambda \in [0,1]$, we give the notion of $F\lambda$-$ACFL$ and give their Pumping lemma. In Sect. 4, the algebraic properties and a hierarchical characterization of $F\lambda$-$ACFL$ are also given. In Sect. 5, we discuss the closure of various operations of $F\lambda$-$ACFL$ and obtain the operations about union, concatenation and Kleene closure are closed of $F\lambda$-$ACFL$, but not closed under the operations intersection, complement, Łukasiewicz addition, Łukasiewicz product and Łukasiewicz implication. Furthermore, we discuss the relationships between $F\lambda$-$ACFL$ and $F\lambda$-ARL and prove that the intersection of a $F\lambda$-$ACFL$ and a $F\lambda$-ARL is also a $F\lambda$-$ACFL$.

2 Preliminaries

Now we introduce some related concepts about fuzzy languages. And we use $\vee$ and $\wedge$ to represent the supremum operation and infimum operation on $[0,1]$, respectively.

Definition 1 ([13]). Let Σ be a nonempty finite set of symbols, Σ^* denote the set of all words of finite length over Σ and ε denote the empty word. For any $\theta \in \Sigma^*$, $|\theta|$ denotes the length of θ. A fuzzy language A over Σ^* is a function from Σ^* to $[0,1]$.

Generally, we denote the set of all fuzzy languages over Σ^* as $FL(\Sigma^*)$.

For a $A \in FL(\Sigma^*)$, we define the set $R(A) = \{A(\theta)|A(\theta) > 0, \theta \in \Sigma^*\}$ and call $R(A)$ is the image set of A. $\forall a \in R(A)$, $A_a = \{\theta \in \Sigma^*|A(\theta) \geq a\}$ is called the a-cut of A.

For fuzzy languages A and B over Σ^* are called to be equal if $A(\theta) = B(\theta)$ for all $\theta \in \Sigma^*$.

We introduce some operations on $FL(\Sigma^*)$, which are union, intersection, complement, concatenation and Kleene closure of language-theoretic.

Definition 2 ([13]). Let $A, B \in FL(\Sigma^*)$.

(1) The union of A and B is denoted as $A \cup B \in FL(\Sigma^*)$, $\forall \theta \in \Sigma^*$,

$$(A \cup B)(\theta) = A(\theta) \vee B(\theta).$$

(2) The intersection of A and B is denoted as $A \cap B \in FL(\Sigma^*)$, $\forall \theta \in \Sigma^*$,

$$(A \cap B)(\theta) = A(\theta) \wedge B(\theta).$$

(3) The complement of A is denoted as $A^c \in FL(\Sigma^*)$,$\forall \theta \in \Sigma^*$,

$$A^c(\theta) = 1 - A(\theta).$$

(4) The concatenation of A and B is denoted as $AB \in FL(\Sigma^*)$, $\forall \theta \in \Sigma^*$,

$$(AB)(\theta) = \bigvee_{\theta=\theta_1\theta_2} (A(\theta_1) \wedge B(\theta_2)).$$

(5) The Kleene closure of A is denoted as $A^* \in FL(\Sigma^*)$, $\forall\theta \in \Sigma^*$,

$$A^*(\theta) = \bigvee_{k=0}^{\infty} A^k(\theta),$$

where $A^0(\theta) = \begin{cases} 1, \theta = \varepsilon \\ 0, \theta \neq \varepsilon \end{cases}$, $A^1 = A$, $A^{k+1} = A^k A$, $k \in N$.

Now we introduce three Łukasiewicz operations on $FL(\Sigma^*)$.

Definition 3 ([13]). Let $A, B \in FL(\Sigma^*)$.

(1) The Łukasiewicz addition of A and B is denoted as $A \oplus B \in FL(\Sigma^*)$, $\forall\theta \in \Sigma^*$,

$$(A \oplus B)(\theta) = (A(\theta) + B(\theta)) \wedge 1.$$

(2) The Łukasiewicz product of A and B is denoted as $A \otimes B \in FL(\Sigma^*)$, $\forall\theta \in \Sigma^*$,

$$(A \otimes B)(\theta) = (A(\theta) + B(\theta) - 1) \vee 0.$$

(3) The Łukasiewicz implication of A and B is denoted as $A \to B \in FL(\Sigma^*)$, $\forall\theta \in \Sigma^*$,

$$(A \to B)(\theta) = (1 - A(\theta) + B(\theta)) \wedge 1.$$

Clearly, we obtain $A \otimes B = (A^c \oplus B^c)^c$ and $A \to B = A^c \oplus B$.

Fuzzy grammars can generate fuzzy languages. Now we give the definition of fuzzy grammars.

Definition 4 ([13]). A fuzzy grammar is a system $G = (N, T, P, S)$, where N and T are non-empty finite sets and $N \cap T = \phi$, the elements of N and T are called nonterminals and terminals, respectively. $S \in N$ is called an initial symbol, P is called a finite set of fuzzy productions. And for a fuzzy production $u \xrightarrow{\rho} v \in P$, $\rho \in (0, 1]$ represents the degree of membership of the generation rule $u \to v$, $\forall u \in (N \cup T)^* N (N \cup T)^*$, $v \in (N \cup T)^*$. We specify that S appears only on the left of fuzzy productions.

We say that $\alpha v \beta$ can be directly derived from $\alpha u \beta$ if $u \xrightarrow{\rho} v \in P$, then $\alpha u \beta \xrightarrow{\rho} \alpha v \beta \in P$ for any $\alpha, \beta \in (N \cup T)^*$, and we write it as $\alpha u \beta \Rightarrow^{\rho} \alpha v \beta$. For $u_1, u_2, \ldots, u_p \in (N \cup T)^*$, if $u_1 \Rightarrow^{\rho_1} u_2 \Rightarrow^{\rho_2} \ldots \Rightarrow^{\rho_{p-1}} u_p$, then we say u_1 derives u_p and write it as $u_1 \Rightarrow^{\rho}_{p-1} u_p$, where $\rho = \rho_1 \wedge \rho_2 \wedge \ldots \wedge \rho_{p-1}$. We use $u \Rightarrow^{\rho}_{*} v$ to express that there exists a positive integer n such that $u \Rightarrow^{\rho}_{n} v$.

The fuzzy language $L(G) : T^* \to [0, 1]$ generated by the fuzzy grammar $G = (N, T, P, S)$ is defined as,

$$L(G)(\theta) = \vee \{\rho \mid S \Rightarrow^{\rho}_{*} \theta\}$$

for all $\theta \in T^*$.

Two fuzzy grammars G_1 and G_2 are said to be equivalent if $L(G_1) = L(G_2)$.

Definition 5 ([13]). Let $G = (N, T, P, S)$ be a fuzzy grammar. If $|u| \leq |v|$ and $u \in N$, for any $u \xrightarrow{\rho} v \in P$, then G is called a fuzzy context-free grammar, and $L(G)$ is called a fuzzy context-free language.

Generally, we denote the set of all fuzzy context-free languages over Σ^* as $FCFL(\Sigma^*)$.

Definition 6 ([13]). Let $G = (N, T, P, S)$ be a $FCFG$. If the fuzzy productions in G have the following form

$$A \xrightarrow{\rho} BC \quad or \quad A \xrightarrow{\rho} a \quad or \quad S \xrightarrow{\rho} \varepsilon$$

$\forall A, B, C \in N$, $a \in T$, $\rho \in [0, 1]$, then G is called a fuzzy Chomsky normal form grammar (shortly, $FCNFG$).

Theorem 1 ([13]). *For any $FCFG$, there exists an equivalent $FCNFG$.*

We give the Pumping lemma in $FCFL(\Sigma^*)$. The content of the syntax tree in the proof of Theorem 2 can be found in [7], so it will not be repeated.

Theorem 2. *Let $A \in FCFL(\Sigma^*)$, there is a positive integer p, $\forall \theta \in \Sigma^*$, if $|\theta| \geq p$, then there exist $u, v, w, x, y \in \Sigma^*$ such that $\theta = uvwxy$ and*

(1) $|vwx| \leq p$,
(2) $|vx| > 0$,
(3) for any positive integer k, $A(uv^kwx^ky) = A(uvwxy)$.

Proof. It follows from Theorem 1 that there exist a $FCNFG$ $G = (N, T, P, S)$ such that $L(G) = A$. For any $\theta \in \Sigma^*$, when t is the longest path in θ's syntax tree, the inequality $|\theta| \leq 2^{t-1}$ holds. In fact, the equal sign holds only if the syntax tree of θ is a full binary tree as shown in Fig. 1, where $h = 2^{t-2}$, g=2^{t-1} in Fig. 1. At this point, the length of each path is t, and there are t non-lead nodes marked as syntax variables and a lead node marked as terminal on each path.

Now take $p = 2^{|N|}$. For $\theta \in \Sigma^*$, $|\theta| \geq p$, we know that there is at least one path in the syntax tree of θ whose length is greater than or equal to $|N| + 1$ and the number of non-leaf nodes in the path is greater than or equal to $|N|+1$. Take the longest path s in the tree, the number of non-leaf nodes in s is greater than or equal to $|N|+1$, and their markers are grammar variables. Since $|N|+1 > |N|$, there must be different nodes among these non-leaf nodes labeled with the same grammar variable. Now take the two nodes in path s that are closest to the leaf, v_1 and v_2, and they are marked with the same grammar variable A. For the sake of certainty, let's say that v_1 is the ancestor node of v_2. Obviously, the length of the path from v_1 to the leaf node is less than or equal to $|N| + 1$. As shown in Fig. 2, assume that

(1) All leaf nodes on the left of node v_1 are marked with a string u from left to right and the degree of membership is ρ_1.

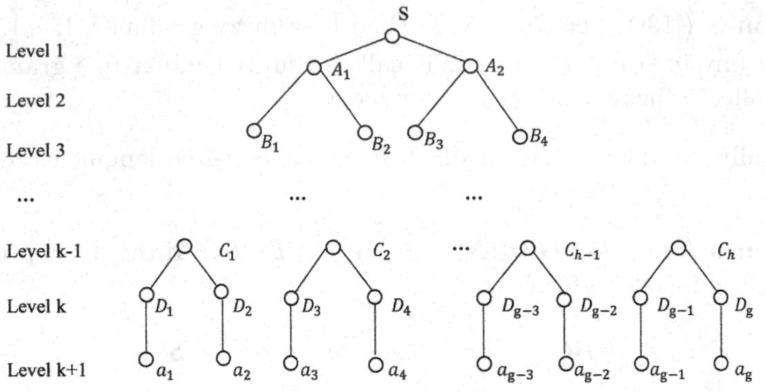

Fig. 1. A full binary tree

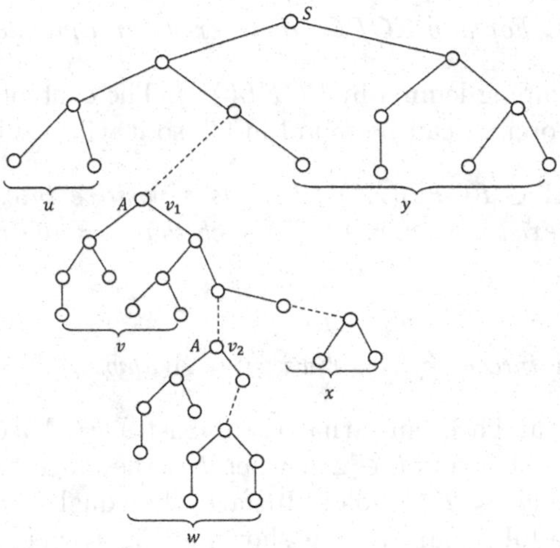

Fig. 2. The derived tree of θ

(2) In the subtree with node v_1 as the root, all leaf nodes on the left of node v_2 are marked with a string v from left to right and the degree of membership is ρ_2.
(3) The result of the subtree with node v_2 as the root is w and the degree of membership is ρ_3.
(4) In the subtree with node v_1 as the root, all the leaf nodes on the right of node v_2 are marked with a string x from left to right and the degree of membership is ρ_4.
(5) All the leaf nodes on the right of node v_1 are marked with a string y from left to right and the degree of membership is ρ_5.

Then $\theta = uvwxy$. Noticed with v_1 as the root of the subtree of the maximum path length less than or equal to $|N| + 1$, so vwx as a result of v_1 meet

$|vwx| \leq 2^{(|N|+1)-1} = 2^{|N|} = p$. And because G is a $FCFG$, v_2 is the off-spring of v_1 and v_2 is marked as A variable, so $|vx| > 0$. Hence $S \Rightarrow_*^{\rho_1 \wedge \rho_5} uAy \Rightarrow_*^{\rho_1 \wedge \rho_2 \wedge \rho_4 \wedge \rho_5} uvAxy \Rightarrow_*^{\rho_1 \wedge \rho_2 \wedge \rho_3 \wedge \rho_4 \wedge \rho_5} uvwxy$. Clearly, for any positive integer k, $A \Rightarrow_*^{\rho_2^k \wedge \rho_4^k} v^k A x^k \Rightarrow_*^{\rho_2 \wedge \rho_3 \wedge \rho_4} v^k w x^k$, where $\rho_2^k = \rho_2 \wedge \rho_2 \wedge \cdots \wedge \rho_2$. So, $S \Rightarrow_*^{\rho_1 \wedge \rho_5} uAy \Rightarrow_*^{\rho_1 \wedge \rho_2 \wedge \rho_4 \wedge \rho_5} uv^k A x^k y \Rightarrow_*^{\rho_1 \wedge \rho_2 \wedge \rho_3 \wedge \rho_4 \wedge \rho_5} uv^k w x^k y$. That is, $S \Rightarrow_*^{\rho} uvwxy$ and $S \Rightarrow_*^{\rho} uv^k w x^k y$, where $\rho = \rho_1 \wedge \rho_2 \wedge \rho_3 \wedge \rho_4 \wedge \rho_5$. Hence, for any positive integer k, $A(uv^k w x^k y) = \vee\{\rho | S \Rightarrow_*^{\rho} uv^k w x^k y\} = \vee\{\rho | S \Rightarrow_*^{\rho} uvwxy\} = A(uvwxy)$.

Lemma 1 ([13]). *Let $A \in FL(\Sigma^*)$, A is a fuzzy context-free language over Σ^* if and only if the image set $R(A)$ of A is a finite set and $A_a \in FCFL(\Sigma^*)$, $\forall a \in R(A)$.*

Lemma 2 ([13]). *The operations union, concatenation and Kleene closure are closed in $FCFL(\Sigma^*)$.*

Lemma 3 ([13]). *The operations intersection and complement are not closed in $FCFL(\Sigma^*)$.*

3 Fuzzy λ-Approximate Context-Free Languages and Their Pumping Lemma

Now we gives the definition of $F\lambda$-$ACFL$ and the Pumping lemma of $F\lambda$-$ACFL$.

Definition 7. Let $A \in FL(\Sigma^*)$ and $\lambda \in [0, 1]$. If there exists a $A' \in FCFL(\Sigma^*)$ such that $|A(\theta) - A'(\theta)| \leq \lambda$ for any $\theta \in \Sigma^*$, then we say that the A' λ-approximates A (or A is λ-approximated by A') and call A a fuzzy λ-approximate context-free language (shortly, $F\lambda$-$ACFL$) over Σ.

We denote the set of all fuzzy λ-approximate context-free languages over Σ^* as $F\lambda$-$ACFL(\Sigma^*)$.

To better discuss the algebraic properties in $F\lambda$-$ACFL(\Sigma^*)$, we first give the Pumping lemma in $F\lambda$-$ACFL(\Sigma^*)$.

Theorem 3. *Let $A \in F\lambda$-$ACFL(\Sigma^*)$, there is a positive integer p, $\forall \theta \in \Sigma^*$, if $|\theta| \geq p$, then there exist $u, v, w, x, y \in \Sigma^*$ such that $\theta = uvwxy$ and*

(1) $|vwx| \leq p$,
(2) $|vx| > 0$,
(3) for any positive integer k, $|A(uv^k w x^k y) - A(uvwxy)| \leq (2\lambda) \wedge 1$.

Proof. Since $A \in F\lambda$-$ACFL(\Sigma^*)$, from the definition of $F\lambda$-$ACFL(\Sigma^*)$, there exists a $A' \in FCFL(\Sigma^*)$ such that $|A(\theta) - A'(\theta)| \leq \lambda$ for any $\theta \in \Sigma^*$. Hence, it is known from Pumping Lemma in $FCFL(\Sigma^*)$ that there is a positive integer p, $\forall \theta \in \Sigma^*$, if $|\theta| \geq p$, then there exist $u, v, w, x, y \in \Sigma^*$ such that $\theta = uvwxy$, $|vwx| \leq p$, $|vx| > 0$ and for any positive integer k, $A'(uv^k w x^k y) = A'(uvwxy)$. So $|A(uv^k w x^k y) - A(uvwxy)| \leq |A(uv^k w x^k y) - A'(uv^k w x^k y)| + |A(uvwxy) - A'(uvwxy)| \leq 2\lambda$. Obviously, for any positive integer k, $|A(uv^k w x^k y) - A(uvwxy)| \leq 1$. Hence, for any positive integer k, $|A(uv^k w x^k y) - A(uvwxy)| \leq (2\lambda) \wedge 1$.

Now we give Example 1 and Example 2 to show that there exist $F\lambda$-$ACFL$ such that the equal sign and the less-than sign in (3) of Theorem 3 hold.

Example 1. Let$A \in FL(\Sigma^*)$ be defined as: $\forall\theta \in \Sigma^*$, $\Sigma = \{a\}$.

$$A(\theta) = \begin{cases} \frac{1}{4}, & \theta = a^m \text{and } m \text{ is prime,} \\ \frac{3}{4}, & \text{otherwise.} \end{cases}$$

Let $\lambda = \frac{1}{4}$. Take $A'(\theta) = \frac{1}{2}$, $\forall\theta \in \Sigma^*$, then A' is a $FCFL(\Sigma^*)$, $\forall\theta \in \Sigma^*$, we have $|A(\theta) - A'(\theta)| \leq \lambda$. So $A \in F\lambda\text{-}ACFL(\Sigma^*)$. We take $\theta = a^{p+d}$, where $p + d$ is prime, then $A(\theta) = \frac{1}{4}$. Let $u = a^s$, $v = a^t$, $w = a^j$, $x = a^{i-t-j}$, $y = a^{p+d-s-i}$, where $0 \leq t + j < i \leq p$. It is easy to prove that $\theta = uvwxy$, $|vwx| = t + j + i - t - j = i \leq p$, $|vx| = t + i - t - j = i - j > 0$. And for any positive integer k, $uv^kwx^ky = a^{(k-1)(i-j)+(p+d)}$. If $k = p + d + 1$, then $uv^kwx^ky = a^{(p+d)(i-j+1)}$ and $A(uv^kwx^ky) = \frac{3}{4}$. Therefore, there exists a positive integer k such that $|A(uv^kwx^ky) - A(uvwxy)| = |\frac{3}{4} - \frac{1}{4}| = \frac{2}{4} = (2\lambda) \wedge 1$.

The Example 1 shows that there exists a $F\lambda$-$ACFL$ such that the equal sign in (3) of Theorem 3 holds.

Example 2. Let $A \in FL(\Sigma^*)$ be defined as: $\forall\theta \in \Sigma^*$, $\Sigma = \{a\}$.

$$A(\theta) = \begin{cases} \frac{1}{3}, & \theta = a^m \text{and } m \text{ is prime,} \\ \frac{2}{3}, & \text{otherwise.} \end{cases}$$

Let $\lambda = \frac{1}{3}$. Take $A'(\theta) = \frac{1}{2}$, $\forall\theta \in \Sigma^*$, then A' is a $FCFL(\Sigma^*)$, $\forall\theta \in \Sigma^*$, we have $|A(\theta) - A'(\theta)| \leq \lambda$. So $A \in F\lambda\text{-}ACFL(\Sigma^*)$. We take $\theta = a^{p+d}$, where $p + d$ is prime, then $A(\theta) = \frac{1}{3}$. Let $u = a^s$, $v = a^t$, $w = a^j$, $x = a^{i-t-j}$, $y = a^{p+d-s-i}$, where $0 \leq t + j < i \leq p$. It is easy to prove that $\theta = uvwxy$, $|vwx| = t + j + i - t - j = i \leq p$, $|vx| = t + i - t - j = i - j > 0$. And for any positive integer k, $uv^kwx^ky = a^{(k-1)(i-j)+(p+d)}$. If $k = p + d + 1$, then $uv^kwx^ky = a^{(p+d)(i-j+1)}$ and $A(uv^kwx^ky) = \frac{2}{3}$. Therefore, there exists a positive integer k such that $|A(uv^kwx^ky) - A(uvwxy)| = |\frac{2}{3} - \frac{1}{2}| = \frac{1}{3} < (2\lambda) \wedge 1$.

The Example 2 shows that there exists a $F\lambda$-$ACFL$ such that the less-than sign in (3) of Theorem 3 holds.

4 Hierarchical Characterization in $F\lambda$-$ACFL(\Sigma^*)$

Now we study the relationships between $F\lambda_1$-$ACFL$ and $F\lambda_2$-$ACFL$ for two real numbers λ_1, $\lambda_2 \in [0, 1]$, and a hierarchical characterization of $F\lambda$-$ACFL$ is given.

Theorem 4 and Theorem 6 can be derived directly from Definition 7.

Theorem 4. *If* $\lambda = 0$, *then* $FCFL(\Sigma^*) = F\lambda\text{-}ACFL(\Sigma^*)$.

Theorem 5. *If* $\lambda \in [\frac{1}{2}, 1]$, *then* $F\lambda\text{-}ACFL(\Sigma^*) = FL(\Sigma^*)$.

The proof is similar to that given in [25]. Therefore, in the following discussion, $\lambda \in (0, \frac{1}{2})$ is always set.

Theorem 6. *For any* $\lambda, \lambda^{'} \in [0,1]$, *if* $\lambda \leq \lambda^{'}$, *then* $F\lambda\text{-}ACFL(\Sigma^*) \subseteq F\lambda^{'}\text{-}ACFL(\Sigma^*)$.

Corollary 1. *For any* $\lambda \in [0,1]$, $FCFL(\Sigma^*) \subseteq F\lambda\text{-}ACFL(\Sigma^*) \subseteq FL(\Sigma^*)$.

Now we give Example 3 to show that $F\lambda\text{-}ACFL$ are different from $FCFL$.

Example 3. Let fuzzy language A over Σ^* be defined as: for any $\theta \in \Sigma^*$, $\Sigma = \{a, b, c\}$.

$$A(\theta) = \begin{cases} \frac{1}{n}, & \theta = a^n b^n c^n, 0 \leq n \leq 5, \\ 0.1, \theta = a^n b^n c^n, n \geq 6, \\ 0, & \text{otherwise.} \end{cases}$$

Let $\lambda = 0.1$. $\forall \theta \in \Sigma^*$,

$$A'(\theta) = \begin{cases} \frac{1}{n}, \theta = a^n b^n c^n, 0 \leq n \leq 5, \\ 0, \text{otherwise.} \end{cases}$$

Obviously, $A' \in FCFL(\Sigma^*)$, $\forall \theta \in \Sigma^*$, we have $|A(\theta) - A'(\theta)| \leq \lambda$, so f is a $F\lambda\text{-}ACFL(\Sigma^*)$. However, $A_{0.1} = \{a^n b^n c^n | n \geq 0\}$ is not a context-free language, it can be obtained from Lemma 1 that A is not a $FCFL(\Sigma^*)$. That is, there is a $A \in FL(\Sigma^*)$ which is a $F\lambda\text{-}ACFL(\Sigma^*)$ but not a $FCFL(\Sigma^*)$.

Theorem 7. *For any* $\lambda, \lambda^{'} \in (0, \frac{1}{2})$, *if* $\lambda < \lambda^{'}$, *then* $F\lambda\text{-}ACFL(\Sigma^*) \subset F\lambda^{'}\text{-}ACFL(\Sigma^*)$.

Proof. Obviously, $F\lambda\text{-}ACFL(\Sigma^*) \subseteq F\lambda^{'}\text{-}ACFL(\Sigma^*)$. If there exists a $A \in FL(\Sigma^*)$ such that $A \in F\lambda^{'}\text{-}ACFL(\Sigma^*)$ but $A \notin F\lambda\text{-}ACFL(\Sigma^*)$, then the theorem holds.

Let $A \in FL(\Sigma^*)$ be defined as, $\forall \theta \in \Sigma^*$, $\Sigma = \{a\}$.

$$A(\theta) = \begin{cases} \frac{1}{2} - \lambda^{'}, \theta = a^m \text{ and } m \text{ is prime,} \\ \frac{1}{2} + \lambda^{'}, \text{otherwise.} \end{cases}$$

There exists a $A' \in FCFL(\Sigma^*)$ which is defined by $A'(\theta) = \frac{1}{2}$ such that A' $\lambda^{'}$-approximate A, $\forall \theta \in \Sigma^*$. Hence, $A \in F\lambda^{'}\text{-}ACFL(\Sigma^*)$. Assume that $A \in F\lambda\text{-}ACFL(\Sigma^*)$. We take $\theta = a^{p+d}$, where $p + d$ is prime, then $A(\theta) = \frac{1}{2} - \lambda^{'}$. Let $u = a^s$, $v = a^t$, $w = a^j$, $x = a^{i-t-j}$, $y = a^{p+d-s-i}$, where $0 \leq t + j < i \leq p$. It is obvious that $\theta = uvwxy$, $|vwx| = i \leq p$, $|vx| = i - j > 0$. And for any positive integer k, $uv^k wx^k y = a^{(k-1)(i-j)+(p+d)}$. If $k = p + d + 1$, then $uv^k wx^k y = a^{(p+d)(i-j+1)}$ and $A(uv^k wx^k y) = \frac{1}{2} + \lambda^{'}$. Therefore, there exists a positive integer k such that $|A(uv^kwx^ky) - A(uvwxy)| = 2\lambda' > 2\lambda$. Hence our assumption is invalid. We have $A \notin F\lambda\text{-}ACFL(\Sigma^*)$. Therefore, $F\lambda\text{-}ACFL(\Sigma^*) \subset F\lambda'\text{-}ACFL(\Sigma^*)$.

Compared with the previous hierarchy in fuzzy languages, fuzzy languages can be divided into fuzzy regular languages, $FCFL(\Sigma^*)$, fuzzy context-related languages and fuzzy recursive enumerable languages. A detail hierarchical characterization in $FCFL(\Sigma^*)$ can be obtained from Theorem 4, Theorem 5 and Theorem 7.

Theorem 8. *If* $0 < \lambda_1 < \lambda_2 < \frac{1}{2} < \lambda_3 < 1$, *then* $FCFL(\Sigma^*) = F0\text{-}ACFL(\Sigma^*) \subset F\lambda_1\text{-}ACFL(\Sigma^*) \subset F\lambda_2\text{-}ACFL(\Sigma^*) \subset F\frac{1}{2}\text{-}ACFL(\Sigma^*) = F\lambda_3\text{-}ACFL(\Sigma^*) = F1\text{-}ACFL(\Sigma^*) = FL(\Sigma^*)$.

5 Closure of Operations in $F\lambda$-$ACFL(\Sigma^*)$

The closure of the operations mentioned above in $F\lambda\text{-}ACFL(\Sigma^*)$ is discussed below.

First we give the Lemma 4 which is an important result in mathematical analysis.

Lemma 4 ([11]). *Let* $A, B : X \to [0, 1]$ *be two functions, then*

(1) $|\bigvee_{\sigma\in X} A(\sigma) - \bigvee_{\sigma\in X} B(\sigma)| \le \bigvee_{\sigma\in X} |A(\sigma) - B(\sigma)|$;
(2) $|\bigwedge_{\sigma\in X} A(\sigma) - \bigwedge_{\sigma\in X} B(\sigma)| \le \bigvee_{\sigma\in X} |A(\sigma) - B(\sigma)|$.

Theorem 9. *For any* $\lambda \in (0, \frac{1}{2})$, *the operations union, concatenation and Kleene closure are closed in* $F\lambda\text{-}ACFL(\Sigma^*)$.

Proof. Let $\lambda \in (0, \frac{1}{2})$. If $A, B \in F\lambda\text{-}ACFL(\Sigma^*)$, then there exist $A', B' \in FCFL(\Sigma^*)$ such that $|A(\theta) - A'(\theta)| \le \lambda$ and $|B(\theta) - B'(\theta)| \le \lambda$ for any $\theta \in \Sigma^*$. Since A', $B' \in FCFL(\Sigma^*)$, thus $A' \cup B'$, $A'B'$, $A'^* \in FCFL(\Sigma^*)$ by Lemma 2.

For the union operation, we have $|(A \cup B)(\theta) - (A' \cup B')(\theta)| = |A(\theta) \vee B(\theta) - A'(\theta) \vee B'(\theta)| \le |A(\theta) - A'(\theta)| \vee |B(\theta) - B'(\theta)| \le \lambda$, $\forall \theta \in \Sigma^*$. So, $A \cup B \in F\lambda\text{-}ACFL(\Sigma^*)$.

For the concatenation operation, then we have $|(AB)(\theta) - (A'B')(\theta)| = |\bigvee_{\theta=\theta_1\theta_2}(A(\theta_1) \wedge B(\theta_2)) - \bigvee_{\theta=\theta_1\theta_2}(A'(\theta_1) \wedge B'(\theta_2))| \le \bigvee_{\theta=\theta_1\theta_2} |A(\theta_1) \wedge B(\theta_2) - A'(\theta_1) \wedge B'(\theta_2)| \le \bigvee_{\theta=\theta_1\theta_2}(|A(\theta) - A'(\theta)| \vee |B(\theta) - B'(\theta)|) \le \lambda$, for any $\theta \in \Sigma^*$. Therefore, $AB \in F\lambda\text{-}ACFL(\Sigma^*)$.

And we have that for any positive integer k, $A^k \in F\lambda\text{-}ACFL(\Sigma^*)$.

For the Kleene closure operation, we have $|A^*(\theta) - A'^*(\theta)| = |\bigvee_{k=0}^{\infty} A^k(\theta) - \bigvee_{k=0}^{\infty} A'^k(\theta)| \le \bigvee_{k=0}^{\infty} |A^k(\theta) - A'^k(\theta)| \le \lambda$ for any $\theta \in \Sigma^*$. Therefore, $A^* \in F\lambda\text{-}ACFL(\Sigma^*)$.

Theorem 10. *For any* $\lambda \in (0, \frac{1}{2})$, *the operations intersection and complement are not closed in* $F\lambda\text{-}ACFL(\Sigma^*)$.

Proof. For the intersection operation, if there exist $A, B \in F\lambda\text{-}ACFL(\Sigma^*)$ such that $A \cap B \notin F\lambda\text{-}ACFL(\Sigma^*)$, then the theorem holds.

Let $\Sigma = \{0, 1, 2\}$, $\lambda \in (0, \frac{1}{2})$, we take $a \in (0, 1 - 2\lambda]$, and $A, B \in F\lambda\text{-}ACFL(\Sigma^*)$ are defined as follows, $\forall \theta \in \Sigma^*$,

$$A(\theta) = \begin{cases} a + 2\lambda, & \theta = 0^n 1^n 2^m, m, n \geq 0, \\ 0, & \text{otherwise.} \end{cases}$$

$$B(\theta) = \begin{cases} a + 2\lambda, & \theta = 0^n 1^m 2^m, m, n \geq 0, \\ 0, & \text{otherwise.} \end{cases}$$

Since A and B are $FCFL(\Sigma^*)$, then $A, B \in F\lambda\text{-}ACFL(\Sigma^*)$, we can obtain that, $\forall \theta \in \Sigma^*$,

$$(A \cap B)(\theta) = \begin{cases} a + 2\lambda, & \theta = 0^n 1^n 2^n, n \geq 0, \\ 0, & \text{otherwise.} \end{cases}$$

We show that $A \cap B \notin F\lambda\text{-}ACFL(\Sigma^*)$.

Assume that $A \cap B \in F\lambda\text{-}ACFL(\Sigma^*)$. We take $\theta = 0^p 1^p 2^p$, then $(A \cap B)(\theta) = a + 2\lambda$. We need to prove that there don't exist $u, v, w, x, y \in \Sigma^*$ such that $\theta = uvwxy$ and $|(A \cap B)(uv^k wx^k y) - (A \cap B)(uvwxy)| \leq 2\lambda$ for any positive integer k. That is, for any $u, v, w, x, y \in \Sigma^*$, we can find a special k such that $|(A \cap B)(uv^k wx^k y) - (A \cap B)(uvwxy)| > 2\lambda$. Since $|vwx| \leq p$, v,w and x together cannot have three letters. Therefore, we now discuss it in two cases below.

(1) When vwx takes only one letter, let $vwx = 0^s (0 < s \leq p)$. we suppose $v = 0^i$, $w = 0^j$, $x = 0^{s-i-j}$, where $s - j > 0$. Then for any positive integer k, $uv^k wx^k y = 0^{p+(k-1)(s-j)} 1^p 2^p$.
If $k = 2$, because $s - j > 0$, $p + (k-1)(s-j) = p + s - j \neq p$. So $(A \cap B)(uv^2 wx^2 y) = 0$. Hence, there exists a positive integer k such that $|(A \cap B)(uv^k wx^k y) - (A \cap B)(uvwxy)| = a + 2\lambda > 2\lambda$.
In a similar way, when $vwx = 1^s$ or $vwx = 2^s$, there exists a positive integer k such that $|(A \cap B)(uv^k wx^k y) - (A \cap B)(uvwxy)| = a + 2\lambda > 2\lambda$.

(2) When vwx takes two letters, let $vwx = 0^h 1^f (h + f \geq 1)$, and then $u = 0^{p-h}$, $y = 1^{p-f} 2^p$.
 (i) If $v = 0^i$, $w = 0^j$, $x = 0^{h-i-j} 1^f$, where $h - j + f > 0$, then $uv^k wx^k y = 0^{p-h+ki+j} (0^{h-i-j} 1^f)^k 1^{p-f} 2^p$.
 If $k = 2$, $uv^2 wx^2 y = 0^{p-h+2i+j} (0^{h-i-j} 1^f)^2 1^{p-f} 2^p$, then $(A \cap B)(uv^2 wx^2 y) = 0$. Hence, there exists a positive integer k such that $|(A \cap B)(uv^k wx^k y) - (A \cap B)(uvwxy)| = a + 2\lambda > 2\lambda$.
 (ii) If $v = 0^i$, $w = 0^{h-i} 1^j$, $x = 1^{f-j}$, where $j > 0$ and $h - j + f > 0$, then $uv^k wx^k y = 0^{p+(k-1)i} 1^{p+(k-1)(f-j)} 2^p$.
 If $k = 2$, because $i + f - j > 0$, $p + i \neq p$ and $p + f - j \neq p$, at least one of them is true, then $(A \cap B)(uv^2 wx^2 y) = 0$. Hence, there exists a positive integer k such that $|(A \cap B)(uv^k wx^k y) - (A \cap B)(uvwxy)| = a + 2\lambda > 2\lambda$.
 (iii) If $v = 0^h 1^i$, $w = 1^j$, $x = 1^{f-i-j}$, where $j > 0$ and $h + f - j > 0$, then $uv^k wx^k y = 0^{p-h} (0^h 1^i)^k 1^{p+(k-1)(f-j)+i} 2^p$.

If $k = 2$, $uv^2wx^2y = 0^{p-h}(0^h1^i)^2 1^{p+f-j+i}2^p$, then $(A \cap B)(uv^2wx^2y) = 0$. Hence, there exists a positive integer k such that $|(A \cap B)(uv^kwx^ky) - (A \cap B)(uvwxy)| = a + 2\lambda > 2\lambda$.
In a similar way, when $vwx = 1^h2^f(h + f \geq 1)$, there exists a positive integer k such that $|(A \cap B)(uv^kwx^ky) - (A \cap B)(uvwxy)| = a + 2\lambda > 2\lambda$.

So the assumption is invalid. Therefore, $A \cap B$ is not a $F\lambda$-$ACFL(\Sigma^*)$. That is to say, the operation intersection is not closed in $F\lambda$-$ACFL(\Sigma^*)$.

According to Theorem 9, union operation is closed in $F\lambda$-$ACFL(\Sigma^*)$. If complement operation is closed in $F\lambda$-$ACFL(\Sigma^*)$, $A \cap B = ((A \cap B)^c)^c = (A^c \cup B^c)^c \in F\lambda$-$ACFL(\Sigma^*)$ for any $A, B \in F\lambda$-$ACFL(\Sigma^*)$, it is contradictory. Therefore, the operation complement is not closed in $F\lambda$-$ACFL(\Sigma^*)$.

Theorem 11. *For any $\lambda \in (0, \frac{1}{2})$, the operations Łukasiewicz addition, Łukasiewicz product and Łukasiewicz implication are not closed in $F\lambda$-$ACFL(\Sigma^*)$.*

Proof. Let $\lambda \in (0, \frac{1}{2})$. For the Łukasiewicz addition operation, if we find a $A \in F\lambda$-$ACFL(\Sigma^*)$, but $A \oplus A \notin \lambda$-$ACFL(\Sigma^*)$, then the theorem holds.

Let fuzzy language A be defined as, $\forall \theta \in \Sigma^*$, $\Sigma = \{a\}$.

$$A(\theta) = \begin{cases} 2\lambda, \ \theta = a^m \text{ and } m \text{ is prime,} \\ 0, \ \text{ otherwise.} \end{cases}$$

For any $\theta \in \Sigma^*$, we take $A'(\theta) = \lambda$, then $A' \in FCFL(\Sigma^*)$ and A' λ-approximate A. So $A \in F\lambda$-$ACFL(\Sigma^*)$. Then we have

$$(A \oplus A)(\theta) = \begin{cases} (4\lambda) \wedge 1, \ \theta = a^m \text{ and } m \text{ is prime,} \\ 0, \qquad\qquad \text{otherwise.} \end{cases}$$

We assume that $A \oplus A \in F\lambda$-$ACFL(\Sigma^*)$, then it satisfies Theorem 3. We take $\theta = a^{p+d}$, where $p + d$ is prime, then $(A \oplus A)(\theta) = (4\lambda) \wedge 1$. Let $u = a^s$, $v = a^t$, $w = a^j$, $x = a^{i-t-j}$, $y = a^{p+d-s-i}$, where $0 \leq t + j < i \leq p$. It is clear that $\theta = uvwxy$, $|vwx| = i \leq p$, $|vx| = i - j > 0$ and for any positive integer k, $uv^kwx^ky = a^{(k-1)(i-j)+(p+d)}$. If $k = p + d + 1$, then $uv^kwx^ky = a^{(d+p)(i-j+1)}$ and $(A \oplus A)(uv^kwx^ky) = 0$. Hence, there exists a positive integer k, such that $|(A \oplus A)(uv^kwx^ky) - (A \oplus A)(uvwxy)| = (4\lambda) \wedge 1 > 2\lambda$. So the assumption is invalid. That is, $A \oplus A \notin F\lambda$-$ACFL(\Sigma^*)$.

Since $\forall A, B \in FL(\Sigma^*)$, we have $A \otimes B = (A^c \oplus B^c)^c$ and $A \to B = A^c \oplus B$, Thus, the operations Łukasiewicz addition, Łukasiewicz product and Łukasiewicz implication are not closed in $F\lambda$-$ACFL(\Sigma^*)$.

Now we discuss the relationships between $F\lambda$-$ACFL(\Sigma^*)$ and $F\lambda$-$ARL(\Sigma^*)$.

Definition 8 ([25]). Let $f \in FL(\Sigma^*)$. If there exists a fuzzy regular language f' such that $|f(\theta) - f'(\theta)| \leq \lambda$, $\forall \theta \in \Sigma^*$, then we say that f' λ-approximate f(or f λ-approximated by f') and call f is a fuzzy λ-approximate regular language over Σ^*.

We usually denote the set of all fuzzy λ-approximate regular languages over Σ^* as $F\lambda\text{-}ARL(\Sigma^*)$.

Theorem 12. *For any* $\lambda \in (0, \frac{1}{2})$, $F\lambda\text{-}ARL(\Sigma^*) \subseteq F\lambda\text{-}ACFL(\Sigma^*)$.

Proof. If $A \in F\lambda\text{-}ARL(\Sigma^*)$, then there exists a fuzzy regular language A' such that $|A(\theta) - A'(\theta)| \leq \lambda$, $\forall\theta \in \Sigma^*$. Because a fuzzy regular language must be a $FCFL(\Sigma^*)$, then $A' \in FCFL(\Sigma^*)$ and $|A(\theta) - A'(\theta)| \leq \lambda$ for any $\theta \in \Sigma^*$. So $A \in F\lambda\text{-}ACFL(\Sigma^*)$. Therefore, $F\lambda\text{-}ARL(\Sigma^*) \subseteq F\lambda\text{-}ACFL(\Sigma^*)$.

Theorem 10 shows that the intersection operation in $F\lambda\text{-}ACFL(\Sigma^*)$ is not closed, but the intersection of a $F\lambda\text{-}ACFL(\Sigma^*)$ and a $F\lambda\text{-}ARL(\Sigma^*)$ is still a $F\lambda\text{-}ACFL(\Sigma^*)$, that is, Theorem 13 holds.

Theorem 13. *For any* $\lambda \in (0, \frac{1}{2})$, *the intersection of a* $F\lambda\text{-}ACFL(\Sigma^*)$ *and a* $F\lambda\text{-}ARL(\Sigma^*)$ *is a* $F\lambda\text{-}ACFL(\Sigma^*)$.

Proof. Let $A \in F\lambda\text{-}ACFL(\Sigma^*)$, $B \in F\lambda\text{-}ARL(\Sigma^*)$. From Definitions 7 and Theorem 4, we know that there exists a $A' \in FCFL(\Sigma^*)$ and a fuzzy regular language B' such that $|A(\theta) - A'(\theta)| \leq \lambda$ and $|B(\theta) - B'(\theta)| \leq \lambda$, $\forall\theta \in \Sigma^*$. And because the intersection of a fuzzy regular language and a $FCFL(\Sigma^*)$ is a $FCFL(\Sigma^*)$ [13], $A' \cap B'$ is still a $FCFL(\Sigma^*)$. So $|(A \cap B)(\theta) - (A' \cap B')(\theta)| = |A(\theta) \wedge B(\theta) - A'(\theta) \wedge B'(\theta)| \leq |A(\theta) - A'(\theta)| \vee |B(\theta) - B'(\theta)| \leq \lambda$, $\forall\theta \in \Sigma^*$. Hence, $A \cap B \in F\lambda\text{-}ACFL(\Sigma^*)$. That is, the intersection of a $F\lambda\text{-}ACFL(\Sigma^*)$ and a $F\lambda\text{-}ARL(\Sigma^*)$ is a $F\lambda\text{-}ACFL(\Sigma^*)$.

6 Conclusion

In this paper, the properties in $F\lambda\text{-}ACFL$ are studied. Firstly, we give the concept of $F\lambda\text{-}ACFL$ and their Pumping lemma. Secondly, we study the relationships between $F\lambda_1\text{-}ACFL$ and $F\lambda_2\text{-}ACFL$ for different real numbers λ_1, $\lambda_2 \in [0, 1]$, and a hierarchical characterization in $F\lambda\text{-}ACFL$ is given. Finally, we discuss the closure of various operations in $F\lambda\text{-}ACFL$ and study the relationships between $F\lambda\text{-}ACFL$ and $F\lambda\text{-}ARL$. We know if a $A \in FL(\Sigma^*)$ is a $F\lambda\text{-}ACFL$, there exists a fuzzy pushdown automaton M such that $|A(\theta) - L(M)(\theta)| \leq \lambda$, for any $\theta \in \Sigma^*$, and we call that M λ-accepts A (or A is λ-accepted by M). However, for a $F\lambda\text{-}ACFL$, the construction method of fuzzy pushdown automata λ-accepting has not been given for the time being, and will be further studied in the future.

Acknowledgements. This work was supported by the National Natural Science Foundation of China [grant numbers 61673250]; the Fundamental Research Funds for the Central Universities [grant number Gk201803008]; and the China CEE 410 University Joint Education Project [grant number 202008].

References

1. Asveld, P.R.J.: Algebraic aspects of fuzzy languages. Theoret. Comput. Sci. **293**, 417–445 (2003)
2. Asveld, P.R.J.: Fuzzy context-free languages-part 1: generalized fuzzy context-free grammars. Theoret. Comput. Sci. **347**(1–2), 167–190 (2005)
3. Asveld, P.R.J.: Fuzzy context-free languages-part 2: recognition and parsing algorithms. Theoret. Comput. Sci. **347**(1–2), 191–213 (2005)
4. Asveld, P.: A fuzzy approach to erroneous inputs in context-free language recognition. Institute of Formal Applied Linguistics Charles University (1995)
5. Bucurescu, I., Pascu, A.: Fuzzy pushdown automata. Comput. Math. **10**, 109–119 (1981)
6. Gaines, B.R., Kohout, L.J.: The logic of automata. Int. J. Gen Syst **2**, 191–208 (1976)
7. Jiang, Z.L., Jiang, S.X.: Formal Languages and Automata Theory. Tsinghua University Press, Beijing (2003). (in Chinese)
8. Lee, E.T., Zadeh, L.A.: Note on fuzzy languages. Inf. Sci. **1**, 421–434 (1969)
9. Li, P.: Pseudo semiring and its application in automata theory. Ph.D. thesis, Shanxi normal university (2010). (in Chinese with English abstract)
10. Li, P., Li, Y.M.: Algebraic properties of la-languages. Inf. Sci. **176**(21), 3232–3255 (2006)
11. Li, Y.M.: Approximation and robustness of fuzzy finite automata. Int. J. Approximate Reasoning **47**, 247–257 (2008)
12. Li, Y.M.: Approximation and universality of fuzzy Turing machines. Sci. China Ser. F: Inf. Sci. **51**(10), 1445–1465 (2008)
13. Li, Y.M., Li, P.: Theory of Fuzzy Computation. Science Press, Beijing (2016). https://doi.org/10.1007/978-1-4614-8379-3.pdf(in Chinese)
14. Li, Y.M., Pedrycz, W.: Fuzzy finite automata and fuzzy regular expressions with membership values in lattice ordered monoids. Fuzzy Sets Syst. **156**, 68–92 (2005)
15. Li, Z.H., Li, P., Li, Y.M.: The relationships among several types of fuzzy automata. Inf. Sci. **176**, 2208–2226 (2006)
16. Peng, J.Y.: The relationship between fuzzy pushdown automata and fuzzy context-free grammars. J. Sichuan Normal Uni. (Nat. Sci. Ed.) **23**(1), 27–30 (2000)
17. Qiu, D.W.: Automata theory based on quantum logic: reversibilities and pushdown automata. Theoret. Comput. Sci. **386**, 38–56 (2007)
18. Santos, E.S.: Context-free fuzzy languages. Inf. Control **26**, 1–11 (1974)
19. Santos, E.S.: Max-product grammars and languages. Inf. Sci. **9**, 1–23 (1975)
20. Wang, Y.B., Li, Y.M.: Approximation property of fuzzy context-free grammars. In: Proceedings of the First Intelligent Computing Conference(ICC2007) (2007)
21. Wang, Y.B., Li, Y.M.: Approximation of fuzzy regular grammars. Fuzzy Syst. Math. **22**(6), 130–134 (2008)
22. Wang, Y.B., Li, Y.M.: Approximation of fuzzy context-free grammars. Inf. Sci. **179**(22), 3920–3929 (2009)
23. Xing, H.: Fuzzy pushdown automata. Fuzzy Sets Syst. **158**, 1437–1449 (2007)
24. Xing, H.Y., Qiu, D.W.: Pumping lemma in context-free grammar theory based on complete residuated lattice-valued logic. Fuzzy Sets Syst. **160**, 1141–1151 (2009)
25. Yang, C., Li, Y.M.: Fuzzy ϵ-approximate regular languages and minimal deterministic fuzzy automata ϵ-accepting them. Fuzzy Sets Syst. **420**, 72–86 (2020)
26. Zadeh, L.A.: Fuzzy sets. Inf. Control **8**(3), 338–353 (1965)

Fuzzy-Classical Linguistic Concept Acquisition Approach Based on Attribute Topology

Kuo Pang[1], Ning Kang[2], Li Zou[3], and Mingyu Lu[1](✉)

[1] Information Science and Technology College, Dalian Maritime University, Dalian 116026, China
{pangkuo_p,lumingyu}@dlmu.edu.cn

[2] Software Engineering Institute, East China Normal University, Shanghai 200062, China

[3] School of Computer Science and Technology, Shandong Jianzhu University, Jinan 250102, China

Abstract. There is a large amount of fuzzy linguistic-valued data in real life. For the complexity of fuzzy linguistic concept acquisition, in this paper, we propose a Fuzzy-classical Linguistic Concept Lattice Acquisition (FLCLC) approach based on attribute topology. Specifically, the fuzzy-classical linguistic concept induced operator is proposed according to the fuzzy linguistic formal context. On this basis, the fuzzy linguistic attribute topology is generated by the fuzzy linguistic coupling relationship between attributes. Furthermore, in order to improve the interpretability of the fuzzy-classical linguistic concept acquisition process, we traverse the weight paths of the coarse fuzzy linguistic attribute topology, and the set of attributes obtained by traversing the paths is the intents of the fuzzy-classical linguistic concept, and the weights on the paths are the extents of the fuzzy-classical linguistic concept, resulting in the fuzzy-classical linguistic concept. Finally, examples are used to demonstrate the effectiveness and practicality of our proposed approach.

Keywords: Fuzzy-classical linguistic concepts · Attribute topology · Concept lattice · Fuzzy linguistic attribute topology

1 Introduction

Formal concept analysis (FCA), introduced by Wille [1] in 1982, is mainly used to describe the relationships between objects and attributes in a formal context. For a given formal context, the concept we obtain consists of two parts: intent and extent, which are used to portray the hierarchical relationships of formal concepts and to analyze the generalization and specialization relationships between concepts [2,3]. The concept lattice has been widely used in knowledge discovery [4,5], machine learning [6–8], data mining [9,10], etc.

The concept lattice is a beneficial tool for data analysis, and with the development of computer technology, the amount of data is increasing. The number and complexity of concepts generated from formal contexts has increased.

Y. Chen and S. Zhang (Eds.): AILA 2022, CCIS 1657, pp. 131–142, 2022.
https://doi.org/10.1007/978-981-19-7510-3_10

Many scholars have conducted in-depth research on the concept lattice construction algorithm [9,22]. For example, Zhang et al. [11] constructed three concept lattices and discussed their properties and relations. Mao et al. [12] defined a weighted graph in the formal context, and the complexity of the algorithm was greatly reduced by constructing a classical-fuzzy concept lattice through the weighted graph. The intent of classical-fuzzy concept is expressed by fuzzy set. Unlike classical-fuzzy concept, the extent of fuzzy-classical concept is expressed by fuzzy set. Zhang et al. [13] use the idea of Attribute Topology (AT) to represent formal contexts and search all concepts using visual paths by decomposing the AT with the top-level attributes as the core. This approach reduces both the complexity of the concept lattice construction algorithm and the redundancy of concept acquisition, while making concept lattice construction more intuitive and visualizable. However, such methods do not take into account fuzzy linguistic information. Inspired by the attribute topology, it is necessary to consider a new method of fuzzy linguistic concept lattice construction.

In daily life, using linguistic values to express information is more in line with human thinking patterns. Since Zadeh proposed fuzzy sets (FSs) [14] and linguistic variables [15], they have been applied by many scholars in evaluation, decision making, and reasoning. Zou et al. [16,17] designed a personalized teaching resource recommendation approach for the linguistic concept lattice with fuzzy object. In order to deal with the incomparable information in fuzzy linguistic values, Yang et al. [18,19] proposed a fuzzy linguistic-valued concept lattice construction and rule extraction method based on the linguistic truth-valued lattice implication algebra. However, due to the inherent uncertainty of fuzzy linguistic values, it is inefficient in constructing linguistic concept lattices.

Through the above analysis, in this paper, we propose a fuzzy-classical linguistic concept lattice construction approach based on attribute topology. The approach not only reduces the complexity of concept acquisition, but also preserves the relatively important concepts in fuzzy-classical linguistic concepts.

The rest of this paper is organized as follows. In Sect. 2, we review some basic notions relevant to FCA, linguistic term set and attribute topology. In Sect. 3, we propose the fuzzy-classical linguistic concept based on the fuzzy linguistic formal context. In Sect. 4, we generate a coarse object-oriented attribute topology based on fuzzy linguistic attribute topology, on this basis, we get the fuzzy-classical linguistic concepts according to the reachable paths. Finally, we conclude the paper with a summary and outlook for further research in Sect. 5.

2 Preliminaries

This section briefly recalls the concepts of linguistic term set [20], attribute topology [13] and FCA [1,2].

Definition 1. *[1] A formal context is a triple* (U, A, I), *where* $U = \{x_1, x_2, \cdots, x_n\}$ *is a set of objects,* $A = \{a_1, a_2, \cdots, a_n\}$ *is a set of attributes, and* I *is a binary relation between* U *and* I. *For* $x_i \in U$ *and* $a_i \in A$, *we write*

$(x_i, a_i) \in I$ as $x_i I a_i$, and say that the object x_i has the attribute a_i. Alternatively, the attribute a_i is possessed by the object x_i.

Definition 2. *[1] Let (U, A, I) be a formal context, a formal concept is a pair (X, B) where $X \subseteq U$ and $B \subseteq A$ such that $X^{\uparrow} = B$ and $X = B^{\downarrow}$. X and B are respectively the extent and intent of (X, B), two operators "$\uparrow$" and "$\downarrow$" can be defined as follows:*

$$(\bullet)^{\uparrow} : 2^U \rightarrow 2^A,$$

$$X^{\uparrow} = \{a_i | a_i \subset A,\ \forall x_i \in X,\ (x_i, a_i) \in I\}, \tag{1}$$

$$(\bullet)^{\downarrow} : 2^A \rightarrow 2^U,$$

$$B^{\downarrow} = \{x_i | x_i \in U,\ \forall a_i \in B,\ (x_i, a_i) \in I\}. \tag{2}$$

$X^{\uparrow}$ denotes the set of attributes common to all objects in X. $B^{\downarrow}$ denotes the set consisting of objects that share all the attributes in B.

We denote the set consisting of all concepts of the formal context (U, A, I) as $L(U, A, I)$, for $(X_1, B_1), (X_2, B_2) \in L(U, A, I)$, the partial order relation "$\leq$" can be defined as follows:

$$(X_1, B_1) \leq (X_2, B_2) \Leftrightarrow X_1 \subseteq X_2\, (B_2 \subseteq B_1), \tag{3}$$

$(L(U, A, I), \leq)$ forms a complete concept lattice in the partial order relation "$\leq$", which is called a concept lattice.

Definition 3. *[13] Let (U, A, I) be a formal context, the adjacency matrix of an attribute topology can be expressed as $T = (V, E)$, where $V = A$ is the vertex set of the T, E is the weight of edges in T, T can be defined as follows:*

$$E(v_i, v_j) = \begin{cases} v_i^*, & i = j \\ \emptyset, & v_i^* \cap v_j^* = \emptyset (i \neq j) \\ \emptyset, & v_i^* \cap v_j^* = v_i^* (i \neq j) \\ v_j^*, & v_i^* \cap v_j^* = v_j^* (i \neq j) \\ v_i^* \cap v_j^*, & Others \end{cases}. \tag{4}$$

According to Definition 3, in the construction process of AT, the following coupling relationships exist between attributes:

1. When $i = j$, $E(v_i, v_j)$ is the set of objects of the attributes belong.
2. When $v_i^* \cap v_j^* = \emptyset (i \neq j)$, v_i and v_j are attributes mutually exclusive from each other.
3. When $v_i^* \cap v_j^* = v_i^* (i \neq j)$, v_j contains v_i, which is represented in the topology as v_j that cannot be reached by v_i, but can be reached by v_j.
4. When $v_i^* \cap v_j^* = v_j^* (i \neq j)$, v_i contains v_j, which is represented in the topology as v_i that is not reachable from v_i.
5. When $E(v_i, v_j) = v_i^* \cap v_j^*$, v_i and v_j are compatible.

Definition 4. *[13] Let (U, A, I) be a formal context, $T = (V, E)$ be an attribute topology generated by (U, A, I), for any $v_i, v_j \in V$, the set of attributes V can be divided into three categories as follows:*

1. *If $E(v_i, v_j) \neq \emptyset$ and $E(v_j, v_i) \neq \emptyset$, then v_i is called the top-level attribute.*
2. *If $E(v_i, v_j) = \emptyset$ and $E(v_j, v_i) \neq \emptyset$, then v_i is called the bottom-level attribute.*
3. *An attribute that is neither a top-level attribute nor a bottom-level attribute is called a transition attribute.*

Definition 5. *[20] Let $S = \{s_0, s_1, \cdots, s_g\}$ be a linguistic term set consisting of an odd number of linguistic terms, where $g+1$ is the granularity of the linguistic term set, then S satisfies the following properties:*

1. *Order: $s_i > s_j \Leftrightarrow i > j$.*
2. *Negative operator: $Neg(s_i) = s_j$, where $j = g - i$.*
3. *Maximal operator: If $i \geq j$, then $Max(s_i, s_j) = s_i$.*
4. *Minimal operator: If $i \geq j$, then $Min(s_i, s_j) = s_j$.*

3 Attribute Topology Representation of Fuzzy Linguistic Formal Context

Definition 6. *[21] A fuzzy linguistic formal context is a triple (U, A, S), where $U = \{u_i \mid i \in 1, 2, \cdots, n\}$ is a set of objects, $A = \{a_j \mid j \in 1, 2, \cdots, m\}$ is a set of attributes, $S = \{s_0, s_1, \cdots, s_g\}$ is the fuzzy linguistic relationship between U and A, i.e., $S \subseteq U \times A$. For any $(x, a) \in U \times A$, such that*

$$S(x, a) = \{s_l \mid l \in 0, 1, \cdots, g\}, \tag{5}$$

where $s_0 \leq s(x, a) \leq s_g$.

In the fuzzy linguistic formal context (U, A, S), if $\varphi(x) = \{s(x, a_1), s(x, a_2), \cdots, s(x, a_m)\}$, then $\varphi(x)$ is called the fuzzy linguistic-valued set of object x on A. If $\varphi(a) = \{s(x_1, a), s(x_2, a), \cdots, s(x_n, a)\}$, then $\varphi(a)$ is called the fuzzy linguistic-valued set of attribute a on U.

Definition 7. *Let (U, A, S) be a fuzzy linguistic formal context, s_ϕ be a fuzzy linguistic threshold and $\varphi(A)$ be the fuzzy linguistic-valued set of attribute set A on U, then two operators "$\triangleright$" and "$\triangleleft$" can be defined as follows:*

$$X^{\triangleright} = \cap \{a \in A \mid \forall x \in X, s(x, a) \geq s_\phi\}, \tag{6}$$

$$B^{\triangleleft} = \cap \{(x, \varphi(a)) \mid x \in U, \forall a \in A, s(x, a) \geq s_\phi\}, \tag{7}$$

$X^{\triangleright}$ represents the attribute set of all objects in X that satisfy the fuzzy linguistic threshold s_ϕ. $B^{\triangleleft}$ represents the set of minimum fuzzy linguistic values corresponding to the attributes that satisfy the fuzzy linguistic threshold s_ϕ.

Definition 8. *Let (U, A, S) be a fuzzy linguistic formal context, a fuzzy-classical linguistic concept is a pair (Y, C) $(Y \subseteq U, C \subseteq A)$ such that $X^{\triangleright} = B$ and $X = B^{\triangleleft}$. Y and C are respectively the extent and intent of (Y, C).*

It should be noted that the fuzzy linguistic threshold s_ϕ can be selected according to the linguistic preference of users or decision makers. The larger the fuzzy linguistic threshold, the less fuzzy-classical linguistic concepts we get.

In the fuzzy linguistic formal context (U, A, S), for any $X_1, X_2 \subseteq U$ and $B_1, B_2 \subseteq A$, we denote $FVLL(U, A, S)$ as the set of all fuzzy-classical linguistic concepts in the fuzzy linguistic formal context (U, A, S). For any $(X_1, B_1), (X_2, B_2) \in FVLL(U, A, S)$, the corresponding partial order relation "$\preceq$" is as follows:

$$(X_1, B_1) \preceq (X_2, B_2) \Leftrightarrow X_1 \subseteq X_2 \,(B_2 \subseteq B_1), \tag{8}$$

where (X_1, B_1) is the fuzzy-classical linguistic subconcept of (X_2, B_2), (X_2, B_2) is the fuzzy-classical linguistic parent concept of (X_1, B_1).

Proposition 1. *Let (U, A, S) be a fuzzy linguistic formal context, $s_\phi (s_0 \le s_\phi \le s_g)$ be a fuzzy linguistic threshold, for any $B, B_1, B_2 \subseteq A$ and $X, X_1, X_2 \subseteq U$, then:*

1. $X_1 \subseteq X_2 \Rightarrow X_2^{\triangleright} \subseteq X_1^{\triangleright}$, $B_1 \subseteq B_2 \Rightarrow B_2^{\triangleleft} \subseteq B_1^{\triangleleft}$.
2. $X \subseteq X^{\triangleright\triangleleft}$, $B \subseteq B^{\triangleleft\triangleright}$.
3. $X^{\triangleright} = X^{\triangleright\triangleleft\triangleright}$, $B^{\triangleleft} = B^{\triangleleft\triangleright\triangleleft}$.
4. $(X_1 \cup X_2)^{\triangleright} = X_1^{\triangleright} \cap X_2^{\triangleright}$, $(B_1 \cup B_2)^{\triangleleft} = B_1^{\triangleleft} \cap B_2^{\triangleleft}$.
5. $(X_1 \cap X_2)^{\triangleright} \supseteq X_1^{\triangleright} \cup X_2^{\triangleright}$, $(B_1 \cap B_2)^{\triangleleft} \supseteq B_1^{\triangleleft} \cup B_2^{\triangleleft}$.

Proof. According to Definition 7 and the partial order relation between fuzzy-classical linguistic concepts, we can easily prove that properties 1, 2 hold. According to properties 1, 2, we have $X^{\triangleright\triangleleft\triangleright} \subseteq X^{\triangleright}$. Suppose $X^{\triangleright} = X$, according to property 2, we get $X^{\triangleright} \subseteq X^{\triangleright\triangleleft\triangleright}$. Therefore, property 3 is proved. Similarly, we have $B^{\triangleleft} = B^{\triangleleft\triangleright\triangleleft}$. Similarly, properties 4, 5 hold.

Example 1. We give a fuzzy linguistic formal context (U, A, S) as shown in Table 1, where $U = \{u_1, u_2, u_3, u_4, u_5\}$ represents the set of objects, $A = \{a_1, a_2, a_3, a_4, a_5\}$ represents the set of attributes, $S = \{s_0 = strongly\ disagree, s_1 = very\ disagree, s_2 = disagree, s_3 = somewhat\ disagree, s_4 = neutral, s_5 = somewhat\ agree, s_6 = agree, s_7 = very\ agree, s_8 = strongly\ agree\}$ represents the fuzzy linguistic relationship between U and A.

We take the fuzzy linguistic threshold $s_\phi = s_4$. According to Definition 7, we can obtain all fuzzy-classical linguistic concepts $FVLL(U, A, S)$ on the fuzzy linguistic formal context (U, A, S) and construct the corresponding fuzzy-classical linguistic concept lattice as shown in Table 2 and Fig. 1.

For example, fuzzy-classical linguistic concept $c_4 : (\{(u_1, s_5), (u_5, s_7)\}, \{a_1, a_5\})$ indicates that objects u_1 and u_5 evaluate attributes a_1 and a_5 to a higher

Table 1. Fuzzy linguistic formal context (U, A, S)

U	a_1	a_2	a_3	a_4	a_5
u_1	s_5	s_1	s_6	s_5	s_6
u_2	s_6	s_3	s_7	s_2	s_3
u_3	s_3	s_5	s_0	s_3	s_4
u_4	s_4	s_2	s_1	s_2	s_0
u_5	s_7	s_6	s_2	s_1	s_8

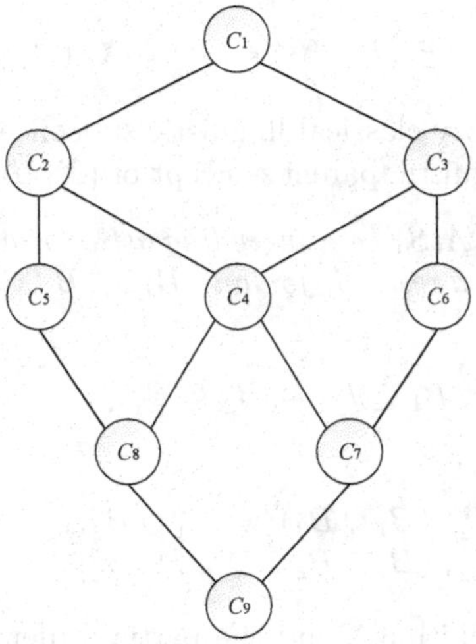

Fig. 1. Fuzzy-classical linguistic concept lattice $FVLL(U, A, S)$

Table 2. All fuzzy-classical linguistic concepts

Index	Extent	Intent
c_1	U	$\emptyset$
c_2	$\{(u_1, s_5), (u_2, s_6), (u_4, s_4), (u_5, s_7)\}$	$\{a_1\}$
c_3	$\{(u_1, s_6), (u_3, s_4), (u_5, s_8)\}$	$\{a_5\}$
c_4	$\{(u_1, s_5), (u_5, s_7)\}$	$\{a_1, a_5\}$
c_5	$\{(u_1, s_5), (u_2, s_6)\}$	$\{a_1, a_3\}$
c_6	$\{(u_3, s_4), (u_5, s_6)\}$	$\{a_2, a_5\}$
c_7	$\{(u_5, s_6)\}$	$\{a_1, a_2, a_5\}$
c_8	$\{(u_1, s_5)\}$	$\{a_1, a_3, a_4, a_5\}$
c_9	$\emptyset$	A

degree. Furthermore, the common fuzzy linguistic evaluation index of object u_1 for attribute a_1 and a_2 is s_5, and the common fuzzy linguistic evaluation index of object u_5 for attribute a_1 and a_2 is s_7.

4 Fuzzy-Classical Linguistic Concept Computation Based on Attribute Topology

4.1 Coarse Fuzzy Linguistic Attribute Topology

Definition 9. *Let (U, A, S) be a fuzzy linguistic formal context, $K = (V, G)$ be the fuzzy linguistic attribute topology of (U, A, S), where $V = A$ is the set of nodes in the fuzzy linguistic attribute topology., and G represents the set of edges in the fuzzy linguistic attribute topology. For $a_i, a_j \in V$, $G(a_i, a_j)$ is expressed as follows:*

$$G(a_i, a_j) = \begin{cases} \emptyset, & a_i^{\triangleleft} \cap a_j^{\triangleleft} = \emptyset \text{ or } a_i^{\triangleleft} \cap a_j^{\triangleleft} = a_i^{\triangleleft} \\ \varphi_T(a_i) \wedge \varphi_T(a_j), & \text{Others} \end{cases} . \tag{9}$$

where

$$\varphi_T(a) = \sum_i^{|U|} \frac{S_T(x_i, a)}{x_i}$$

represents the fuzzy linguistic-valued set for which attribute a satisfies the fuzzy linguistic threshold T on U.

In Eq. 9, the weight between any two attributes in the fuzzy linguistic attribute topology is $\mathcal{K}(a_i, a_j) = \varphi_T(a_i) \wedge \varphi_T(a_j)$, which represents the linguistic membership degree of attributes a_i and a_j to objects that they have in common.

Definition 10. *Let (U, A, S) be a fuzzy linguistic formal context, $K = (V, G)$ be the fuzzy linguistic attribute topology of (U, A, S), for $a_i \in V$, $\nexists a_j \in V$, if $G(a_j, a_i) \neq \emptyset$, both have $G(a_i, a_j) = \emptyset$, then a_i is called the top-level attribute.*

Definition 11. *Let (U, A, S) be a fuzzy linguistic formal context, $K = (V, G)$ be the fuzzy linguistic attribute topology of (U, A, S), for $a_i \in V$, if a_i is a top-level attribute, a self-loop o_i is added to that node and the weight is noted as $\mathcal{K}(o_i) = \varphi_T(a_i)$. The fuzzy linguistic attribute topology $K'(V, \overline{G})$ considering the self-loop of top-level attribute nodes is called coarse fuzzy linguistic attribute topology.*

Definition 12. *Let (U, A, S) be a fuzzy linguistic formal context, $K'(V, \overline{G})$ be the coarse fuzzy linguistic attribute topology of (U, A, S), for $a \in A$, if a is considered to be self-loop, then a is called the initial attribute, and the set of initial attributes is denoted as H.*

Theorem 1. *Suppose (U, A, S) is the fuzzy linguistic formal context, $K'(V, \overline{G})$ is the coarse fuzzy linguistic attribute topology of (U, A, S), if the top-level attribute a_i contains a self-loop, then $(\varphi_T(a_i), a_i)$ is a fuzzy-classical linguistic concept.*

Proof. According to Definition 10, since the top-level attribute a_i contains a self-loop, there is no arbitrary attribute a_j, satisfying $\varphi_T(a_i) \subseteq \varphi_T(a_j)$. Therefore, $(\varphi_T(a_i), a_i)$ is a fuzzy-classical linguistic concept.

Definition 13. *Let (U, A, S) be a fuzzy linguistic formal context, $K'(V, \overline{G})$ be the coarse fuzzy linguistic attribute topology of (U, A, S), $T = \{a, a_1, \cdots, a_l\}$ is the path starting with attribute a, where $a_i \in T$ and $a_i \in H$. If $\mathcal{K}(T) = \mathcal{K}(a, a_1) \cap \mathcal{K}(a_1, a_2) \cap \cdots \cap \mathcal{K}(a_{l-1}, a_l) \neq \emptyset$, then T is called the reachable path.*

Theorem 2. *Suppose (U, A, S) is the fuzzy linguistic formal context, $K'(V, \overline{G})$ is the coarse fuzzy linguistic attribute topology of (U, A, S), if T is a reachable path, then $(\mathcal{K}(T), T)$ is a fuzzy-classical linguistic concept.*

Proof. According to Definition 13, we have $\mathcal{K}(T) = \mathcal{K}(a, a_1) \cap \mathcal{K}(a_1, a_2) \cap \cdots \cap \mathcal{K}(a_{l-1}, a_l) \neq \emptyset$. According to the inclusion relation of fuzzy linguistic attribute topology, we get $\mathcal{K}(a_{l-1}, a_l) \subset \cdots \subset \mathcal{K}(a_1, a_2) \subset \mathcal{K}(a, a_1)$. Therefore, the intent of the top-level attribute contains the intent of all node attributes, so all fuzzy-classical linguistic concepts can be obtained from the top-level attribute, i.e., $(\mathcal{K}(T), T)$ is a fuzzy-classical linguistic concept.

4.2 Algorithm and Illustrations

According to the above analysis, in order to generate fuzzy-classical linguistic concepts, we first need to determine whether the node attributes corresponding to the fuzzy linguistic attribute topology are equal to the weights on the path. Then we traverse each node attribute in the coarse fuzzy linguistic attribute topology and determine whether it is a reachable path to obtain fuzzy-classical linguistic concepts. We give the Fuzzy-Classical Linguistic Concepts Acquisition (FCLCA) approach based on the attribute topology as follows:

Input: the fuzzy linguistic formal context (U, A, S).

Output: all fuzzy-classical linguistic concepts $FVLL(U, A, S)$.

Step 1: Construct the adjacency matrix of fuzzy linguistic attribute topology (V, G) by Eq. 9.

Step 2: Generate a fuzzy linguistic attribute topology according to Definition 9.

Step 3: For $a_i \in V$, if a_i is the top-level attribute, a self-loop o_i is added to this node to generate a coarse fuzzy linguistic attribute topology $K'(V, \overline{G})$.

Step 4: If the nodes consider the self-loop, the fuzzy-classical linguistic concept $(\varphi_T(a_i), a_i)$ is obtained.

Step 5: Using a_i as the path starting point, find the maximal path T. If $\mathcal{K} \neq \emptyset$, then use $\mathcal{K}(T)$ as an extent of T to generate the fuzzy-classical linguistic concept $(\mathcal{K}(T), T)$.

Step 6: Output all fuzzy-classical linguistic concepts $FVLL(U, A, S)$.

Example 2. (Continued Example 1) We take the fuzzy linguistic formal context (U, A, S) shown in Table 1 as an example. According to Step 1, we can derive the adjacency matrix of the fuzzy linguistic attribute topology $G(a_i, a_j)$ as follows:

$$G(a_i,a_j)=\begin{bmatrix} \frac{s_5}{u_1}+\frac{s_6}{u_2}+\frac{s_4}{u_4}+\frac{s_7}{u_5} & \frac{s_6}{u_5} & \frac{s_5}{u_1}+\frac{s_6}{u_2} & \frac{s_5}{u_1} & \frac{s_5}{u_1}+\frac{s_7}{u_5} \\ \frac{s_6}{u_5} & \frac{s_5}{u_3}+\frac{s_6}{u_5} & \varnothing & \varnothing & \varnothing \\ \varnothing & \varnothing & \frac{s_6}{u_1}+\frac{s_7}{u_2} & \frac{s_5}{u_1} & \frac{s_5}{u_4} \\ \varnothing & \varnothing & \varnothing & \frac{s_5}{u_1} & \varnothing \\ \frac{s_5}{u_1}+\frac{s_7}{u_5} & \frac{s_4}{u_3}+\frac{s_6}{u_5} & \frac{s_5}{u_4} & \frac{s_5}{u_1} & \frac{s_6}{u_1}+\frac{s_4}{u_3}+\frac{s_8}{u_5} \end{bmatrix},$$

similar to the Zadeh representation for fuzzy sets, we use the Zadeh representation for fuzzy linguistic values.

We can generate a fuzzy linguistic attribute topology (V, G) based on the fuzzy linguistic adjacency matrix $G(a_i, a_j)$ as shown in Fig. 2. It represents the relationship between attributes.

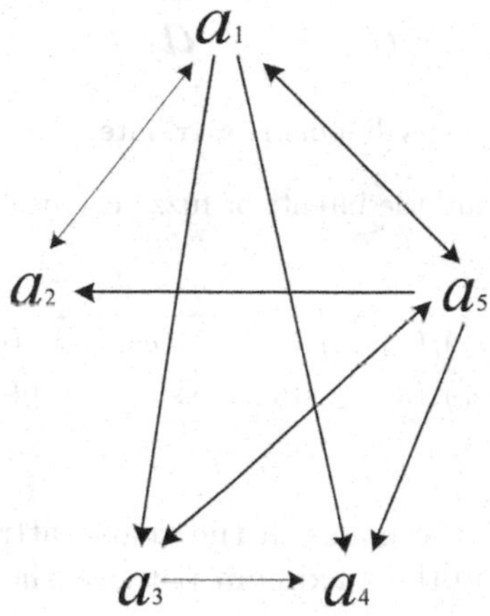

Fig. 2. Fuzzy linguistic attribute topology (V, G)

According to the fuzzy linguistic attribute topology (V, G), there is a one-way edge connected between attributes a_1 and a_3, indicating that there are common objects between attributes a_1 and a_3 and the set of objects owned by attribute a_1 contains the set of objects owned by attribute a_3, corresponding to a fuzzy linguistic weight $\frac{s_5}{u_1}+\frac{s_6}{u_2}$. There is no edge connection between attributes a_2 and a_3, indicating that the attributes are mutually exclusive and the weight between them is $\emptyset$, so attributes a_2 and a_3 cannot constitute a fuzzy-classical linguistic concept.

According to step 3, we can obtain the coarse fuzzy linguistic attribute topology $K'(V, \overline{G})$ as shown in Fig. 3.

According to Fig. 3, attributes a_1 and a_5 are considered self-loop. Therefore, according to Theorem 1, the fuzzy linguistic weights corresponding to attributes a_1 and a_5 are the extents of fuzzy-classical linguistic concepts, and attributes a_1 and a_5 are the intents of fuzzy-classical linguistic concepts, as shown in Table 3.

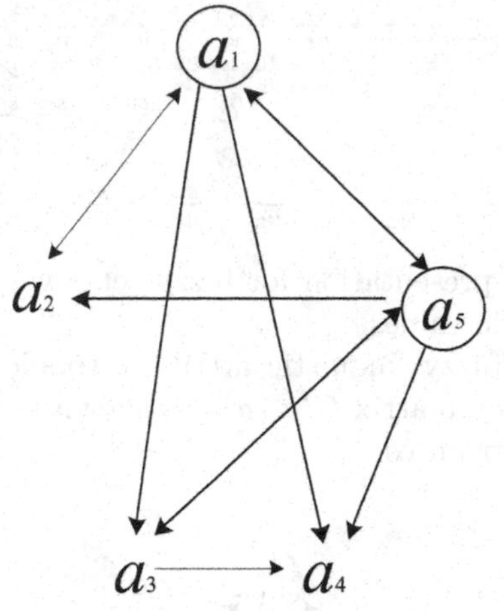

Fig. 3. Coarse fuzzy linguistic attribute topology $K'(V, \overline{G})$

Table 3. a_1 and a_5 are the intents of fuzzy-classical linguistic concepts

Extent	Intent
$\{(u_1, s_5), (u_2, s_6), (u_4, s_4), (u_5, s_7)\}$	$\{a_1\}$
$\{(u_1, s_6), (u_3, s_4), (u_5, s_8)\}$	$\{a_5\}$

We sequentially select the nodes in the coarse attribute topology graph as path starting points to find the maximum path, as shown in Table 4.

Table 4. The path corresponding to attribute a_1

$\mathcal{K}(T)$	T
$\{(u_1, s_5), (u_5, s_7)\}$	$\{a_1, a_5\}$
$\{(u_1, s_5), (u_2, s_6)\}$	$\{a_1, a_3\}$
$\{(u_5, s_6)\}$	$\{a_1, a_2, a_5\}$
$\{(u_1, s_5)\}$	$\{a_1, a_3, a_4, a_5\}$

According to Tables 4 and 5, when using attribute a_1 as the starting point of the path, we can obtain the fuzzy-classical linguistic concept intent set as follows:

$$\{\{(u_1, s_5), (u_5, s_7)\}, \{(u_1, s_5), (u_2, s_6)\}, \{(u_5, s_6)\}, \{(u_1, s_5)\}\}.$$

Similarly, when taking attribute a_5 as the starting point of the path, we can get the intent set of fuzzy-classical linguistic concepts as follows:

$$\{\{(u_3, s_4), (u_5, s_6)\}\}.$$

Table 5. The path corresponding to attribute a_5

$\mathcal{K}(T)$	T
$\{(u_3, s_4), (u_5, s_6)\}$	$\{a_2, a_5\}$

In the above paths, the redundant fuzzy-classical linguistic concepts generated on different paths have been removed, so they are the same as the fuzzy-classical linguistic obtained in Example 1.

5 Conclusions

In this paper, we propose a fuzzy-classical linguistic concept acquisition approach based on attribute topology, combining fuzzy linguistic formal context and concept lattice. Our proposed FCLCA approach reduces redundant fuzzy-classical linguistic concepts while retaining relatively important fuzzy linguistic concept information. In the process of fuzzy-classical linguistic concept acquisition, the reachable paths of coarse fuzzy linguistic attribute topology is visualized, which improves the interpretability of the FCLCA approach and is more beneficial for people to select different fuzzy linguistic concepts according to different downstream tasks. We verify the correctness and effectiveness of our proposed approach by examples.

In the future, due to the complexity and uncertainty of the fuzzy linguistic formal context, it is necessary for us to consider a new approach to deal with fuzzy linguistic values. In addition, the fuzzy-classical linguistic concept acquisition approach we proposed is still in the theoretical stage. In the big data environment, we can try to apply it to different downstream tasks.

Acknowledgements. This work was supported by the National Natural Science Foundation of P.R. China (Nos. 61772250, 62176142), Foundation of Liaoning Educational Committee (No. LJ2020007) and Special Foundation for Distinguished Professors of Shandong Jianzhu University.

References

1. Wille, R.: Restructuring lattice theory: an approach based on hierarchies of concepts. In: Ordered Sets. D Reidel, vol. 83, pp. 314–339 (1982)
2. Ganter, B., Wille, R.: Mathematical Foundations. Springer, Heidelberg (1999). https://doi.org/10.1007/978-3-642-59830-2
3. Priss, U.: Formal concept analysis in information science. Annu. Rev. Inf. Sci. Technol. **40**(1), 521–543 (2006)
4. Zhi, H., Li, J.: Granule description based knowledge discovery from incomplete formal contexts via necessary attribute analysis. Inf. Sci. **485**, 347–361 (2019)
5. Medina, J., Navareño, P., Ramírez-Poussa, E.: Knowledge implications in multi-adjoint concept lattices. In: Cornejo, M.E., Kóczy, L.T., Medina-Moreno, J., Moreno-García, J. (eds.) Computational Intelligence and Mathematics for Tackling

Complex Problems 2. SCI, vol. 955, pp. 155–161. Springer, Cham (2022). https://doi.org/10.1007/978-3-030-88817-6_18
6. Li, J., Huang, C., Qi, J., Qian, Y., Liu, W.: Three-way cognitive concept learning via multi-granularity. Inf. Sci. **378**, 244–263 (2017)
7. Yan, E., Yu, C., Lu, L., Tang, C.: Incremental concept cognitive learning based on three-way partial order structure. Knowl.-Based Syst. **220**, 106898 (2021)
8. Mi, Y., Shi, Y., Li, J., Liu, W., Yan, M.: Fuzzy-based concept learning method: exploiting data with fuzzy conceptual clustering. IEEE Trans. Cybern. **52**, 582–593 (2020)
9. Hao, F., Yang, Y., Min, G., Loia, V.: Incremental construction of three-way concept lattice for knowledge discovery in social networks. Inf. Sci. **578**, 257–280 (2021)
10. Bayhan, S.: Textural dependency and concept lattices. Int. J. Approx. Reason. **136**, 36–65 (2021)
11. Chen, D.G., Zhang, W.X., Yeung, D., Tsang, E.C.C.: Rough approximations on a complete completely distributive lattice with applications to generalized rough sets. Inf. Sci. **176**(13), 1829–1848 (2006)
12. Mao, H., Zheng, Z.: The construction of fuzzy concept lattice based on weighted complete graph. J. Intell. Fuzzy Syst. **36**(2), 1–9 (2019)
13. Zhang, T., Li, H., Liu, M., Rong, M.: Incremental concept-cognitive learning based on attribute topology. Int. J. Approx. Reason. **118**, 173–189 (2020)
14. Zadeh, L.A.: Fuzzy sets. Inf. Control **8**(3), 338–353 (1965)
15. Zadeh, L.A.: The concept of a linguistic variable and its application to approximate reasoning-I. Inf. Sci. **8**(3), 199–249 (1975)
16. Zou, L., Pang, K., Song, X., Kang, N., Liu, X.: A knowledge reduction approach for linguistic concept formal context. Inf. Sci. **524**, 165–183 (2020)
17. Pang, K., Kang, N., Chen, S., Zheng, H.: Collaborative filtering recommendation algorithm based on linguistic concept lattice with fuzzy object. In: 2019 IEEE 14th International Conference on Intelligent Systems and Knowledge Engineering (ISKE), pp. 57–63. IEEE (2019)
18. Yang, L., Xu, Y.: A decision method based on uncertainty reasoning of linguistic truth-valued concept lattice. Int. J. Gen. Syst. **39**(3), 235–253 (2010)
19. Yang, L., Wang, Y., Li, H.: Research on the disease intelligent diagnosis model based on linguistic truth-valued concept lattice. Complexity **2021**, 1–11 (2021)
20. Herrera, F., Herrera-Viedma, E., Martıinez, L.: A fusion approach for managing multi-granularity linguistic term sets in decision making. Fuzzy Sets Syst. **114**(1), 43–58 (2000)
21. Cui, H., Yue, G., Zou, L., Liu, X., Deng, A.: Multiple multidimensional linguistic reasoning algorithm based on property-oriented linguistic concept lattice. Int. J. Approx. Reason. **131**, 80–92 (2021)
22. Zou, L., He, T., Dai, J.: A new parallel algorithm for computing formal concepts based on two parallel stages. Inf. Sci. **586**, 514–524 (2022)

Generalized Convex Combinations of T-Norms on Bounded Lattices

Yi-Qun Zhang and Hua-Wen Liu(✉)

School of Mathematics, Shandong University, Jinan 250100, People's Republic of China
hw.liu@sdu.edu.cn

Abstract. In this work, we first extend the concept of (α, T_0)-migrative property of binary operations, study some analytical properties of them and then obtain new t-norms on bounded lattices in terms of generalized convex combinations of T_W and t-norms. Conditions for the generalized convex combinations to be t-norms again are investigated.

Keywords: T-norms · Migrativity · Bounded lattices · Generalized convex combinations

1 Introduction

T-norms, investigated in the study of probabilistic metric spaces [27,33], are widely used in fuzzy preference modelling, fuzzy sets, fuzzy decision making and statistics [1,22]. Nowadays, researchers pay more attention to t-norms on bounded lattices instead of those on the real unit interval [7,8,10,11,16,18,24, 32,36], which have more practical applications [23].

The so-called migrativity was introduced by Durante and Sarkoci [9] when they studied the convex combinations of two t-norms and this property was considered first in [26] as a problem to find examples of such t-norms that differ from the t-norms T_a provided, for all $a \in]0,1[$ and $u, v \in [0,1]$,

$$T_a(u,v) = \begin{cases} \min(u,v) & \max(u,v) = 1 \\ auv & \text{otherwise,} \end{cases}$$

which has been solved in [5]. This property has received more and more attention in the cases of t-norms [12–15,30], aggregation functions [6], semicopulas, quasi-copulas and copulas [25], uninorms and nullnorms [31], overlap functions [37] and so on.

As an interesting method to obtain new logic operators from given logic operators, the convex combination of t-norms on the real unit interval was widely studied and it turned out that there are many cases that the non-trivial convex combinations of two different discontinuous t-norms are still t-norms

Supported by Shandong University.

Y. Chen and S. Zhang (Eds.): AILA 2022, CCIS 1657, pp. 143–153, 2022.
https://doi.org/10.1007/978-981-19-7510-3_11

[9,19,29]. Jenei [19] clarified that several examples can be given when taking into account convex combinations of discontinuous t-norms. What's more, Ouyang [29] claimed that the convex combination of a continuous t-norm T and the drastic product T_D is a t-norm if and only if the continuous t-norm T is strict and the additive generator f of T satisfies an additional condition.

Recently, based on the idea given in [28], Karaçal et al. [21] extended the definition of the convex combination for t-norms on the real unit interval to bounded lattices, which is called the generalized convex combination. Since there are no definitions of addition, subtraction and multiplication on lattices, they replaced the addition, subtraction and multiplication in the equation by t-conorms, negations and t-norms, respectively. Inspired by the idea given in [21] and the work in [9,35], we can extend the α-migrative property on $[0,1]$ to the bounded lattices and investigate the generalized convex combination of several kinds of t-norms on bounded lattices.

The rest of this paper is formed as follows. First, we recall some basic notions of lattices, several operators on bounded lattices, the generalized convex combination and some of their properties In Sect. 2. In Sect. 3, we first extend the definition of (α, T_0)-migrativity and study several properties of it. Then we investigate the case when the generalized convex combinations of such migrative t-norms are still t-norms, which can be divided into two parts: the generalized convex combination of T_W and $(\alpha, T_\wedge)$-migrative t-norms, and the generalized convex combination of T_W and (α, T_0)-migrative t-norms, where T_0 is a $\vee$-distributive t-norm.

2 Preliminaries

For some basic notions about t-norms on $[0,1]^2$, the reader can refer to [22]. In this section, we will give some useful concepts about lattices as well as t-norms on bounded lattices. For more detailed information about partially ordered sets and lattices we recommend [3,4].

Definition 1. *[4] Let $(M, \leq)$ be a partially ordered set and N be a subset of M. An element $x \in M$ is said to be an upper bound of N if $x \geq y$ for all $y \in N$. An upper bound p of N is said to be the least upper bound if $p \leq x$ for all upper bounds x of N and we write $p = \bigvee N$. The definitions of the lower bound and the greatest lower bound can be obtained dually.*

Definition 2. *[3] Let $(L, \leq)$ be a partially ordered set, if any two elements $x, y \in L$ have a greatest lower bound, which is denoted by $x \wedge y$, and a least upper bound, which is denoted by $x \vee y$, then L is a lattice. If a lattice $(L, \leq)$ has a bottom element $\perp$ and a top element $\top$, then $(L, \leq)$ is called a bounded lattice.*

In general, we write the top element and bottom element as 1_L and 0_L. We denote the bounded lattice $(L, \leq)$ with a top element 1_L and a bottom element 0_L by $(L, \leq, 0_L, 1_L)$. For m, n in a lattice, we use $m \| n$ to denote that m and n are incomparable and we use the notation $m \nparallel n$ to denote that m and n are comparable.

Definition 3. *[3] Given a bounded lattice $(L, \leq, 0_L, 1_L)$, and $m, n \in L$ with $m \leq n$, a subinterval $[m, n]$ of L is a sublattice of L defined as*

$$[m, n] = \{x \in L \mid m \leq x \leq n\}.$$

Similarly, $]m, n] = \{x \in L \mid m < x \leq n\}$, $[m, n[= \{x \in L \mid m \leq x < n\}$ and $]m, n[= \{x \in L \mid m < x < n\}$.

Definition 4. *[2] Let $(L, \leq, 0_L, 1_L)$ be a bounded lattice. A binary function T(resp. S): $L^2 \to L$ is said to be a t-norm (resp. t-conorm) if it is associative, commutative, has a neutral element 1_L (resp. 0_L) and increasing with respect to both variables.*

If we replace the condition that T has a neutral element 1_L by the inequality $T(u, v) \leq u \wedge v$ for all $u, v \in L$, which can be called boundary condition, then we call T a t-subnorm.

Example 1. [20] Given a bounded lattice $(L, \leq, 0_L, 1_L)$. T_W and $T_\wedge$ given below are two common t-norms on the bounded lattice L: for all $u, v \in L$,

$$T_W(u, v) = \begin{cases} v & u = 1_L \\ u & v = 1_L \\ 0_L & \text{otherwise,} \end{cases}$$

$$T_\wedge(u, v) = u \wedge v.$$

Dually, $S_\vee$ given below is a t-conorm on bounded lattice L: for all $u, v \in L$,

$$S_\vee(u, v) = u \vee v.$$

Recently, Sun and Liu [34] gave the theorem of t-norms with additive generators on bounded lattices, which took a step forward in our study about lattice-valued aggregation functions. As a helpful tool to figure out more properties of t-norms on bounded lattices, here we post the theorem as follows:

Theorem 1. [34] *Let $(L, \leq)$ be a bounded lattice with a bottom element 0_L and a top element 1_L, and $t : L \to [0, \infty]$ be an injective decreasing function with $t(1_L) = 0$. If the conditions below stand:*

(i) for all $u, v \in L$ function t fulfills

$$t(u) + t(v) \in Ran(t) \cup [t(0_L), \infty],$$

(ii) if $t(v)$ and $t(u)$ have at least one same summand in $Ran(t)$, then $u \nparallel v$,

then the following function $T : L^2 \to L$ is a t-norm. And we call t as an additive generator of T:

$$T(u, v) = t^{(-1)}(t(u) + t(v)).$$

Besides, from the Proposition 3.11 in [34] we know that for an arbitrary t-norm T on a bounded lattice $(L, \le, 0_L, 1_L)$ with its additive generator t, T is strict increasing if and only if $t(0_L) = \infty$.

Finally, we recommend the concept of the generalized convex combination of t-norms on bounded lattices.

Definition 5. *[17] Let $(L, \le, 0_L, 1_L)$ be a bounded lattice. A operation $n : L \to L$ is said to be a negation if it is decreasing, $n(0_L) = 1_L$ and $n(1_L) = 0_L$.*

If $L = [0,1]$, the classical (standard) negation n_C on L is defined as $n_C(u) = 1 - u$, for all $u \in [0,1]$.

Definition 6. *[21] Let $(L, \le, 0_L, 1_L)$ be a bounded lattice, T, T_1, T_2 be t-norms, n be a negation on L, S be a t-conorm, $r \in L$ and $S(r, n(r)) = 1_L$. The generalized convex combination of the t-norms T_1 and T_2 is given as follows:*

$$K_{r,n}^{T,S}(u,v) = S(T(r, T_1(u,v)), T(n(r), T_2(u,v)))$$

for all $u, v \in L$.

If we take $L = [0,1]$, $S = S_L$, $T = T_P$ and $n = n_C$, then the function $K_{r,n}^{T,S}$ defined in Definition 6 is equivalent to the definition of the convex combination of two t-norms T_1 and T_2 on $[0,1]$.

Besides, it is clear that $K_{r,n}^{T,S}$ satisfies the monotonicity and the commutativity.

3 The Generalized Convex Combinations of T-Norms and T_W on Bounded Lattices

In the following, we mainly focus on the generalized convex combinations of α-migrative t-norms and T_W on bounded lattices. Recall that in [9], a binary function $T : [0,1]^2 \to [0,1]$ is called α-migrative if $T(\alpha u, v) = T(u, \alpha v)$ holds for every u, v in $[0,1]$. Meanwhile, they proved that the convex combination of a α-migrative t-norm and T_D is still a t-norm.

Since the equation above can also be written as $T(T_P(\alpha, u), v) = T(u, T_P(\alpha, v))$, Fodor and Rudas [13] replaced T_P by a t-norm T_0 on the left-hand side, and on the right-hand side by a possibly different t-norm T_1 as follows:

$$T(T_0(\alpha, u), v) = T(u, T_1(\alpha, v)). \tag{1}$$

And they proved that the α-partial functions of T_0, T and T_1 must coincide on (α, u), i.e. $T(\alpha, u) = T_0(\alpha, u) = T_1(\alpha, u)$. Consequently, it is sufficient to consider $T_0(\alpha, u)$ on both sides in Eq. (1), which leads to the notion of α-migrativity with respect to T_0. To be more specific, we show the definition below.

Definition 7. *[13] Let T_0 be a fixed t-norm and $\alpha \in]0,1[$. A binary function $T : [0,1]^2 \to [0,1]$ is called α-migrative with respect to T_0 (shortly: (α, T_0)-migrative) if, for all $u, v \in [0,1]$,*

$$T(T_0(\alpha, u), v) = T(u, T_0(\alpha, v)).$$

Returning to our study and considering the generalization of the convex combination of such t-norms and T_W on bounded lattices, the first thing we need to do is giving the definition of (α, T_0)-migrativity on bounded lattices.

Definition 8. *Let $(L, \leq, 0_L, 1_L)$ be a bounded lattice, T_0 be a fixed t-norm on L and $\alpha \in L \setminus \{0_L, 1_L\}$. A binary operation $T : L^2 \to L$ is called (α, T_0)-migrative if, for all $u, v \in L$, $T(T_0(\alpha, u), v) = T(u, T_0(\alpha, v))$ holds.*

In the next theorem, we show that (α, T_0)-migrative t-norms on bounded lattices still have the same characterizations in [13], which is simple but very helpful.

Theorem 2. *Let $(L, \leq, 0_L, 1_L)$ be a bounded lattice, T_0 be a fixed t-norm on L and $\alpha \in L \setminus \{0_L, 1_L\}$. Then for a t-norm $T : L^2 \to L$, the following statements coincide.*

(i) T is (α, T_0)-migrative;
(ii) T_0 is (α, T)-migrative, i.e., for all $u, v \in L$: $T_0(T(\alpha, u), v) = T_0(u, T(\alpha, v))$;
(iii) $T(\alpha, u) = T_0(\alpha, u)$, for all $u \in L$.

Proof. It is clear that statements (i) and (iii) are equivalent. Thus, we only need to prove that (ii) and (iii) are equivalent.

$(ii) \Rightarrow (iii)$: take $v = 1$, then (ii) implies (iii).

$(iii) \Rightarrow (ii)$: from the associativity of T_0, we have

$$T_0(T_0(\alpha, u), v) = T_0(u, T_0(\alpha, v))$$
$$\Leftrightarrow T_0(T(\alpha, u), v) = T_0(u, T(\alpha, v)),$$

then (iii) implies (ii).

Thanks to the representation of t-norms by additive generators on bounded lattices, we can step into the following property of (α, T_0)-migrative t-norms on bounded lattices.

Proposition 1. *Let $(L, \leq, 0_L, 1_L)$ be a bounded lattice, T_0 be a fixed t-norm on L, $\alpha \in L \setminus \{0_L, 1_L\}$ and T be a strict increasing t-norm with additive generator $t : L \to [0, \infty]$. Then T is (α, T_0)-migrative if and only if for all $u, v \in L$,*

$$t(T_0(\alpha, u)) - t(u) = t(T_0(\alpha, v)) - t(v). \tag{2}$$

Proof. Since t is the additive generator of T, we have $T(u, v) = t^{(-1)}(t(u) + t(v))$ for all $u, v \in L$. Besides, the (α, T_0)-migrativity of T means that

$$T(T_0(\alpha, u), v) = T(u, T_0(\alpha, v))$$
$$\Leftrightarrow t^{(-1)}(t(T_0(\alpha, u)) + t(v)) = t^{(-1)}(t(u) + t(T_0(\alpha, v)))$$

holds for all $u, v \in L$. Since T is a strict monotone t-norm on bounded lattice L, from Proposition 3.11 in [34] we know that $t(0_L) = \infty$, i.e. for all $u, v \in L$, $t(u) + t(v) \in Ran(t)$. Clearly, t is injective and $t(T_0(\alpha, u)) + t(v)$, $t(u) + t(T_0(\alpha, v)) \in Ran(t)$, we have

$$t(T_0(\alpha, u)) + t(v) = t(u) + t(T_0(\alpha, v)),$$

the equation above is equivalent to Eq. (2).

3.1 The Generalized Convex Combinations of (α, $T_\wedge$)-migrative T-Norms and T_W

Now we investigate the case when the generalized convex combinations of (α, $T_\wedge$)-migrative t-subnorms and T_W are still t-norms. First we need to introduce the following essential lemma.

Lemma 1. *Let $(L, \leq, 0_L, 1_L)$ be a bounded lattice, T be an (α, $T_\wedge$)-migrative t-subnorm and $\alpha \in L \setminus \{0_L, 1_L\}$. Then the function $T_\alpha : L^2 \to L$ defined by $T_\alpha(u, v) = \alpha \wedge T(u, v)$ for all $u, v \in L$ is a t-subnorm.*

Proof. It is immediate to gain that T_α satisfies the commutativity, monotonicity and clearly, it satisfies $T_\alpha(u, v) \leq \alpha \wedge u \wedge v \leq u \wedge v$. Therefore, the only thing we need to check is the associativity of T_α. For any $u, v, w \in L$, we have the following equation:

$$\begin{aligned}T_\alpha(T_\alpha(u,v),w) &= T_\alpha(\alpha \wedge T(u,v),w) = \alpha \wedge T(\alpha \wedge T(u,v),w)\\ &= \alpha \wedge T(T(u,v),\alpha \wedge w) = \alpha \wedge T(u,T(v,\alpha \wedge w))\\ &= \alpha \wedge T(u,T(\alpha \wedge v,w)) = \alpha \wedge T(T(u,\alpha \wedge v),w)\\ &= \alpha \wedge T(T(\alpha \wedge u,v),w) = \alpha \wedge T(\alpha \wedge u,T(v,w))\\ &= \alpha \wedge T(u,\alpha \wedge T(v,w))\\ &= T_\alpha(u,T_\alpha(v,w)).\end{aligned}$$

In conclusion, T_α is a t-subnorm.

After obtaining this useful result, we tend to study the convex combination of two different t-norms on bounded lattices based on this lemma.

Theorem 3. *Let $(L, \leq, 0_L, 1_L)$ be a distributive bounded lattice, $\alpha \in L$, n be a negation on L, $K_{\alpha,n}^{T,S}$ be the generalized convex combination of t-norms T_1 and T_2 such that T is a t-norm and S is a t-conorm. If $T = T_\wedge$, $S = S_\vee$, T_1 be a (α, $T_\wedge$)-migrative t-norm and $T_2 = T_W$, then $K_{\alpha,n}^{T_\wedge,S_\vee}$ is a t-norm.*

Proof. From the conditions given, $K_{\alpha,n}^{T,S}$ can be written as follows:

$$K_{\alpha,n}^{T_\wedge,S_\vee}(u,v) = (\alpha \wedge T_1(u,v)) \vee (n(\alpha) \wedge T_W(u,v))$$

for any $u, v \in L$. It is immediate to gain that $K_{\alpha,n}^{T_\wedge,S_\vee}$ satisfies commutativity and monotonicity. Noticing that when $u = 1_L$, $K_{\alpha,n}^{T_\wedge,S_\vee}(u, v) = (\alpha \wedge v) \vee (n(\alpha) \wedge v) = (\alpha \vee n(\alpha)) \wedge v = v$ since L is a distributive lattice and $\alpha \vee n(\alpha) = 1_L$, which means that the boundary condition also holds. Thus, we just need to check whether $K_{\alpha,n}^{T_\wedge,S_\vee}$ is associate. By simple computations, we know that

$$K_{\alpha,n}^{T_\wedge,S_\vee}(u,v) = \begin{cases} v & u = 1_L,\\ u & v = 1_L,\\ \alpha \wedge T_1(u,v) & \text{otherwise.}\end{cases}$$

Lemma 1 ensures that $a \wedge T_1(u, v)$ is associative. Hence, $K_{\alpha,n}^{T_\wedge,S_\vee}$ is a t-norm.

Remark 1. Theorem 2.30 in [21] has an equivalent form expressed by (α, $T_\wedge$)-migrativity, which is more concise and given as follows:

Let $(L, \leq, 0_L, 1_L)$ be a distributive bounded lattice, $\alpha \in L$, n be a negation on L, $K_{\alpha,n}^{T,S}$ be the generalized linear combination of t-norms T_1 and T_2 such that T is a t-norm and S is a t-conorm. If $T = T_\wedge$, $S = S_\vee$, $T_1 = T_2$ are ($\alpha \vee n(\alpha)$, $T_\wedge$)-migrative t-subnorms, then $K_{\alpha,n}^{T_\wedge,S_\vee}$ is a t-subnorm.

3.2 The Generalized Convex Combinations of (α, T_0)-migrative T-Norms and T_W

Since (α, $T_\wedge$)-migrativity is quite particular, now we tend to replace $T_\wedge$ by T_0, which is a fixed t-norm, to find out whether we can reach the same conclusion. First, Lemma 1 can be generalized naturally.

Lemma 2. *Let $(L, \leq, 0_L, 1_L)$ be a bounded lattice, α be in $L \setminus \{0_L, 1_L\}$, T be an (α, T_0)-migrative t-subnorm and T_0 be a fixed t-subnorm on L. Then the mapping $T_\alpha : L^2 \to L$ defined by $T_\alpha(u, v) = T_0(\alpha, T(u, v))$ is a t-subnorm.*

Proof. It is obvious that T_α satisfies the commutativity and the monotonicity and clearly, it satisfies $T_\alpha(u, v) \leq \alpha \wedge u \wedge v \leq u \wedge v$. Therefore, the only thing we need to check is the associativity of T_α. For all $u, v, w \in L$, we can obtain the following equation:

$$\begin{aligned}
T_\alpha(T_\alpha(u,v),w) &= T_\alpha(T_0(\alpha,T(u,v)),w) = T_0(\alpha,T(T_0(\alpha,T(u,v)),w)) \\
&= T_0(\alpha,T(T(u,v),T_0(\alpha,w))) = T_0(\alpha,T(u,T(v,T_0(\alpha,w)))) \\
&= T_0(\alpha,T(u,T(T_0(\alpha,v),w))) = T_0(\alpha,T(T(u,T_0(\alpha,v)),w)) \\
&= T_0(\alpha,T(T(T_0(\alpha,u),v),w)) = T_0(\alpha,T(T_0(\alpha,u),T(v,w))) \\
&= T_0(\alpha,T(u,T_0(\alpha,T(v,w)))) \\
&= T_\alpha(u,T_\alpha(v,w)).
\end{aligned}$$

In conclusion, T_α is a t-subnorm.

Based on this lemma, we investigate the conditions when $T_\wedge$ is translated into T_0 and find that most of the conditions still remain while T_0 need to be $\vee$-distributive.

Theorem 4. *Let $(L, \leq, 0_L, 1_L)$ be a bounded lattice, $\alpha \in L$, n be a negation on L, T_0 be a $\vee$-distributive t-norm on L, $K_{\alpha,n}^{T,S}$ be the generalized convex combination of t-norms T_1 and T_2 such that T is a t-norm and S is a t-conorm,. If $T = T_0$, $S = S_\vee$, T_1 be a (α, T_0)-migrative t-norm and $T_2 = T_W$, then $K_{\alpha,n}^{T_0,S_\vee}$ is a t-norm.*

Proof. From the conditions given, $K_{\alpha,n}^{T,S}$ can be written as follows:

$$K_{\alpha,n}^{T_0,S_\vee}(u,v) = T_0(\alpha,T_1(u,v)) \vee T_0(n(\alpha),T_W(u,v))$$

for any $u, v \in L$. It is immediate to gain that $K_{\alpha,n}^{T_0,S_\vee}$ satisfies commutativity and monotonicity. Noticing that when $u = 1_L$, $K_{\alpha,n}^{T_0,S_\vee}(u,v) = (T_0(\alpha,v)) \vee (T_0(n(\alpha),v)) = T_0(\alpha \vee n(\alpha), v) = T_0(1,v) = v$. This is because the t-norm T_0 is $\vee$-distributive and $\alpha \vee n(\alpha) = 1_L$. From this, the boundary condition still holds. Thus, we just need to check whether $K_{\alpha,n}^{T_0,S_\vee}$ is associative. By simple computations we can obtain that

$$K_{\alpha,n}^{T_0,S_\vee}(u,v) = \begin{cases} v & u = 1_L, \\ u & v = 1_L, \\ T_0(\alpha, T_1(u,v)) & \text{otherwise.} \end{cases}$$

Lemma 2 ensures that $T_0(\alpha, T_1(u,v))$ is associative. Hence, $K_{a,n}^{T_0,S_\vee}$ is also a t-norm.

In addition, the above theorem can be generalized into n-ary situation. Let $\alpha_{T_0}^{(n)} = T_0(\underbrace{\alpha, \alpha \ldots \alpha}_{\text{n}})$ for any $n \in \mathbb{N}$, then by induction, $T_0(\alpha_{T_0}^{(n)}, T(u,v))$ in Lemma 2 is still a t-subnorm. To show this, we give the proof when $n = 2$. In order to prove that $T_0(\alpha_{T_0}^{(2)}, T(u,v))$ is still a t-subnorm, the difficulty lies in the associativity. Since we have

$$\begin{aligned} T(T_0(T_0(\alpha,\alpha),T(u,v)),w) &= T(T_0(\alpha,T_0(\alpha,T(u,v))),w) = T(T_0(\alpha,T(u,v)),T_0(\alpha,w)) \\ &= T(T(u,v),T_0(\alpha,T_0(\alpha,w))) = T(u,T(v,T_0(\alpha,T_0(\alpha,w)))) \\ &= T(u,T(T_0(\alpha,v),T_0(\alpha,w))) = T(T(u,T_0(\alpha,v)),T_0(\alpha,w)) \\ &= T(T(T_0(\alpha,u),v),T_0(\alpha,w)) = T(T(T_0(\alpha,T_0(\alpha,u)),v),w) \\ &= T(T(T_0(T_0(\alpha,\alpha),u),v),w) = T(T_0(T_0(\alpha,\alpha),u),T(v,w)) \\ &= T(T_0(\alpha,T_0(\alpha,u)),T(v,w)) = T(T_0(\alpha,u),T_0(\alpha,T(v,w))) \\ &= T(u,T_0(\alpha,T_0(\alpha,T(v,w)))) \\ &= T(u,T_0(T_0(\alpha,\alpha),T(v,w))). \end{aligned}$$

Therefore, we have $T_0(T_0(\alpha,\alpha), T(T_0(T_0(\alpha,\alpha),T(u,v)),w)) = T_0(T_0(\alpha,\alpha),T(u, T_0(T_0(\alpha,\alpha),T(v,w))))$, i.e., $T_\alpha(T_\alpha(u,v),w) = T_\alpha(u,T_\alpha(y,w))$, where $T_\alpha(u,v) = T_0(T_0(\alpha,\alpha),T(u,v))$. The cases when $n \geq 3$ can be obtained by induction.

Consequently, we can provide the following corollary.

Corollary 1. *Let $(L, \leq, 0_L, 1_L)$ be a bounded lattice, n be a negation on L, T_0 be a $\vee$-distributive t-norm on L, $\alpha \in L$, $m \in \mathbb{N}$ and $K_{\alpha_{T_0}^{(m)},n}^{T,S}$ be the generalized convex combination of t-norms T_1 and T_2 such that T is a t-norm and S is a t-conorm. If $T = T_0$, $S = S_\vee$, T_1 be a (α, T_0)-migrative t-subnorm and $T_2 = T_W$, then $K_{\alpha_{T_0}^{(m)},n}^{T_0,S_\vee}$ is a t-norm.*

Proof. By simple computations we have that

$$K_{\alpha_{T_0}^{(n)},n}^{T_0,S_\vee}(u,v) = \begin{cases} v & u = 1_L, \\ u & v = 1_L, \\ T_0(\alpha_{T_0}^{(m)}, T_1(u,v)) & \text{otherwise.} \end{cases}$$

It is easy to see that $K^{T_0,S_\vee}_{\alpha^{(m)}_{T_0},n}$ is a t-norm.

4 Conclusion

In this work, we mainly investigated the conditions for the generalized convex combination of (α, $T_\wedge$)-migrative or (α, T_0)-migrative t-norms and T_W to be t-norms again. Besides, the notion of (α, T_0)-migrativity was given and some of its properties were researched. There is still a lot of work to do in the future. For example, we will investigate the (α, T_0)-migrativity, the generalized linear combination of t-norms as well as the generalized convex combination of other aggregation operators on bounded lattices.

Acknowledgment. This work was supported by the National Natural Science Foundation of China (Nos. 12071259 and 11531009) and the National Key R&D Program of China (No. 2018YFA0703900).

References

1. Alsina, C., Frank, M.J., Schweizer, B.: Associative Functions. Triangular Norms and Copulas. World Scientific, Hackensack (2006)
2. Bedregal, B.C., Santos, H.S., Callejas-Bedregal, R.: T-norms on bounded lattices: t-norm morphisms and operators. In: IEEE International Conference on Fuzzy Systems (2006). https://doi.org/10.1109/FUZZY.2006.1681689
3. Birkhoff, G.: Lattice Theory, 3rd edn. American Mathematical Society Colloquium Publications, Rhode Island (1973)
4. Blyth, T.S.: Lattices and Ordered Algebraic Structures. Springer, Cham (2005)
5. Budinčević, M., Kurilić, M.S.: A family of strict and discontinuous triangular norms. Fuzzy Sets Syst. **95**, 381–384 (1998). https://doi.org/10.1016/S0165-0114(96)00284-9
6. Bustince, H., Montero, J., Mesiar, R.: Migrativity of aggregation functions. Fuzzy Sets Syst. **160**, 766–777 (2009). https://doi.org/10.1016/j.fss.2008.09.018
7. Çaylı, G.D.: On a new class of t-norms and t-conorms on bounded lattices. Fuzzy Sets Syst. **332**, 129–143 (2018). https://doi.org/10.1016/j.fss.2017.07.015
8. De Baets, B., Mesiar, R.: Triangular norms on product lattices. Fuzzy Sets Syst. **104**, 61–75 (1999). https://doi.org/10.1016/S0165-0114(98)00259-0
9. Durante, F., Sarkoci, P.: A note on the convex combinations of triangular norms. Fuzzy Sets Syst. **159**, 77–80 (2008). https://doi.org/10.1016/j.fss.2007.07.005
10. El-Zekey, M., Medina, J., Mesiar, R.: Lattice-based sums. Inf. Sci. **223**, 270–284 (2013). https://doi.org/10.1016/j.ins.2012.10.003
11. Ertuğrul, Ü., Karaçal, F., Mesiar, R.: Modified ordinal sums of triangular norms and triangular conorms on bounded lattices. Int. J. Intell. Syst. **30**, 807–817 (2015). https://doi.org/10.1002/int.21713
12. Fodor, J., Rudas, I.J.: On continuous triangular norms that are migrative. Fuzzy Sets Syst. **158**, 1692–1697 (2007). https://doi.org/10.1016/j.fss.2007.02.020
13. Fodor, J., Rudas, I.J.: An extension of the migrative property for triangular norms. Fuzzy Sets Syst. **168**(1), 70–80 (2011). https://doi.org/10.1016/j.fss.2010.09.020

14. Fodor, J., Rudas, I.J.: Migrative t-norms with respect to continuous ordinal sums. Inf. Sci. **181**(21), 4860–4866 (2011). https://doi.org/10.1016/j.ins.2011.05.014
15. Fodor, J., Klement, E.P., Mesiar, R.: Cross-migrative triangular norms. Int. J. Intell. Syst. **27**(5), 411–428 (2012). https://doi.org/10.1002/int.21526
16. Goguen, J.: L-fuzzy sets. J. Math. Anal. Appl. **18**, 145–174 (1967)
17. Grabisch, M., Marichal, J., Mesiar, R., Pap, E.: Aggregation Functions. Cambridge University Press. In: 2008 6th International Symposium on Intelligent Systems and Informatics, pp. 1–7, Subotica (2008)
18. Jenei, S., De Baets, B.: On the direct decomposability of t-norms on product lattices. Fuzzy Sets Syst. **139**, 699–707 (2003). https://doi.org/10.1016/S0165-0114(03)00125-8
19. Jenei, S.: On the convex combination of left-continuous t-norms. Aequationes Math. **72**, 47–59 (2006). https://doi.org/10.1007/s00010-006-2840-z
20. Karaçal, F., Kesicioğlu, M.N.: A T-partial order obtained from t-norms. Kybernetika **47**, 300–314 (2011)
21. Karaçal, F., Kesicioğlu, M.N., Ertuğrul, Ü.: Generalized convex combination of triangular norms on bounded lattices. Int. J. Gen Syst **49**(3), 277–301 (2020). https://doi.org/10.1080/03081079.2020.1730358
22. Klement, E.P., Mesiar, R., Pap, E.: Triangular Norms. Kluwer Academic Publishers, Dordrecht (2000)
23. Mayor, G., Torrens, J.: Triangular norm on discrete settings. In: Klement, E.P., Mesiar, R., (eds.) Logical, Algebraic, Analytic, and Probabilistic Aspects of Triangular Norms, pp. 189–230. Elsevier, Amsterdam (2005). https://doi.org/10.1016/B978-044451814-9/50007-0
24. Medina, J.: Characterizing when an ordinal sum of t-norms is a t-norm on bounded lattices. Fuzzy Sets Syst. **202**, 75–88 (2012). https://doi.org/10.1016/j.fss.2012.03.002
25. Mesiar, R., Bustince, H., Fernandez, J.: On the α-migrativity of semicopulas quasi-copulas, and copulas. Inf. Sci. **180**, 1967–1976 (2010). https://doi.org/10.1016/j.ins.2010.01.024
26. Mesiar, R., Novák, V. (eds.). Open Problems. Tatra Mountains Mathematical Publications, vol. 6, pp. 12–22 (1995)
27. Menger, K.: Statistical metrics. In: Proceedings of the National Academy of Sciences of the U.S.A, vol. 8, pp. 535–537 (1942). https://doi.org/10.1073/pnas.28.12.535
28. Ouyang, Y., Fang, J.: Some observations about the convex combinations of continuous triangular norms. Nonlinear Anal. **68**, 3382–3387 (2008). https://doi.org/10.1016/j.na.2007.03.027
29. Ouyang, Y., Fang, J., Li, G.: On the convex combination of TD and continuous triangular norms. Inf. Sci. **177**, 2945–2953 (2007). https://doi.org/10.1016/j.ins.2007.01.023
30. Ouyang, Y.: Generalizing the migrativity of continuous t-norms. Fuzzy Sets Syst. **211**, 73–87 (2013). https://doi.org/10.1016/j.fss.2012.03.008
31. Qiao, J., Hu, B.Q.: On the migrativity of uninorms and nullnorms over overlap and grouping functions. Fuzzy Sets Syst. **346**, 1–54 (2018). https://doi.org/10.1016/j.fss.2017.11.012
32. Saminger, S.: On ordinal sums of triangular norms on bounded lattices. Fuzzy Sets Syst. **157**, 1403–1416 (2006). https://doi.org/10.1016/j.fss.2005.12.021
33. Schweizer, B., Sklar, A.: Statistical metric spaces. Pac. J. Math. **10**, 313–334 (1960). https://doi.org/10.2140/pjm.1960.10.313

34. Sun, X.-R., Liu, H. W.: The additive generators of t-norms and t-conorms on bounded lattices. Fuzzy Sets Syst. **408**, 13–25 (2021). https://doi.org/10.1016/j.fss.2020.04.005
35. Wang, H.W.: Constructions of overlap functions on bounded lattices. Int. J. Approx. Reason. **125**, 203–217 (2020). https://doi.org/10.1016/j.ijar.2020.07.006
36. Zhang, D.: Triangular norms on partially ordered sets. Fuzzy Sets Syst. **153**(2), 195–209 (2005). https://doi.org/10.1016/j.fss.2005.02.001
37. Zhou, H., Yan, X.: Migrativity properties of overlap functions over uninorms. Fuzzy Sets Syst. **403**, 10–37 (2021). https://doi.org/10.1016/j.fss.2019.11.011

A Transformation Model for Different Granularity Linguistic Concept Formal Context

Ning Kang[1], Kuo Pang[2(✉)], Li Zou[3], and Meiqiao Sun[4]

[1] Software Engineering Institute, East China Normal University, Shanghai 200062, China
[2] Information Science and Technology College, Dalian Maritime University, Dalian 116026, China
pangkuo_p@dlmu.edu.cn
[3] School of Computer Science and Technology, Shandong Jianzhu University, Jinan 250102, China
[4] School of Computer and Information Technology, Liaoning Normal University, Dalian 116081, China

Abstract. Due to certain differences in people's knowledge level and personal preferences, for the description of the same thing phenomenon, different people usually choose different granularity of linguistic term sets to make judgments, so it is necessary to propose a transformation method for linguistic concept formal context with different granularity. In this paper, we firstly define the normalized distance between multi-granularity linguistic formal contexts and linguistic terms for different granularity linguistic concept formal context. Then, the normalized distance from linguistic terms to intermediate linguistic terms is kept constant to realize the transformation for different granularity linguistic concept formal context. Finally, under different threshold constraints, we achieve dynamic transformation of formal contexts of linguistic values by adjusting the thresholds. The transformation method process is reversible and it can avoid information loss.

Keywords: Multi-granularity fuzzy linguistic formal context · Granularity transformation · Concept lattice

1 Introduction

As an important knowledge discovery tool, Formal Concept Analysis (FCA) was proposed by Wille in 1982 [1]. In FCA, the formal context accurately describes the only defined binary relationship between objects and attributes, and concepts are composed of intent and extent, and all concepts form a concept lattice through partial order relations. The concept lattice has been widely used in various fields such as knowledge engineering [2], data mining [3], and pattern recognition [4].

In the classical formal context, we can know with certainty that the relationship between the object and the attribute. However, in real life, in some uncertain environments, decision makers are accustomed to using linguistic values to represent uncertain information. In order to describe the fuzzy relationships between objects and attributes,

Y. Chen and S. Zhang (Eds.): AILA 2022, CCIS 1657, pp. 154–165, 2022.
https://doi.org/10.1007/978-981-19-7510-3_12

the model of Bêlohlávek [5] expressed uncertainty information directly in terms of Fuzzy Sets (FSs), and established fuzzy concept lattice on the basis of fuzzy formal context, which provides a theoretical basis for Fuzzy Formal Concept Analysis (FFCA). In the uncertain environment, people always use linguistic values to describe objects. Inspired by Computing with Words (CW) [6], scholars have integrated linguistic value information into FCA and made great progress. Pang et al. [7] proposed a personalized recommendation algorithm based on linguistic concept lattice for the problem of fuzzy interpretation of recommendations in recommender systems. Zou et al. [8, 9] proposed linguistic-valued formal context based on linguistic truth-valued lattice implication algebra, and constructed the corresponding linguistic concept lattice with the trust degree, on this basis, giving decision rule extraction and linguistic concept reduction approaches for linguistic concept lattice.

Inspired by Granular Computing (GrC) and Rough Set (RS) [10], more and more scholars have conducted research on multi-granularity [11, 12]. As an important research content in FCA, multi-granularity FCA has attracted many experts and scholars. For example, Chu et al. [13] quantitatively described the uncertainty decision problem using RS and FCA approaches in a multi-granularity rough set framework. Hu et al. [14] studied the generation of concept lattice under Multi-Granularity Formal Context (MGFC) and gave a series of related algorithms, on this basis, proposed the approach of interconversion of formal concepts under different granularity.

Through the above analysis of fuzzy linguistic value and FCA, we are inspired to introduce the idea of multi-granularity into the fuzzy linguistic concept lattice, and propose the transformation models for different granularity linguistic concept formal context. The model achieves transformation of different granularity linguistic concept formal context.

The organizational structure of the paper is as follows. Section 2 reviews some notions relevant to FCA, Linguistic Term Set (LTS) and linguistic-valued formal context. Section 3 proposes the transformation model for different granularity linguistic concept formal context. Section 4 gives an example of linguistic granularity transformation. Finally in Sect. 5, we give the conclusions and future prospects of this paper.

2 Preliminaries

This section briefly reviews some notions related to LTS and linguistic-valued formal context.

Definition 1. [15] Let $S = \{s_\alpha | \alpha = -\tau, \ldots, -1, 0, 1, \ldots, \tau\}$ be a linguistic term set, where τ is a positive integer, then S satisfies the following properties:

(1) Order: $s_k \le s_l$ if and only if $k \le l$,
(2) Reversibility: $\mathrm{Neg}(s_{-k}) = s_k$. In particular, $\mathrm{Neg}(s_0) = s_0$,
(3) Boundedness: For $s_i \in S$, $s_{-\tau} \le s_i \le s_\tau$.

Then S is the set of linguistic terms with symmetric subscripts.

Definition 2. [8] A fuzzy linguistic-valued information system is defined as (P, Q, I, f), where $P = \{x_1, x_2, ..., x_m\}$ is an object set, $Q = \{q_1, q_2, ..., q_n\}$ is an attribute set, $I = \cup\{s_k | k \in -\tau, ..., -1, 0, 1, ..., \tau\}$ is a fuzzy linguistic values set $f : P \times Q \to I$ is an information function to represent the fuzzy linguistic-valued relationship.

It is worth mentioning that all fuzzy linguistic values in I are contained in the preset LTS.

Definition 3. ([8]) A linguistic concept formal context is a triplet (U, L_{s_α}, I), where $U = \{x_1, x_2, ..., x_n\}$ is an object set, $L_{s_\alpha} = \{l_{s_\alpha}^{x_i} | i = 1, 2, ..., n, \alpha = -\tau, ..., 0, ..., \tau\}$ is a set of linguistic concepts, I is used to characterize whether a certain object can be described in terms of attribute-linguistic relations, i.e., $I \subseteq U \times L_{s_\alpha}$. $(x, l_{s_\alpha}^{x_i}) \in I$ means the object can be described by $l_{s_\alpha}^{x_i}$, $(x, l_{s_\alpha}^{x_i}) \notin I$ means the object cannot be described by $l_{s_\alpha}^{x_i}$.

In the (U, L_{s_α}, I), for $X \subseteq U$ and $B_{s_\alpha} \subseteq L_{s_\alpha}$, a pair of operators "$\Rightarrow$" and "$\Leftarrow$" are defined as follows:

$$X^{\Rightarrow} = \{l_{s_\alpha} \in L_{s_\alpha} | \forall x \in X, (x, l_{s_\alpha}^{i}) \in I\},$$
$$B_{s_\alpha}^{\Leftarrow} = \{x \in U | \forall l_{s_\alpha}^{i} \in B_{s_\alpha}, (x, l_{s_\alpha}^{i}) \in I\}.$$

Definition 4. [8] Let (U, L_{s_α}, I) be a linguistic concept formal context, for $X \subseteq U$ and $B_{s_\alpha} \subseteq L_{s_\alpha}$, if there exist $X^{\Rightarrow} = B_{s_\alpha}$ and $B_{s_\alpha}^{\Rightarrow} = X$, a pair (X, B_{s_α}) is called linguistic concept knowledge, X and B_{s_α} are called the extent and intent of the linguistic concept knowledge.

Let $(X_1, B_{s_\alpha}^1)$ and $(X_2, B_{s_\alpha}^2)$ be two linguistic concept knowledge, the partial order relation "$\leq$" between $(X_1, B_{s_\alpha}^1)$ and $(X_2, B_{s_\alpha}^2)$ can be defined as follows:

$$(X_1, B_{s_\alpha}^1) \leq (X_2, B_{s_\alpha}^2) \Leftrightarrow X_1 \subseteq X_2 (\Leftrightarrow B_{s_\alpha}^2 \subseteq B_{s_\alpha}^1),$$

then the complete lattice $LCKL(U, L_{s_\alpha}, I)$ is a linguistic concept lattice.

3 Transformation Models for Different Granularity Linguistic Concept Formal Context

3.1 Multi-granularity Linguistic-Valued Formal Context

Definition 5. Let (U, L_{s_α}, I) be a linguistic concept formal context, $Il_{s_\alpha}^i$ denotes the object described by the linguistic concept $l_{s_\alpha}^i$. For $B \subseteq L_{s_\alpha}$, if $U = \bigcup_{l_{s_\alpha}^i \in B} Il_{s_\alpha}^i$ $(\cap Il_{s_\alpha}^i = \emptyset)$, then B is a single granularity class linguistic concept block of (U, L_{s_α}, I).

We can find that each attribute in a Fuzzy Linguistic-valued Information System (FLIS) can constitute a single granularity class linguistic concept block. According to the properties of the Linguistic Concept Formal Context (LCFC) that all linguistic terms describing each attribute necessarily constitute a single granularity class linguistic concept block.

Definition 6. Let $(U, L^1_{s_\alpha}, I_1)$ and $(U, L^2_{s_\alpha}, I_2)$ be two single granularity linguistic formal context, the sets of single granularity class linguistic concept blocks are $B_1 = \{B_{11}, B_{12}, \cdots, B_{1n}\}$ and $B_2 = \{B_{21}, B_{22}, \cdots, B_{1m}\}$, respectively. If the merging of linguistic concepts in a single granularity class linguistic concept block $B_{1k} \in B_1$ can generate $B_{2k} \in B_2$, then B_{1k} is a specialized linguistic concept of B_{2k} and B_{2k} is a generalized linguistic concept of B_{1k}, i.e., the granularity of B_{1k} is finer than that of B_{2k}, which is recorded as $B_{1k} \succ_l B_{2k}$.

Definition 7. Let $(U, L^i_{s_\alpha}, I_i)(i \in \{1, 2, \cdots, n\})$ be n single-granularity linguistic concept formal contexts, $B_i = \{B_{i1}, B_{i2}, \cdots, B_{in}\}$ be the set of class linguistic concept blocks of $L^i_{s_\alpha}$, where $B_{ik} = \{B_{1k}, B_{2k}, \cdots, B_{pk}\}(k \in \{1, 2, \cdots, n\})$ is a class linguistic concept block at different granularities. If $(U, L^1_{s_\alpha}, I_1) \prec_l (U, L^2_{s_\alpha}, I_2) \prec_l \cdots \prec_l (U, L^n_{s_\alpha}, I_n)$, then $\rho = \bigcup_{i=1}^{n} (U, L^i_{s_\alpha}, I_i)$ is a multi-granularity linguistic concept formal context.

Specially, supposing that $S = \{s_\alpha | \alpha = -\tau, \ldots, -1, 0, 1, \ldots, \tau\}$ is a LTS, we use S to describe different attributes, and the resulting (U, L_{s_α}, I) is called the $(2\tau + 1)$-granularity linguistic concept formal context.

Example 1. We take a FLIS (P, Q, I, f) as shown in Table 1, where $P = \{x_1, x_2, \ldots, x_8\}$ denotes eight students in a high school, the attribute set $Q = \{a, b, c\}$ denotes students' examination subject scores, a represents language scores, b represents mathematics scores, and c is English scores, The linguistic terms set $S = \{s_\alpha | \alpha = -\tau, \ldots, -1, 0, 1, \ldots, \tau\}$ is a binary relation between P and Q, i.e., $I \subseteq P \times Q$. When $\tau = 3$, the linguistic terms set represents $S = \{s_{-3} =$ very low, $s_{-2} =$ low, $s_{-1} =$ a little low, $s_0 =$ medium, $s_1 =$ a little high, $s_2 =$ high, $s_3 =$ very high$\}$.

In order to construct the seven-granularity linguistic concept lattice, the FLIS (P, Q, I, f) in Table 1 is scaled to the LCFC in Table 2. Fuzzy linguistic relations for attributes that none of the objects have are omitted in Table 2, such as $a_{s_{-2}}$, etc. In Table 2, 1 means that the object has the linguistic concept and 0 means that the object does not have the linguistic concept.

Table 1. Linguistic-valued information system (P, Q, I, f)

P	a	b	c
x_1	s_{-3}	s_0	s_0
x_2	s_{-1}	s_{-1}	s_{-2}
x_3	s_0	s_2	s_3
x_4	s_1	s_2	s_{-1}
x_5	s_{-1}	s_0	s_{-1}
x_6	s_2	s_1	s_1
x_7	s_{-1}	s_1	s_1
x_8	s_0	s_{-1}	s_{-2}

According to Table 2, we obtain the following Linguistic Concept Knowledge (LCK) by calculation in Table 3.

The number of LCK generated in the seven-granularity linguistic formal context is seventeen, and the LCK with the largest number of objects in the extents is $(x_2, x_5, x_7\}, \{a_{s_{-1}}\})$, and the number of objects in the extents is three. The seven-granularity conceptual hierarchy is constructed based on the partial order relationship between LCK, and the resulting linguistic concept lattice $LCKL(U, L_{s_\alpha}, I)$ is shown in Fig. 1.

Table 2. Seven-granularity linguistic concept formal context (P, Q, I, f)

	$a_{s_{-3}}$	$a_{s_{-1}}$	a_{s_0}	a_{s_1}	a_{s_2}	$b_{s_{-1}}$	b_{s_0}	b_{s_1}	b_{s_2}	$c_{s_{-2}}$	$c_{s_{-1}}$	c_{s_0}	c_{s_1}	c_{s_3}
x_1	1	0	0	0	0	0	1	0	0	0	0	1	0	0
x_2	0	1	0	0	0	1	0	0	0	1	0	0	0	0
x_3	0	0	1	0	0	0	0	0	1	0	0	0	0	1
x_4	0	0	0	1	0	0	0	0	1	0	1	0	0	0
x_5	0	1	0	0	0	0	1	0	0	0	1	0	0	0
x_6	0	0	0	0	1	0	0	1	0	0	0	0	1	0
x_7	0	1	0	0	0	0	0	1	0	0	0	0	1	0
x_8	0	0	1	0	0	1	0	0	0	1	0	0	0	0

Table 3. Seven-granularity linguistic concept

#1	$(P, \emptyset)$
#2	$(\{x_2, x_5, x_7\}, \{a_{s_{-1}}\})$
#3	$(\{x_6, x_7\}, \{b_{s_1}, c_{s_1}\})$
#4	$(\{x_4, x_5\}, \{c_{s_{-1}}\})$
#5	$(\{x_3, x_8\}, \{a_{s_0}\})$
#6	$(\{x_1, x_5\}, \{b_{s_0}\})$
#7	$(\{x_2, x_8\}, \{b_{s_{-1}}, c_{s_{-2}}\})$
#8	$(\{x_3, x_4\}, \{b_{s_2}\})$
#9	$(\{x_1\}, \{a_{s_{-3}}, b_{s_0}, c_{s_0}\})$
#10	$(\{x_2\}, \{a_{s_{-1}}, b_{s_{-1}}, c_{s_{-1}}\})$
#11	$(\{x_3\}, \{a_{s_0}, b_{s_2}, c_{s_3}\})$
#12	$(\{x_4\}, \{a_{s_1}, b_{s_2}, c_{s_{-1}}\})$

(continued)

Table 3. (*continued*)

#1	$(P, \emptyset)$
#13	$(\{x_5\}, \{a_{s_{-1}}, b_{s_0}, c_{s_{-1}}\})$
#14	$(\{x_6\}, \{a_{s_2}, b_{s_1}, c_{s_1}\})$
#15	$(\{x_7\}, \{a_{s_{-1}}, b_{s_1}, c_{s_1}\})$
#16	$(\{x_8\}, \{a_{s_0}, b_{s_{-1}}, c_{s_{-2}}\})$
#17	$(\emptyset, L_{s_\alpha})$

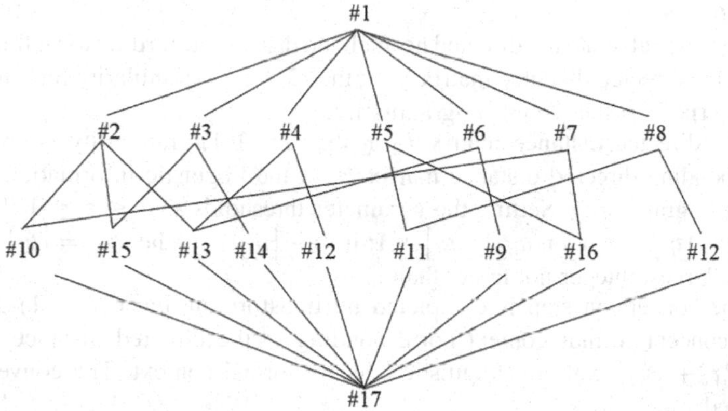

Fig. 1. Seven-granularity linguistic concept lattice

3.2 Transformation Models for Different Granularity Linguistic Concept formal Context

Definition 8. Let $S = \{s_a | a = -\tau, ..., -1, 0, 1, ..., \tau\}$ be a finite set of fully ordered linguistic terms consisting of an odd number of linguistic terms, the directed distance from a linguistic term s_a to s_0 is defined as a, the directed normalized distance is defined as $\frac{a}{\tau}$ $(-1 \leq \frac{a}{\tau} \leq 1)$.

Theroem 1. The directed normalized distance $\frac{a}{\tau}$ from a linguistic term s_a to an intermediate linguistic term s_0 satisfies the following properties:

(1) $\frac{a}{\tau} = 0$ if and only if $a = 0$;
(2) $\frac{a}{\tau} = 1$ if and only if $a = \tau$; $\frac{a}{\tau} = -1$ if and only if $a = -\tau$;
(3) Boundedness: $-1 \leq \frac{a}{\tau} \leq 1$.

When $-1 \leq \frac{a}{\tau} \leq 0$, it means that s_a is said to be symmetrically distributed on the left side of s_0. When $0 \leq \frac{a}{\tau} \leq 1$, it means that s_0 is symmetrically distributed on the right side of s_a.

Definition 9. In the linguistic-valued information system (P, Q, I, f), for $x_i \in U$, $a_j \in M$, $s_k \in S$, the corresponding directed distance matrix of $<U, M, I, f>$ as follows:

$$T = (a_{ij})_{m \times n},$$

where a_j denotes the directed distance between the binary fuzzy linguistic relations x_i to a_j between s_k and s_0.

The method for transforming a $(2\tau_1 + 1)$-granularity linguistic concept formal context into a $(2\tau_2 + 1)$-granularity linguistic concept formal context are as follows.

Step 1: We calculate the directed distance matrix T_1 corresponding to (P, Q, I, f) from the $(2\tau_1 + 1)$-granularity linguistic-valued information system (P, Q, I, f) according to Definition 6.
Step 2: Then, we calculate the directed normalized distance according to Definition 5 and converting the directed distance matrix T_1 at the $(2\tau_1 + 1)$-granularity into the directed distance matrix T_2 at the $(2\tau_2 + 1)$-granularity;
Step 3: The directed distance matrix T_2 at the $(2\tau_2 + 1)$-granularity is converted to the corresponding directed distance matrix T_3 of the linguistic information system at the $(2\tau_2 + 1)$-granularity. Setting the parameter thresholds ε, $0 \le \varepsilon \le 1$. If $t_{ij} \in T_2$, $u_{ij} \in T_3$,$t_{ij} - [t_{ij}] \ge \varepsilon$, then $u_{ij} = [t_{ij}] + 1$; if $t_{ij} - [t_{ij}] < \varepsilon$, then $u_{ij} = [t_{ij}]$. Here $[t_{ij}]$ denotes the largest integer not larger than t_{ij};
Step 4: The conversion step is completed by transforming the $(2\tau_2 + 1)$-granularity linguistic concept formal context corresponding to the directed distance matrix T_3 into the $(2\tau_2 + 1)$-granularity linguistic concept formal context. The conversion step is completed.

In the above process, T_1,T_2 satisfy the relation $T_2 = \frac{\tau_2}{\tau_1} T_1$, i.e., the directed distance matrices T_1,T_2 can be converted to each other, which ensures that there is no data loss in the process of converting different granularity fuzzy linguistic-valued information systems to each other.

4 Example Analysis and Illustration

We take the seven-granularity LCFC in Table 1 as an example, and the seven-granularity LCFC is converted into a 3-granularity LCFC, and the steps are as follows.

Step 1: We calculate the corresponding directed distance matrix T_1 from the seven-granularity FLIS (P, Q, I, f).

$$T_1 = \begin{bmatrix} -3 & 0 & 0 \\ -1 & -1 & -2 \\ 0 & 2 & 3 \\ 1 & 2 & -1 \\ -1 & 0 & -1 \\ 2 & 1 & 1 \\ -1 & 1 & 1 \\ 0 & -1 & -2 \end{bmatrix}$$

Step 2: Then, we calculate the directed normalized distance by transforming the directed distance matrix T_1 at seven-granularity into the directed normalized distance matrix T_2.

$$T_2 = \begin{bmatrix} -1 & 0 & 0 \\ -\frac{1}{3} & -\frac{1}{3} & -\frac{2}{3} \\ 0 & \frac{2}{3} & 1 \\ \frac{1}{3} & \frac{2}{3} & -\frac{1}{3} \\ -\frac{1}{3} & 0 & -\frac{1}{3} \\ \frac{2}{3} & \frac{1}{3} & \frac{1}{3} \\ -\frac{1}{3} & \frac{1}{3} & \frac{1}{3} \\ 0 & -\frac{1}{3} & -\frac{2}{3} \end{bmatrix}$$

Step 3: Next, we set the threshold $\varepsilon = 0.5$ and transform the directed normalized distance matrix T_2 at seven-granularity to the corresponding matrix T_2 at three-granularity (P, Q, I, f).

$$T_3 = \begin{bmatrix} -1 & 0 & 0 \\ 0 & 0 & -1 \\ 0 & 1 & 1 \\ 0 & 1 & 0 \\ 0 & 0 & 0 \\ 1 & 0 & 0 \\ 0 & 0 & 0 \\ 0 & 0 & -1 \end{bmatrix}$$

Step 4: Transforming the matrix T_3 corresponding to the seven-granularity LCFC into a three-granularity LCFC.

When we set the threshold value $\varepsilon = 0.5$, the seven-granularity linguistic context in Table 1 is transformed to the three-granularity linguistic context as shown in Table 4.

Table 4. Three-granular linguistic formal context

U	a	b	c
x_1	s_{-1}	s_0	s_0
x_2	s_0	s_0	s_{-1}
x_3	s_0	s_1	s_1
x_4	s_0	s_1	s_0
x_5	s_0	s_0	s_0
x_6	s_1	s_0	s_0

(continued)

Table 4. *(continued)*

U	a	b	c
x_7	s_0	s_0	s_0
x_8	s_0	s_0	s_{-1}

In order to reduce the scale of the linguistic concept lattice $LCKL(U, L_{s_\alpha}, I)$, Table 4 was scaled into the corresponding LCFC as shown in Table 5.

We calculate the linguistic concept knowledge of Table 5 as shown in Table 6.

Table 5. Three-granular linguistic concept formal context

U	$a_{s_{-1}}$	a_{s_0}	a_{s_1}	b_{s_0}	b_{s_1}	$c_{s_{-1}}$	c_{s_0}	c_{s_1}
x_1	1	0	0	1	0	0	1	0
x_2	0	1	0	1	0	1	0	0
x_3	0	1	0	0	1	0	0	1
x_4	0	1	0	0	1	0	1	0
x_5	0	1	0	1	0	0	1	0
x_6	0	0	1	1	0	0	1	0
x_7	0	1	0	1	0	0	1	0
x_8	0	1	0	1	0	1	0	0

Table 6. Three-granularity linguistic concept

#1	$(U, \emptyset)$
#2	$(\{x_1, x_2, x_5, x_6, x_7, x_8\}, \{b_{s_0}\})$
#3	$(\{x_2, x_3, x_4, x_5, x_7, x_8\}, \{a_{s_0}\})$
#4	$(\{x_1, x_4, x_5, x_6, x_7\}, \{c_{s_0}\})$
#5	$(\{x_4, x_5, x_7\}, \{a_{s_0}, c_{s_0}\})$
#6	$(\{x_1, x_5, x_6, x_7\}, \{b_{s_0}, c_{s_0}\})$
#7	$(\{x_2, x_5, x_7, x_8\}, \{a_{s_0}, b_{s_0}\})$
#8	$(\{x_3, x_4\}, \{a_{s_0}, b_{s_1}\})$
#9	$(\{x_5, x_7\}, \{a_{s_0}, b_{s_0}, c_{s_0}\})$
#10	$(\{x_2, x_8\}, \{a_{s_0}, b_{s_0}, c_{s_{-1}}\})$
#11	$(\{x_1\}, \{a_{s_{-1}}, b_{s_0}, c_{s_0}\})$

(continued)

Table 6. (*continued*)

#1	$(U, \emptyset)$
#12	$(\{x_3\}, \{a_{s0}, b_{s1}, c_{s1}\})$
#13	$(\{x_4\}, \{a_{s0}, b_{s1}, c_{s0}\})$
#14	$(\{x_6\}, \{a_{s1}, b_{s0}, c_{s0}\})$
#15	$(\emptyset, L_{s_a})$

Based on the partial order relationship between linguistic concepts, we construct a three-granularity lattice of linguistic concepts as shown in Fig. 2.

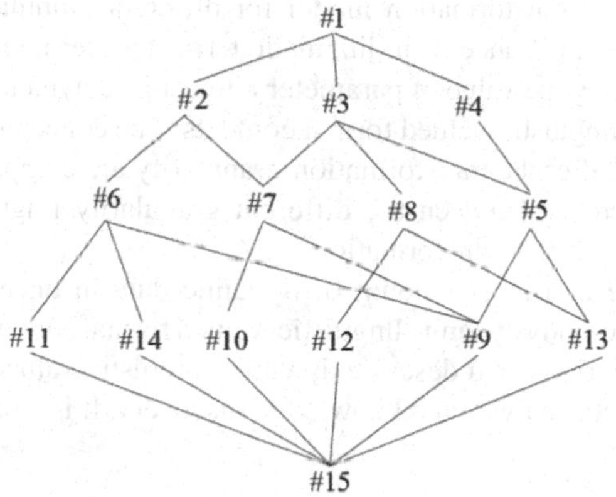

Fig. 2. Three-granular linguistic concept lattice

Since linguistic term s_a is distributed symmetrically around the intermediate linguistic term s_0, the same linguistic formal context is in the description of different granular linguistic terms, the meaning of the intermediate linguistic term s_0 is unchanged. Therefore, we can compare the linguistic concepts generated by different granular linguistic formal context. $(\{x_1, x_5\}, \{b_{s_0}\})$ and $(\{x_3, x_8\}, \{a_{s_0}\})$ are the concepts generated in the seven-granularity linguistic formal context.$(\{x_1, x_2, x_5, x_6, x_7, x_8\}, \{b_{s_0}\})$ and $(\{x_2, x_3, x_4, x_5, x_7, x_8\}, \{a_{s_0}\})$ are the concepts generated in the three-granularity linguistic formal context. Compared with the seven-granularity linguistic formal context is thinner than the granularity size of the three-granularity linguistic formal context.

When the intent is the same, the number of extents in the concept of the three-granularity linguistic formal context is more extended than the number of extents generated in the background of the seven-granularity linguistic formal context. The extent x_1 of the generated concept in the three-granularity linguistic formal context and the extent x_2 of the generated concept in the seven-granularity linguistic formal context satisfy the inclusion relationship $x_2 \subseteq x_1$. For example, $\{x_1, x_5\} \subseteq \{x_1, x_2, x_5, x_6, x_7, x_8\}$ and $\{x_3, x_8\} \subseteq \{x_2, x_3, x_4, x_5, x_7, x_8\}$. The transformation to the attribute from fine granularity to coarse granularity, and judging the fact that the attributes are divided into the basis of the object to increase the fact that the same object is increased. For example, the number of non-empty linguistic concepts in this example is fifteen in the seven-granularity linguistic formal context, and the number of non-empty linguistic concepts in the three-granularity linguistic formal context is thirteen. The maximum number of objects in the extent is three in the LCK corresponding to the seven-granularity linguistic formal context, and the maximum number of objects in the extent is six in the LCK corresponding to the three-granularity linguistic formal context.

Generally speaking, the finer the linguistic granularity of the linguistic formal context, the more the amount of language concept knowledge calculated, and the less the number of linguistic concept extents.

5 Conclusions

This paper proposes a transformation model for different granularity LCFC. Firstly, we keep the normalized distance from linguistic terms to intermediate linguistic terms unchanged, by adjusting the value of parameter ε to achieve dynamic transformation of different granularity linguistic-valued formal contexts. The concepts and concept lattice generated before and after the transformation granularity are compared to illustrate the connection and difference between the different granularity linguistic-valued formal contexts before and after the transformation.

In real life, there is a large amount of dynamic data in uncertain environments, therefore, we intend to study dynamic linguistic-valued formal contexts, fuse information from different time periods, and describe dynamic linguistic-valued formal contexts in stages for dynamic data analysis and knowledge discovery. It is worth studying and has significance.

Acknowledgments. This work is partially supported by the National Natural Science Foundation of P.R. China (No. 61772250, 62176142), Foundation of Liaoning Educational Committee (No. LJ2020007) and Special Foundation for Distinguished Professors of Shandong Jianzhu University.

References

1. Wille, R.: Restructuring lattice theory: an approach based on hierarchies of concepts. In: Rival, I. (ed.) Ordered Sets, pp. 445–470. Springer, Dordrecht (1982)
2. Poelmans, J., Ignatov, D.I., Kuznetsov, S.O., Dedene, G.: Formal concept analysis in knowledge processing: a survey on applications. Expert Syst. Appl. An Int. J. **40**(16), 6538–6560 (2013)
3. Kaytoue, M., Kuznetsov, S.O., Napoli, A., Duplessis, S.: Mining gene expression data with pattern structures in formal concept analysis. Inform. Sci. **181**(10), 1989–2001 (1989)
4. Adriana, M., de Farias, G., Cintra, M.E., Felix, A.C., Cavalcante, D.L.: Definition of strategies for crime prevention and combat using fuzzy clustering and formal concept analysis. Int. J. Unc. Fuzz. Knowl.-Based Syst. **26**(03), 429–452 (2018). https://doi.org/10.1142/S0218488518500216
5. Baixeries, J., Kaytoue, M., Napoli, A.: Characterizing functional dependencies in formal concept analysis with pattern structures. Ann. Math. Artif. Intell. **72**(1–2), 129–149 (2014). https://doi.org/10.1007/s10472-014-9400-3

6. Zadeh, L.A., Kacprzyk, J.: Computing with Words in Information/Intelligent Systems, vol. 2. Physica-Verlag, HD (1999)
7. Pang, K., Kang, N., Chen, S., Zheng, H.: Collaborative filtering recommendation algorithm based on linguistic concept lattice with fuzzy object. In: 2019 IEEE 14th International Conference on Intelligent Systems and Knowledge Engineering (ISKE), pp. 57–63. IEEE (2019)
8. Zou, L., Pang, K., Song, X., Kang, N., Liu, X.: A knowledge reduction approach for linguistic concept formal context. Inform. Sci. **524**, 165–183 (2020)
9. Zou, L., Kang, N., Che, L., Liu, X.: Linguistic-valued layered concept lattice and its rule extraction. Int. J. Mach. Learn. Cybern. **13**(1), 83–98 (2021). https://doi.org/10.1007/s13042-021-01351-3
10. Yao, Y.Y.: Concept lattices in rough set theory. In: IEEE Annual Meeting of the Fuzzy Information, 2004. Processing NAFIPS '04, vol. 2, pp. 796–801 (2004)
11. Wu, W.Z., Chen, Y., Xu, Y.H.: Optimal granularity selections in consistent incomplete multi-granular labeled decision systems. Pattern Recogn. Artif. Intell. **29**(2), 108–115 (2016)
12. Shao, M.-W., Lv, M.-M., Li, K.-W., Wang, C.-Z.: The construction of attribute (object)-oriented multi-granularity concept lattices. Int. J. Mach. Learn. Cybern. **11**(5), 1017–1032 (2019). https://doi.org/10.1007/s13042-019-00955-0
13. Chu, X., et al.: Multi-granularity dominance rough concept attribute reduction over hybrid information systems and its application in clinical decision-making. Inf. Sci. **597**, 274–299 (2022)
14. Hu, Q., Qin, K.Y.: The construction of multi-granularity concept lattices. J. Intell. Fuzzy Syst. **39**(3), 2783–2790 (2020)
15. Liao, H., Zeshui, X., Zeng, X.-J.: Distance and similarity measures for hesitant fuzzy linguistic term sets and their application in multi-criteria decision making. Inform. Sci. **271**, 125–142 (2014). https://doi.org/10.1016/j.ins.2014.02.125

Paraconsistent Rough Set Algebras

Hao Wu(✉)

Institute of Logic and Cognition, Sun Yat-Sen University, Guangzhou, China
wuhao43@mail2.sysu.edu.cn

Abstract. Paraconsistent Pawlakian rough sets and paraconsistent covering based rough sets are introduced for modeling and reasoning about inconsistent information. Topological quasi-Boolean algebras are shown to be algebras for paraconsistent rough sets. We also give two sequent calculi as the modal systems for these paraconsistent rough sets.

Keywords: Rough set · Paraconsistency · Quasi-boolean algebra · Sequent calculus

1 Introduction

Rough sets by Pawlak [14,15] was proposed as an approach to imprecise knowledge about objects in the field of knowledge representation. Knowledge and data differ in the way that the former is organized while the latter is loosely scattered. Pawlakian knowledge is based on the notion of *classification*. A knowledge base is understood as a relational structure $(U, \mathbf{R})$ where $U \neq \varnothing$ is a set of objects and $\mathbf{R}$ is a family of equivalence relations over U. Hence the logic for approximate reasoning in Pawlakian rough sets is indeed a multimodal logic **S5**. Later works focus on the generalization of Pawlak's rough set theory by extending the equivalence relation to similarity relation [18], altering the equivalence relation to arbitrary binary relations [21,22] or by replacing the partitions by coverings [6,23]. Interactions between rough set theory and modal logic are presented in [9,16].

In many practical scenarios, classifications of objects are given by a set of attributes. A usual assumption is that each attribute determines a set of objects. Given an object x in the universe and an attribute ϕ, either $\phi(x)$ holds or not. This is usually called the *consistency assumption*. Thus in standard rough set systems we make the lower and upper approximations of a given set X of objects. However, in practice, there exist datebases which are *inconsistent* in the sense that there are contradictions or conflict. For example, a toy is classified into both round objects and square objects. In such a case, we need a *paraconsistent* rough set theory which can be used to deal with inconsistent information.

This work was supported by Chinese National Funding of Social Sciences (Grant no. 18ZDA033).

Y. Chen and S. Zhang (Eds.): AILA 2022, CCIS 1657, pp. 166–179, 2022.
https://doi.org/10.1007/978-981-19-7510-3_13

Previous work bridging paraconsistent logic and rough set theory can be found in [12,20] where basic notions such as set, approximations and similarity relation are allowed to have four values. Later, paraconsistency was treated as membership function, set containment and set operations in [13,19]. Four-valued logic was employed as the semantics to express approximate reasoning. However, bilattice in Belnap's logic was discarded since Belnapian truth ordering was considered counterintuitive. Therefore, only knowledge ordering was retained in their framework and truth ordering was changed into a linear order in their approach.

Here, we consider an attribute ϕ as a pair of sets of objects $\langle \phi^+, \phi^- \rangle$ where ϕ^+ is the set of all objects having the attribute ϕ and ϕ^- is the set of all objects lacking the attribute ϕ. Then $\phi^+ \cap \phi^-$ consists of those objects with inconsistent information, and the objects outside $\phi^+ \cup \phi^-$ have no information with respect to the attribute ϕ. Similar idea can be found in the Belnap-Dunn four-valued logic [3,4]. We will introduce the notion of *polarity* in knowledge base and approximations of polarity. In such a way we obtain new paraconsistent rough sets. A polarity is simply a pair of sets of objects $\langle X, Y \rangle$ which are candidates for approximations. The following figure shows four cases of a polarity $\langle X, Y \rangle$ in a universe U of objects: Objects in $X \cap Y$ have inconsistent information, and

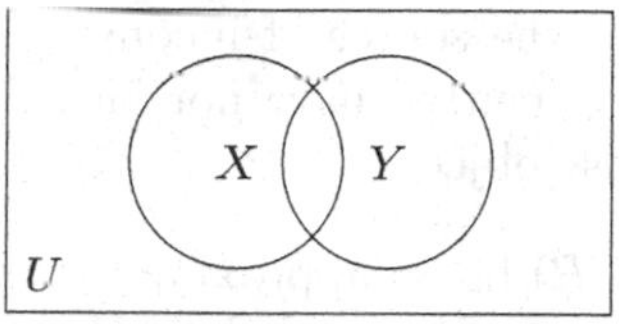

Fig. 1. Polarity in a universe

those outside of $X \cup Y$ have no information. If $X \cap Y \neq \varnothing$, we say that $\langle X, Y \rangle$ is inconsistent; and if $X \cup Y \neq U$, we say that $\langle X, Y \rangle$ is incomplete. In the standard rough set theory, a set of objects X can be viewed as a polarity $\langle X, X^c \rangle$ where X^c is the complement of X in U. It is clearly both consistent and complete.

Quasi-Boolean algebras and topological quasi-Boolean algebras are algebras for rough sets. Algebras for Pawlakian rough sets and covering based rough sets can be found in previous work such as [1,2,7,11]. Recent development on the interrelations betwenn rough sets and logic is presented in [9,17]. We construct the algebras for paraconsistent Pawlakian rough set and paraconsistent covering based rough sets respectively. Jonsson-Tarskian style [5] duality results are also provided.

In the present work, we shall introduce two types of paraconsistent rough sets. One is the Pawlakian, and the other is covering based. Lower and upper approximations in each type will be defined. Then paraconsistent rough set algebras for each type of paraconsistent rough sets will be proposed and representation theorems for these algebras will be proved. Finally we establish two sequent calculi for reasoning in these paraconsistent rough sets.

2 Paraconsistent Rough Sets

In this section, we introduce paraconsistent rough sets which are variants of the standard rough sets. We shall define the upper and lower approximations, and prove some basic properties of them.

Definition 1. A (Pawlakian) *approximation space* is a structure $\mathcal{K} = (U, R)$ where $U \neq \varnothing$ is the set of objects and R is an equivalence relation on U. For every $x \in U$, let $R(x) = \{y \in U : xRy\}$, i.e., the equivalence class of x.

A *polarity* in U is a paraconsistent pair of sets $\langle X, Y\rangle$ with $X, Y \subseteq U$ and $X \cap Y \neq \varnothing$. The set of all polarities in U is denoted by $\mathfrak{P}(U)$ which is exactly the product $\mathcal{P}(U) \times \mathcal{P}(U)$ of power sets of U. We use capital letters G, H etc. for polarities. For each $G \in \mathfrak{P}(U)$, if $G = \langle X, Y\rangle$, we write $G^+ = X$ and $G^- = Y$.

Let $\mathcal{K} = (U, R)$ be an approximation space. A polarity G is *consistent* in $\mathcal{K}$ if $G^+ \cap G^- = \varnothing$; and G is *complete* in $\mathcal{K}$ if $G^+ \cup G^- = U$. In the standard Pawlakian rough set theory, a set of objects X to be approximated can be viewed as a consistent and complete polarity $\langle X, X^c\rangle$ where $X^c = U \setminus X$ is the complement of X in U. Each polarity G can be viewed as an attribute or property of objects. The set G^+ stands for the set of all objects having the attribute G, and G^- for the set of all objects lacking G. Then objects in $G^+ \cap G^-$ both have G and do not have G. This part is the source of inconsistent information. Objects in the complement $(G^+ \cup G^-)^c$ neither have nor lack the attribute G, namely no information is given for these objects.

Definition 2. Let $\mathcal{K} = (U, R)$ be an approximation space and $G, H \in \mathfrak{P}(U)$ be polarities. The operations $\sim$, $\sqcap$ and $\sqcup$ are defined as follows:

$$
\begin{aligned}
{\sim}G &= \langle G^-, G^+\rangle \\
G \sqcap H &= \langle G^+ \cap H^+, G^- \cup H^-\rangle \\
G \sqcup H &= \langle G^+ \cup H^+, G^- \cap H^-\rangle.
\end{aligned}
$$

The binary relation $\sqsubseteq$ on $\mathfrak{P}(U)$ is defined by setting $G \sqsubseteq H$ if and only if $G^+ \subseteq H^+$ and $H^- \subseteq G^-$. We define the following sets:

$$
\begin{aligned}
\underline{G}^+ &= \{x \in U : R(x) \subseteq G^+\}, & \underline{G}^- &= \{x \in U : R(x) \cap G^- \neq \varnothing\}, \\
\overline{G}^+ &= \{x \in U : R(x) \cap G^+ \neq \varnothing\}, & \overline{G}^- &= \{x \in U : R(x) \subseteq G^-\}.
\end{aligned}
$$

The *lower approximation* of G is defined as the polarity $\underline{G} = \langle \underline{G}^+, \underline{G}^-\rangle$, and the *upper approximation* of G is defined as the polarity $\overline{G} = \langle \overline{G}^+, \overline{G}^-\rangle$.

For every approximation space $\mathcal{K} = (U, R)$, it is clear that $\sqsubseteq$ is a partial order on the set $\mathfrak{P}(U)$ of all polarities. Moreover, for all $G, H \in \mathfrak{P}(U)$, $G = H$ if and only if $G \sqsubseteq H$ and $H \sqsubseteq G$.

Example 1. Let $\mathcal{K} = (U, R)$ be the approximation space where $U = \{x_1, x_2, x_3, x_4, x_5\}$ and R is an equivalence relation with classification $\{\{x_1\}, \{x_2, x_4\}, \{x_3, x_5\}\}$. Let $G^+ = \{x_1, x_2\}$ and $G^- = \{x_2, x_3, x_5\}$. Note that the polarity $G = \langle G^+, G^- \rangle$ is inconsistent and incomplete. Note that $R(x_1) = \{x_1\}$, $R(x_2) = R(x_4) = \{x_2, x_4\}$ and $R(x_3) = R(x_5) = \{x_3, x_5\}$. Then $\underline{G}$ and $\overline{G}$ are calculated as follows:

$$\underline{G}^+ = \{x_1\};\ \underline{G}^- = \{x_2, x_3, x_4, x_5\};\ \overline{G}^+ = \{x_1, x_2, x_4\};\ \overline{G}^- = \{x_3, x_5\}$$

Note that $\underline{G}$ and $\overline{G}$ are consistent and complete.

Proposition 1. *Let $\mathcal{K} = (U, R)$ be an approximation space and $G, H \in \mathfrak{P}(U)$ be polarities. Then the following conditions hold:*

(1) $\sim(G \sqcap H) = \sim G \sqcup \sim H$.
(2) $\sim(G \sqcap H) = \sim Q \sqcup \sim H$.
(3) $\sim\sim G = G$.
(4) $\langle \varnothing, U \rangle \sqsubseteq G$ *and* $G \sqsubseteq \underline{\langle U, \varnothing \rangle}$.
(5) $\underline{\langle U, \varnothing \rangle} = \langle U, \varnothing \rangle$ *and* $\overline{\langle \varnothing, U \rangle} = \langle \varnothing, U \rangle$.
(6) *if* $G \sqsubseteq H$, *then* $\underline{G} \sqsubseteq \underline{H}$ *and* $\overline{G} \sqsubseteq \overline{H}$.
(7) $\underline{G \sqcap H} = \underline{G} \sqcap \underline{H}$ *and* $\overline{G \sqcup H} = \overline{G} \sqcup \overline{H}$.
(8) $\sim\underline{G} = \overline{\sim G}$ *and* $\sim\overline{G} = \underline{\sim G}$.
(9) $\underline{G} \sqcap \overline{H} \sqsubseteq \overline{G \sqcap H}$.

Proof. For (6), assume $G \sqsubseteq H$. Suppose $x \in \underline{G}^+$. Then $R(x) \subseteq G^+$. By the assumption, $G^+ \subseteq H^+$. Hence $x \in \underline{H}^+$. Suppose $x \in \underline{H}^-$. Then $R(x) \cap H^- \neq \varnothing$. By the assumption, $H^- \subseteq G^-$. Then $R(x) \cap G^- \neq \varnothing$ and so $x \in \underline{G}^-$. Hence $\underline{G} \sqsubseteq \underline{H}$. Similarly $\overline{G} \sqsubseteq \overline{H}$. For (7), clearly $G \sqcap H \sqsubseteq G$ and $C \sqcap H \sqsubseteq H$. By (6), $\underline{G \sqcap H} \sqsubseteq \underline{G}$ and $\underline{G \sqcap H} \sqsubseteq \underline{H}$. Hence $\underline{G \sqcap H} \sqsubseteq \underline{G} \sqcap \underline{H}$. Conversely, $\underline{G} \sqcap \underline{H} = \langle \underline{G}^+ \cap \underline{H}^+, \underline{G}^- \cup \underline{H}^- \rangle$ and $\underline{G \sqcap H} = \langle \underline{G^+ \cap H^+}, \underline{G^- \cup H^-} \rangle$. Clearly $\underline{G}^+ \cap \underline{H}^+ \subseteq \underline{G^+ \cap H^+}$ and $\underline{G^- \cup H^-} \subseteq \underline{G}^- \cup \underline{H}^-$. Hence $\underline{G} \sqcap \underline{H} \sqsubseteq \underline{G \sqcap H}$. Then $\underline{G \sqcap H} = \underline{G} \sqcap \underline{H}$. Similarly $\overline{Q \sqcup H} = \overline{G} \sqcup \overline{H}$. Other items are shown directly. □

Proposition 2. *Let $\mathcal{K} = (U, R)$ be an approximation space and $G \in \mathfrak{P}(U)$. Then* (1) $\underline{G} \sqsubseteq G \sqsubseteq \overline{G}$; (2) $\underline{G} \sqsubseteq \underline{\underline{G}}$ *and* $\overline{\overline{G}} \sqsubseteq \overline{G}$; *and* (3) $\overline{G} \sqsubseteq \underline{(\overline{G})}$ *and* $G \sqsubseteq \underline{(\overline{G})}$.

Proof. (1) Assume $x \in \underline{G}^+$. Then $R(x) \subseteq G^+$. Since R is an equivalence relation, we have $x \in R(x)$ and so $x \in G^+$. Hence $\underline{G}^+ \subseteq G^+$. Assume $y \in G^-$. By the reflexivity of R, we have $y \in R(y)$. Then $R(y) \cap G^- \neq \varnothing$. Hence $y \in \underline{G}^-$. Then $G^- \subseteq \underline{G}^-$. It follows that $\overline{G} \sqsubseteq G$. Similarly $G \sqsubseteq \overline{G}$.

(2) Assume $x \in \underline{G}^+$. Then $R(x) \subseteq G^+$. Suppose $y \in R(x)$. Let $z \in R(y)$. By the transitivity of R, we have $z \in R(x)$. Then $z \in G^+$. Hence $R(y) \subseteq G^+$, i.e., $y \in \underline{G}^+$. Then $R(x) \subseteq \underline{G}^+$, i.e., $x \in \underline{\underline{G}}^+$. It follows that $\underline{G}^+ \subseteq \underline{\underline{G}}^+$. Assume $x \in \underline{\underline{G}}^-$. Then $R(x) \cap \underline{G}^- \neq \varnothing$. Let $y \in R(x)$ and $y \in \underline{G}^-$. Then $R(y) \cap G^- \neq \varnothing$. Let $z \in R(y)$ and $z \in G^-$. By the transitivity of R, we have $z \in R(x)$. Then $R(x) \cap G^- \neq \varnothing$. Hence $x \in \underline{G}^-$. Then $\underline{\underline{G}}^- \sqsubseteq \underline{G}^-$. Similarly $\overline{\overline{G}} \sqsubseteq \overline{G}$.

(3) Assume $x \in \overline{G}^+$. Then $R(x) \cap G^+ \neq \varnothing$. Let $y \in R(x)$ and $y \in G^+$. Suppose $z \in R(x)$. Since R is an equivalence relation, we have $y \in R(z)$. Then $R(z) \cap G^+ \neq \varnothing$, i.e., $z \in \overline{G}^+$. Hence $R(x) \subseteq \overline{G}^+$, i.e., $x \in \underline{(\overline{G})}^+$. Then $\overline{G}^+ \subseteq \underline{(\overline{G})}^+$. Assume $x \in \underline{(\overline{G})}^-$. Then $R(x) \cap \overline{G}^- \neq \varnothing$. Let $y \in R(x)$ and $y \in \overline{G}^-$. Then $R(y) \subseteq G^-$. Suppose $z \in R(x)$. Since R is an equivalence relation, $z \in R(y)$. Then $z \in G^-$. Hence $R(x) \subseteq G^-$, i.e., $x \in \overline{G}^-$. It follows that $\underline{(\overline{G})}^- \subseteq \overline{G}^-$. Moreover, by (1), we have $G \sqsubseteq \overline{G} \sqsubseteq \underline{(\overline{G})}$. □

Now we continue by introducing the paraconsistent covering based rough sets which are defined based on covering frames by a neighborhood function N.

Definition 3. Given a nonempty set of objects U, a *covering* of U is a nonempty collection $\mathscr{C} = \{C_i \subseteq U \mid i \in I\}$ such that $\bigcup \mathscr{C} = U$.

A *covering frame* is a pair $\mathscr{F} = (U, \mathscr{C})$ where $U \neq \varnothing$ and $\mathscr{C}$ is a covering of U. The *neighborhood function* $N : U \to \mathcal{P}(U)$ is defined by setting

$$N(x) = \bigcap \{C \in \mathscr{C} \mid x \in C\}.$$

For every polarity $G \in \mathfrak{P}(U)$, we define the following sets:

$$\Box_N G^+ = \{x \in U : N(x) \subseteq G^+\}, \qquad \Box_N G^- = \{x \in U : N(x) \cap G^- \neq \varnothing\},$$
$$\Diamond_N G^+ = \{x \in U : N(x) \cap G^+ \neq \varnothing\}, \quad \Diamond_N G^- = \{x \in U : N(x) \subseteq G^-\}.$$

The *lower N-approximation* of G is defined as $\Box_N G = \langle \Box_N G^+, \Box_N G^- \rangle$, and the *upper N-approximation* of G is defined as $\Diamond_N G = \langle \Diamond_N G^+, \Diamond_N G^- \rangle$.

An **S4**-frame is a pair $\mathfrak{F} = (U, R)$ where $U \neq \varnothing$ is a nonempty set of objects and $R \subseteq U \times U$ is a preorder, i.e., a reflexive and transitive relation on U. For every $x \in U$, let $R(x) = \{y \in U : xRy\}$ be the set of all objects related with x in $\mathfrak{F}$.

Example 2. Let $U = \{x_1, x_2, x_3, x_4, x_5\}$ and $R = \{\langle x_i, x_j \rangle \in U \times U : i \leq j\}$. Then we have an **S4** frame $\mathfrak{F} = (U, R)$. Consider the polarity $G = \langle G^+, G^- \rangle$ where $G^+ = \{x_1, x_2\}$ and $G^- = \{x_2, x_3, x_5\}$. Then we calculate $\Box_N G^+ = \varnothing, \Box_N G^- = U, \Diamond_N G^+ = \{x_1, x_2\}$ and $\Diamond_N G^- = \{x_5\}$. Note that $\Box_N G$ is consistent and complete, and $\Diamond_N G$ is consistent and incomplete.

Proposition 3. *Let $\mathscr{F} = (U, \mathscr{C})$ be a covering frame and $G, H \in \mathfrak{P}(U)$ be polarities. Then the following conditions hold:*

(1) *if $G \sqsubseteq H$, then $\Box_N G \sqsubseteq \Box_N H$ and $\Diamond_N G \sqsubseteq \Diamond_N H$.*
(2) *$\Box_N G \sqcap H = \Box_N G \sqcap \Box_N H$ and $\Diamond_N (Q \sqcup H) = \Diamond_N G \sqcup \Diamond_N H$.*
(3) *$\sim\Box_N G = \Diamond_N {\sim} G$ and $\sim\Diamond_N G = \Box_N {\sim} G$.*
(4) *$\Box_N G \sqcap \Diamond_N H \sqsubseteq \Diamond_N (G \sqcap H)$.*
(5) *$\Box_N G \sqsubseteq G \sqsubseteq \Diamond_N G$.*
(6) *$\Box_N G \sqsubseteq \Box_N \Box_N G$ and $\Diamond_N \Diamond_N G \sqsubseteq \Diamond_N G$.*

Proof. Items (1)–(4) are shown by the definition.

(5) Assume $x \in \Box_N G^+$. Then $N(x) \subseteq G^+$. Clearly $x \in N(x)$. Then $x \in G^+$. Hence $\Box_N G^+ \subseteq G^+$. Assume $y \in G^-$. Then $y \in N(y)$ and so $N(y) \cap G^- \neq \varnothing$. Hence $y \in \Box_N G^-$. Then $G^- \subseteq \Box_N G^-$. Hence $\Box_N G \sqsubseteq G$. Similarly $G \sqsubseteq \Diamond_N G$.

(6) We show $\Box_N G \sqsubseteq \Box_N \Box_N G$. Assume $x \in \Box_N G^+$. Then $N(x) \subseteq G^+$. Suppose $y \in N(x)$. Let $z \in N(y)$. Then $z \in N(x)$ and so $z \in G^+$. Hence $y \in \Box_N G^+$. Then $N(x) \subseteq \Box_N G^+$, i.e., $x \in \Box_N \Box_N G^+$. Hence $\Box_N G^+ \subseteq \Box_N \Box_N G^+$. Now assume $x \in \Box_N \Box_N G^-$. Then $N(x) \cap \Box_N G^- \neq \varnothing$. Let $y \in N(x)$ and $y \in \Box_N G^-$. Then $N(y) \cap G^- \neq \varnothing$. Let $z \in N(y)$ and $z \in G^-$. Then $z \in N(x)$. Hence $N(x) \cap G^- \neq \varnothing$, i.e., $x \in \Box_N G^-$. Then $\Box_N \Box_N G^- \subseteq \Box_N G^-$. Similarly $\Diamond_N \Diamond_N G \sqsubseteq \Diamond_N G$. □

3 Paraconsistent Rough Set Algebras

In this section, we show that algebras for paraconsistent Pawlakian rough sets are partition topological quasi-Boolean algebras, and algebras for paraconsistent covering based rough sets are topological quasi-Boolean algebras.

An algebra $\mathbb{A} = (A, \cdot, +, ', 0, 1)$ is a *quasi-Boolean algebra* (qBa) if $(A, \cdot, +, 0, 1)$ is a bounded distributive lattice(cf. [8, Definition 2.12]) and for all $a, b \in A$:

(1) $(a \cdot b)' = a' + b'$.
(2) $(a + b)' = a' \cdot b'$.
(3) $a'' = a$
(4) $0' = 1$ and $1' = 0$.

Note that for simplicity, we will abbreviate ab for $a \cdot b$ hereafter. Quasi Boolean algebras are often used as the fundamental part of rough algebras. If we add modal operators to a qBa satisfying additional conditions, we obtain various rough algebras [2].

Definition 4. An algebra $\mathbb{A} = (A, \cdot, +, ', 0, 1, \Box)$ is a *topological quasi-Boolean algebra* (tqBa) if $(A, \cdot, +, ', 0, 1)$ is a quasi-Boolean algebra, $\Box$ is a unary operations on A such that for all $a, b \in A$:

($\mathrm{K}_\Box$) $\Box(ab) = \Box a \Box b$
($\mathrm{N}_\Box$) $\Box 1 = 1$
($\mathrm{T}_\Box$) $\Box a \leq a$
($4_\Box$) $\Box a \leq \Box\Box a$

where the lattice order $\leq$ on A is defined by setting $a \leq b$ if and only if $ab = b$. Let **tqBa** be the variety of all topological quasi-Boolean algebras.

A *partition topological quasi-Boolean algebra* (tqBa5) is a topological quasi-Boolean algebra $\mathbb{A} = (A, \cdot, +, ', \Box, 0, 1)$ such that for all $a \in A$:

($5_\Box$) $\Diamond a \leq \Box \Diamond a$

where $\Diamond$ is the unary operation on A defined by $\Diamond a := (\Box a')'$. Let **tqBa5** be the variety of all partition topological quasi-Boolean algebras.

Fact 1. *Let* $\mathbb{A} = (A, \cdot, +, ', 0, 1, \Box)$ *be a tqBa. For all* $a \in A$, (1) $\Diamond(a+b) = \Diamond a + \Diamond b$; (2) $\Diamond 0 = 0$; (3) $a \leq \Diamond a$ *and* (4) $\Diamond\Diamond a \leq \Diamond a$.

Now we define the dual algebras of approximation space and covering frame respectively.

Definition 5. The *dual algebra* of an approximation space $\mathcal{K} = (U, R)$ is defined as $\mathcal{K}^* = (\mathfrak{P}(U), \sqcap, \sqcup, \sim, \underline{(.)}, \langle\varnothing, U\rangle, \langle U, \varnothing\rangle)$ where $\mathfrak{P}(U)$ is the above defined polarities on U. The operation $\underline{(.)}$ stands for taking the lower approximation of quasicomplement. The *dual algebra* of a covering frame $\mathcal{F} = (U, \mathscr{C})$ is defined as $\mathcal{F}^\sharp = (\mathfrak{P}(U), \sqcap, \sqcup, \sim, \Box_N, \langle\varnothing, U\rangle, \langle U, \varnothing\rangle)$ where $\mathfrak{P}(U)$ is the above defined polarities on U.

By Proposition 1 (8), in the dual algebra $\mathcal{K}^*$ of an approximation space, we can define $\overline{G} := \sim(\underline{\sim G})$. By Proposition 3 (3), in the dual algebra $\mathcal{F}^\sharp$ of a covering frame $\mathscr{F}$, we can define $\Diamond_N G := \sim\Box_N\sim G$.

Proposition 4. *Let* $\mathcal{K} = (U, R)$ *be an approximation space and* $\mathcal{F} = (U, \mathscr{C})$ *be a covering frame. Then* (1) $\mathcal{K}^*$ *is a tqBa5; and* (2) $\mathcal{F}^\sharp$ *is a tqBa.*

Proof. It suffices to show that both algebras defined satisfy the properties of tqBa5 and tqBa respectively. For (1), by Proposition 1 and Proposition 2, $\mathcal{K}^*$ is a tqBa5. For (2), by Proposition 3, $\mathcal{F}^\sharp$ is a tqBa. $\Box$

Let $\mathbb{A} = (A, \cdot, +, ', \Box, 0, 1)$ be a tqBa. A *filter* in $\mathbb{A}$ is a subset $u \subseteq A$ such that the following conditions hold for all $a, b \in A$:

(1) $ab \in u$ for all $a, b \in u$.
(2) $a \in u$ and $a \leq b \in A$ imply $b \in u$.

A filter u in $\mathbb{A}$ is *proper* if $0 \notin u$. A proper filter u in $\mathbb{A}$ is *prime* if $a + b \in u$ implies $a \in u$ or $b \in u$. Let $U(A)$ be the set of all prime filters in $\mathbb{A}$. Let $\mathbb{A} = (A, \cdot, +, ', \Box, 0, 1)$ be a tqBa. A subset $\varnothing \neq X \subseteq A$ has the *finite meet property*, if $a_1 \ldots a_n \neq 0$ for all $a_1, \ldots, a_n \in X$. It is well-known that, by Zorn's lemma, every subset $\varnothing \neq X \subseteq A$ with the finite meet property can be extended to a proper filter, and every proper filter can be extended to a prime filter.

Fact 2. *Let* $\mathbb{A} = (A, \cdot, +, ', \Box, 0, 1)$ *be a tqBa. For all* $u \in U(A)$ *and* $a, b \in u$,

(1) $ab \in u$ *if and only* $a, b \in u$; *and*
(2) $a + b \in u$ *if and only if* $a \in u$ *or* $b \in u$.

Moreover, if $\mathbb{A}$ *is a tqBa5, then* $a \leq \Box\Diamond a$ *and* $\Diamond\Box a \leq \Box a$.

Definition 6. The *dual space* of a tqBa5 $\mathbb{A} = (A, \cdot, +, ', \Box, 0, 1)$ is defined as the structure $\mathbb{A}_* = (U(A), R^A)$ where R^A is defined as follows:

$$uR^A v \text{ if and only if } \{a : \Box a \in u\} \subseteq v.$$

Note that $uR^A v$ if and only if $\{\Diamond b : b \in v\} \subseteq u$.

Lemma 1. *For every tqBa5 $\mathbb{A}$, the dual space $\mathbb{A}_*$ is an approximation space.*

Proof. Let $\mathbb{A} = (A, \cdot, +, {}', \Box, 0, 1)$ be a tqBa5 and $\mathbb{A}_* = (U(A), R^A)$. It suffices to show that R_A is an equivalence relation. By ($\mathrm{T}_\Box$), R^A is reflexive. By ($4_\Box$), R^A is transitive. Suppose uR^Av and $a \in v$. Then $\Diamond a \in u$. By ($5_\Box$), $\Diamond a \leq \Box\Diamond a$ and so $\Box\Diamond a \in u$. Then $\Diamond a \in u$. Hence vR^Au. □

Lemma 2. *Let $\mathbb{A} = (A, \cdot, +, {}', \Box, 0, 1)$ be a tqBa5 and $u \in U(A)$. If $\Box a \notin u$, then there exists $v \in R_A(u)$ with $a \notin v$.*

Proof. Assume $\Box a \notin u$. Consider the set $X = \{b : \Box b \in u\}$ which is clearly closed under meet. Then $a \notin X$. Now we show that X has the finite meet property. Suppose not. Then $\Box 0 \in u$. Clearly $\Box 0 \leq 0$ and so $0 \in u$ which contradicts $u \in U(A)$. Then there exists a prime filter $v \in U(A)$ with uR_Av and $a \notin v$.

Let $\mathbb{A}$ and $\mathbb{B}$ be tqBas. A function $f : A \to B$ is an *embedding* from $\mathbb{A}$ to $\mathbb{B}$ if f is an injective homomorphism, i.e., for all $a, b \in A$, $f(ab) = f(a)f(b)$; $f(a+b) = f(a)+f(b)$; $f(a') = f(a)'$; $f(0) = 0$ and $f(1) = 1$; and $f(\Box a) = \Box f(a)$. We say that $\mathbb{A}$ is *embedded* into $\mathbb{B}$ if there is an embedding from $\mathbb{A}$ to $\mathbb{B}$.

Theorem 1. *Every tqBa5 $\mathbb{A}$ is embedded into $(\mathbb{A}_*)^*$.*

Proof. Let $\mathbb{A} = (A, \cdot, +, {}', \Box, 0, 1)$ be a tqBa5 and $\mathfrak{P}(U(A)) = \mathcal{P}(U(A)) \times \mathcal{P}(U(A))$. Clearly $(\mathbb{A}_*)^* = (\mathfrak{P}(U(A)), \sqcap, \sqcup, \underline{(.)}, \langle\varnothing, U(A)\rangle, \langle U(A), \varnothing\rangle)$. We define the function $f : A \to \mathfrak{P}(U(A))$ by $f(a) = \langle\pi^+(a), \pi^-(a)\rangle$ where

$$\pi^+(a) = \{u \in U(A) : a \in u\};\ \pi^-(a) = \{v \in U(A) : a' \in v\}.$$

Now we show that f is injective. Suppose $a \neq b$. Without loss of generality, suppose $a \not\leq b$. By Zorn's lemma, there exists a prime filter $u \in U(A)$ such that $a \in u$ and $b \notin u$. This implies that $u \in \pi^+(a)$ and $v \notin \pi^+(b)$. Hence $f(a) \neq f(b)$.

Next we show the function f preserves operations:

(1) We have $\pi^+(ab) = \{u \in U(A) : ab \in u\} = \{u \in U(A) : a \in u\} \cap \{u \in U(A) : b \in u\} = \pi^+(a) \cap \pi^+(b)$. Moreover, $\pi^-(ab) = \{u \in U(A) : (ab)' \in u\} = \{u \in U(A) : a' + b' \in u\} = \{u \in U(A) : a' \in u\} \cup \{u \in U(A) : b' \in u\} = \pi^-(a) \cup \pi^-(b)$. Hence $f(ab) = f(a) \sqcap f(b)$.

(2) We have $\pi^+(a+b) = \{u \in U(A) : a+b \in u\} = \{u \in U(A) : a \in u\} \cup \{u \in U(A) : b \in u\} = \pi^+(a) \cup \pi^+(b)$. Moreover, $\pi^-(a+b) = \{u \in U(A) : (a+b)' \in u\} = \{u \in U(A) : a'b' \in u\} = \{u \in U(A) : a' \in u\} \cap \{u \in U(A) : b' \in u\} = \pi^-(a) \cap \pi^-(b)$. Hence $f(a+b) = f(a) \sqcup f(b)$.

(3) We have $\pi^+(a') = \{u \in U(A) : a' \in u\} = \pi^-(a)$. Moreover, $\pi^-(a') = \{u \in U(A) : a'' \in u\} = \{u \in U(A) : a \in u\} = \pi^+(a)$. Hence $f(a') = {\sim}f(a)$.

(4) We have $\pi^+(\Box a) = \{u \in U(A) : \Box a \in u\}$. Now we show $\pi^+(\Box a) = \underline{\pi^+(a)} = \{u \in U(A) : R_A(u) \subseteq \pi^+(a)\}$. Suppose $u \in \pi^+(\Box a)$. Then $\Box a \in u$. If $v \in R_A(u)$, then $a \in v$ and so $v \in \pi^+(a)$. Hence $R_A(u) \subseteq \pi^+(a)$. Suppose $R(u) \subseteq \pi^+(a)$. For a contradiction, assume $\Box a \notin u$. By Lemma 2, there exists $v \in R_A(u)$ with $a \notin v$. Then $v \in \pi^+(a)$, i.e., $a \in v$ which contradicts $a \notin v$. Hence $\pi^+(\Box a) = \underline{\pi^+(a)}$. Similarly we have $\pi^+(\Box a) = \underline{\pi^-(a)}$. Hence $f(\Box a) = \underline{f(a)}$. □

Definition 7. Let $\mathbb{A} = (A, \cdot, +, ', \Box, 0, 1)$ be a tqBa. The binary relation $Q_A \subseteq U(A) \times U(A)$ is defined as follows:

$$uQ_A v \text{ if and only if } \{a : \Box a \in u\} \subseteq v.$$

Note that $uQ_A v$ if and only if $\{\Diamond b : b \in v\} \subseteq u$. The function $N_A : U(A) \to \mathcal{P}(U(A))$ is defined by $N_A(u) = \bigcup\{C \in \mathscr{C}_A : u \in C\}$. Note that $N_A(u) = Q_A(u)$. The *dual frame* of a tqBa $\mathbb{A}$ is defined as the structure $\mathbb{A}_\sharp = (U(A), \mathscr{C}_A)$ where $\mathscr{C}_A = \{Q_A(u) : u \in U(A)\}$.

Lemma 3. *For every tqBa $\mathbb{A}$, the dual frame $\mathbb{A}_\sharp$ is a covering frame.*

Proof. Let $\mathbb{A} = (A, \cdot, +, ', \Box, 0, 1)$ be a tqBa. It suffices to show $\bigcup \mathscr{C}_A = U(A)$. Clearly $\bigcup \mathscr{C}_A \subseteq U(A)$. Let $u \in U(A)$. By $(\mathrm{T}_\Box)$, $u \in Q(u)$. Hence $u \in \bigcup \mathscr{C}_A$. □

Lemma 4. *Let $\mathbb{A} = (A, \cdot, +, ', \Box, 0, 1)$ be a tqBa and $u \in U(A)$. If $\Box a \notin u$, then there exists $v \in Q_A(u)$ with $a \notin v$.*

Proof. The proof is similar to Lemma 2 and details are omitted.

Theorem 2. *Every tqBa $\mathbb{A}$ is embeddable into $(\mathbb{A}_\sharp)^\sharp$.*

Proof. Let $\mathbb{A} = (A, \cdot, +, ', \Box, 0, 1)$ be a tqBa and $\mathfrak{P}(U(A)) = \mathcal{P}(U(A)) \times \mathcal{P}(U(A))$. Clearly $(\mathbb{A}_\sharp)^\sharp = (\mathfrak{P}(U(A)), \sqcap, \sqcup, \Box_{N_A}, \langle\varnothing, U(A)\rangle, \langle U(A), \varnothing\rangle)$. We define the function $f : A \to \mathfrak{P}(U(A))$ by $f(a) = \langle \pi^+(a), \pi^-(a)\rangle$ where $\pi^+(a) = \{u \in U(A) : a \in u\}$ and $\pi^-(a) = \{v \in U(A) : a' \in v\}$. Like the proof of Theorem 1, the function f is an embedding. □

4 Logics for Paraconsistent Rough Sets

In this section, we introduce two sequent calculi for Pawlakian paraconsistent rough sets and covering based rough sets respectively.

Definition 8. Let $\mathbb{V} = \{p_i : i \in \omega\}$ be a denumerable set of variables. The set of all formulas Fm is defined inductively as follows:

$$Fm \ni \phi ::= p \mid \bot \mid \neg\phi \mid (\phi_1 \wedge \phi_2) \mid (\phi_1 \vee \phi_2) \mid \Box\phi$$

where $p \in \mathbb{V}$. We define $\top := \neg\bot$ and $\Diamond\phi := \neg\Box\neg\phi$. A *sequent* is an expression $\Gamma \Rightarrow \phi$ where Γ is a finite multiset of formulas and ϕ is a formula. For every finite multiset of formulas Γ, let $\Box\Gamma = \{\Box\phi : \phi \in \Gamma\}$. If $\Gamma = \varnothing$, then $\Box\varnothing = \varnothing$.

Definition 9. The sequent calculus **G4** consists of the following axiom schemata and inference rules:

(1) Axiom schemata:

$$(\mathrm{Id})\ \phi, \Gamma \Rightarrow \phi \quad (\bot)\ \bot, \Gamma \Rightarrow \phi \quad (\top)\ \Gamma \Rightarrow \top$$

$$(T_\Box)\ \Box\phi \Rightarrow \phi \quad (4_\Box)\ \Box\phi \Rightarrow \Box\Box\phi$$

(2) Logical rules:

$$\frac{\phi, \psi, \Gamma \Rightarrow \chi}{\phi \wedge \psi, \Gamma \Rightarrow \chi}\,(\wedge\mathrm{L}) \qquad \frac{\Gamma \Rightarrow \psi \quad \Gamma \Rightarrow \chi}{\Gamma \Rightarrow \psi \wedge \chi}\,(\wedge\mathrm{R})$$

$$\frac{\phi, \Gamma \Rightarrow \chi \quad \psi, \Gamma \Rightarrow \chi}{\phi \vee \psi, \Gamma \Rightarrow \chi}\,(\wedge\mathrm{L}) \qquad \frac{\Gamma \Rightarrow \psi_i}{\Gamma \Rightarrow \psi_1 \wedge \psi_2}\,(\vee\mathrm{R})(i = 1, 2)$$

$$\frac{\neg\phi, \Gamma \Rightarrow \chi \quad \neg\psi, \Gamma \Rightarrow \chi}{\neg(\phi \wedge \psi), \Gamma \Rightarrow \chi}\,(\neg\wedge\mathrm{L}) \qquad \frac{\Gamma \Rightarrow \neg\psi_i}{\Gamma \Rightarrow \neg(\psi_1 \wedge \psi_2)}\,(\neg\wedge\mathrm{R})(i = 1, 2)$$

$$\frac{\neg\phi, \neg\psi, \Gamma \Rightarrow \chi}{\neg(\phi \vee \psi), \Gamma \Rightarrow \chi}\,(\neg\vee\mathrm{L}) \qquad \frac{\Gamma \Rightarrow \neg\psi_1 \quad \Gamma \Rightarrow \neg\psi_2}{\Gamma \Rightarrow \neg(\psi_1 \wedge \psi_2)}\,(\neg\vee\mathrm{R})$$

$$\frac{\phi, \Gamma \Rightarrow \chi}{\neg\neg\phi, \Gamma \Rightarrow \chi}\,(\neg\neg\mathrm{L}) \qquad \frac{\Gamma \Rightarrow \psi}{\Gamma \Rightarrow \neg{\sim}\psi}\,(\neg\neg\mathrm{R})$$

(3) Contraposition:

$$\frac{\varphi \Rightarrow \psi}{\neg\psi \Rightarrow \neg\varphi}\,(\mathrm{CP})$$

(4) Modal Rules:

$$\frac{\Gamma \Rightarrow \psi}{\Box\Gamma \Rightarrow \Box\psi}\,(\mathrm{K})$$

(5) Cut rule:

$$\frac{\Gamma \Rightarrow \psi \quad \psi, \Delta \Rightarrow \chi}{\Gamma, \Delta \Rightarrow \chi}\,(\mathrm{Cut})$$

The sequent calculus **G5** is obtained from **G4** by adding the following axiom:

$$(5)\ \Diamond\phi \Rightarrow \Box\Diamond\phi.$$

A *derivation* in a sequent calculus is a finite tree of sequents in which each node is either an axiom or derived from child node(s) by a rule. The height of a derivation is defined as the maximal length of branches in it. For $\mathbf{G} \in \{\mathbf{G4}, \mathbf{G5}\}$, let $\mathbf{G} \vdash \Gamma \Rightarrow \psi$ denote that the sequent $\Gamma \Rightarrow \psi$ is derivable in $\mathbf{G}$. A formula ϕ is $\mathbf{G}$-equivalent to ψ (notation: $\mathbf{G} \vdash \phi \Leftrightarrow \psi$) if $\mathbf{G} \vdash \phi \Rightarrow \psi$ and $\mathbf{G} \vdash \psi \Rightarrow \phi$.

Proposition 5. *The following hold:*

(1) $\mathbf{G4} \vdash \neg\top \Leftrightarrow \bot$ *and* $\mathbf{G4} \vdash \phi \Leftrightarrow \neg\neg\phi$.
(2) $\mathbf{G4} \vdash \neg(\phi \wedge \psi) \Leftrightarrow \neg\phi \vee \neg\psi$ *and* $\mathbf{G4} \vdash \neg(\phi \vee \psi) \Leftrightarrow \neg\phi \wedge \neg\psi$.

(3) $\mathbf{G4} \vdash \neg\Box\phi \Leftrightarrow \Diamond\neg\psi$ *and* $\mathbf{G4} \vdash \neg\Diamond\phi \Leftrightarrow \Box\neg\phi$.
(4) *if* $\mathbf{G4} \vdash \phi \Rightarrow \psi$, *then* $\mathbf{G4} \vdash \Box\phi \Rightarrow \Box\psi$ *and* $\mathbf{G4} \vdash \Diamond\phi \Rightarrow \Diamond\psi$.
(5) $\mathbf{G4} \vdash \Box(\phi \wedge \psi) \Leftrightarrow \Box\phi \wedge \Box\psi$ *and* $\mathbf{G4} \vdash \Diamond(\phi \vee \psi) \Leftrightarrow \Diamond\phi \vee \Diamond\psi$.
(6) $\mathbf{G4} \vdash \Box\phi \wedge \Diamond\psi \Rightarrow \Diamond(\phi \wedge \psi)$.
(7) $\mathbf{G4} \vdash \phi \Rightarrow \Diamond\phi$.
(8) $\mathbf{G4} \vdash \Box\phi \Leftrightarrow \Box\Box\phi$ *and* $\mathbf{G4} \vdash \Diamond\phi \Leftrightarrow \Diamond\Diamond\phi$.
(9) $\mathbf{G5} \vdash \phi \Rightarrow \Box\Diamond\phi$ *and* $\mathbf{G5} \vdash \Diamond\Box\phi \Leftrightarrow \Box\phi$.

Proof. We show $\mathbf{G4} \vdash \neg(\phi \wedge \psi) \Leftrightarrow \neg\phi \vee \neg\psi$. We have the following derivations:

$$\dfrac{\dfrac{\dfrac{\phi, \psi \Rightarrow \phi}{\phi \wedge \psi \Rightarrow \phi}\,(\wedge\mathrm{L})}{\neg\phi \Rightarrow \neg(\phi \wedge \psi)}\,(\mathrm{CP}) \qquad \dfrac{\dfrac{\phi, \psi \Rightarrow \psi}{\phi \wedge \psi \Rightarrow \psi}\,(\wedge\mathrm{L})}{\neg\psi \Rightarrow \neg(\phi \wedge \psi)}\,(\mathrm{CP})}{\neg\phi \vee \neg\psi \Rightarrow \neg(\phi \wedge \psi)}\,(\vee\mathrm{L})$$

$$\dfrac{\dfrac{\neg\phi \Rightarrow \neg\phi}{\neg\phi \Rightarrow \neg\phi \vee \neg\psi}\,(\vee\mathrm{R}) \qquad \dfrac{\neg\psi \Rightarrow \neg\psi}{\neg\psi \Rightarrow \neg\phi \vee \neg\psi}\,(\vee\mathrm{R})}{\neg(\phi \wedge \psi) \Rightarrow \neg\phi \vee \neg\psi}\,(\neg\wedge\mathrm{R})$$

We show $\mathbf{G4} \vdash \Box(\phi \wedge \psi) \Leftrightarrow \Box\phi \wedge \Box\psi$. We have the following derivations:

$$\dfrac{\dfrac{\dfrac{\phi, \psi \Rightarrow \phi}{\phi \wedge \psi \Rightarrow \phi}\,(\wedge\mathrm{L})}{\Box(\phi \wedge \psi) \Rightarrow \Box\phi}\,(\mathrm{K}) \qquad \dfrac{\dfrac{\phi, \psi \Rightarrow \phi}{\phi \wedge \psi \Rightarrow \phi}\,(\wedge\mathrm{L})}{\Box(\phi \wedge \psi) \Rightarrow \Box\phi}\,(\mathrm{K})}{\Box(\phi \wedge \psi) \Rightarrow \Box\phi \wedge \Box\psi}\,(\wedge\mathrm{R}) \qquad \dfrac{\dfrac{\dfrac{\phi, \psi \Rightarrow \phi \quad \phi, \psi \Rightarrow \psi}{\phi, \psi \Rightarrow \phi \wedge \psi}\,(\wedge\mathrm{R})}{\Box\phi, \Box\psi \Rightarrow \Box(\phi \wedge \psi)}\,(\mathrm{K})}{\Box\phi \wedge \Box\psi \Rightarrow \Box(\phi \wedge \psi)}\,(\wedge\mathrm{L})$$

The remaining items are shown regularly. □

An assignment in a tqBa $\mathbb{A}$ is a function $\theta : \mathbb{V} \to A$. A sequent $\Gamma \Rightarrow \phi$ is *valid* in a tqBa $\mathbb{A}$ (notation: $\mathbb{A} \models \Gamma \Rightarrow \phi$), if $\bigwedge_{\psi \in \Gamma} \theta(\psi) \leq \theta(\phi)$. The notation $\mathbf{tqBa} \models \Gamma \Rightarrow \phi$ denote that $\Gamma \Rightarrow \phi$ is valid in all tqBas, and the notation $\mathbf{tqBa5} \models \Gamma \Rightarrow \phi$ denote that $\Gamma \Rightarrow \phi$ is valid in all tqBa5s.

Theorem 3. *For every sequent* $\Gamma \Rightarrow \phi$, (1) $\mathbf{G4} \vdash \Gamma \Rightarrow \phi$ *if and only if* $\mathbf{tqBa} \models \Gamma \Rightarrow \phi$*; and* (2) $\mathbf{G5} \vdash \Gamma \Rightarrow \phi$ *if and only if* $\mathbf{tqBa} \models \Gamma \Rightarrow \phi$.

Proof. This is shown by the standard Lindenbaum-Tarski method. The soundness part is shown directly by induction on the height of a derivation of $\Gamma \Rightarrow \phi$ in a sequent calculus. For $\mathbf{G} \in \{\mathbf{G4}, \mathbf{G5}\}$, the binary relation $\equiv_{\mathbf{G}}$ on the set of all formulas Fm is defined by setting

$$\phi \equiv_{\mathbf{G}} \psi \text{ if and only if } \mathbf{G} \vdash \phi \Rightarrow \psi \text{ and } \mathbf{G} \vdash \psi \Rightarrow \phi.$$

One can easily show that $\equiv_{\mathbf{G}}$ is a congruence relation on Fm. Let $Fm/_{\mathbf{G}} = \{[\phi] : \phi \in Fm\}$ where $[\phi] = \{\psi \in Fm : \phi \equiv_{\mathbf{G}} \psi\}$ is the equivalence class of ϕ. The Lindenbaum-Tarski algebra for $\mathbf{G}$ is defined as $\mathfrak{L}_{\mathbf{G}} = (Fm/_{\mathbf{G}}, \cdot, +, {}', [\bot], [\top])$ where we have

$$[\phi][\psi] = [\phi \wedge \psi],\ [\phi] + [\psi] = [\phi \vee \psi] \text{ and } [\phi]' = [\sim\phi].$$

Obviously $\mathfrak{L}_{\mathbf{G4}}$ is a tqBa and $\mathfrak{L}_{\mathbf{G5}}$ is a tqBa5. Suppose $\mathbf{G} \nvdash \Gamma \Rightarrow \phi$. Let θ be the assignment in $\mathfrak{L}_{\mathbf{G}}$ such that $\theta(p) = [p]$ for each $p \in \mathbb{V}$. By induction on the complexity of a formula χ, we have $\theta(\chi) = [\chi]$. Hence $\mathfrak{L}_{\mathbf{G}} \not\models \Gamma \Rightarrow \phi$. □

Let $\mathcal{K} = (U, R)$ be an approximation space and $\mathscr{F} = (U, \mathscr{C})$ be a covering frame. A sequent $\Gamma \Rightarrow \phi$ is *valid* in $\mathcal{K}$ (notation: $\mathcal{K} \models \Gamma \Rightarrow \phi$), if $\mathcal{K}^* \models \Gamma \Rightarrow \phi$. Let $\mathbf{AS} \models \Gamma \Rightarrow \phi$ denote that $\Gamma \Rightarrow \phi$ is valid in all approximation spaces. A sequent $\Gamma \Rightarrow \phi$ is *valid* in $\mathscr{F}$ (notation: $\mathscr{F} \models \Gamma \Rightarrow \phi$), if $\mathscr{F}^\sharp \models \Gamma \Rightarrow \phi$. Let $\mathbf{CF} \models \Gamma \Rightarrow \phi$ denote that $\Gamma \Rightarrow \phi$ is valid in all covering frames.

Corollary 1. *For every sequent* $\Gamma \Rightarrow \phi$, (1) $\mathbf{G4} \vdash \Gamma \Rightarrow \phi$ *if and only if* $\mathbf{AS} \models \Gamma \Rightarrow \phi$*; and* (2) $\mathbf{G5} \vdash \Gamma \Rightarrow \phi$ *if and only if* $\mathbf{CF} \models \Gamma \Rightarrow \phi$.

Proof. It follows immediately from Theorem 3, Theorem 1 and Theorem 2. □

5 Concluding Remarks

The present work contributes new paraconsistent Pawlakian rough sets and paraconsistent covering based rough sets by introducing approximations of polarities in a universe of objects. Moreover, topological quasi-Boolean algebras are shown to be algebras for paraconsistent covering based rough sets, and partition topological quasi-Boolean algebras are shown to be algebras for paraconsistent Palwakian rough sets. Finally, we present sequent calculi as modal systems for these paraconsistent rough sets. There are some problems which need to be explored further. One problem is about the applications of these paraconsistent rough sets in practical scenarios. The other problem is extendeding the approach taken in the present paper to other types of rough sets. For example, as in [9,10], we can consider the connections between Kripke structures, covering frames and algebras for various paraconsistent rough sets.

Ackonwledgements. The author thanks the anonymous reviewers for their helpful comments and suggestions.

References

1. Banerjee, M.: Rough sets and 3-valued lukasiewicz logic. Fundamenta Informaticae **31**(3–4), 213–220 (1997). https://doi.org/10.3233/FI-1997-313401
2. Banerjee, M., Chakraborty, M.K.: Rough sets through algebraic logic. Fundamenta Informaticae **28**(3–4), 211–221 (1996). https://doi.org/10.3233/FI-1996-283401
3. Belnap, N.D.: A useful four-valued logic. In: Dunn, J.M., Epstein, G. (eds.) Modern uses of Multiple-Valued Logic, pp. 5–37. Springer, Dordrecht (1977). https://doi.org/10.1007/978-94-010-1161-7_2
4. Belnap, N.D.: How a computer should think. In: New Essays on Belnap-Dunn Logic. SL, vol. 418, pp. 35–53. Springer, Cham (2019). https://doi.org/10.1007/978-3-030-31136-0_4

5. Blackburn, P., De Rijke, M., Venema, Y.: Modal Logic. Cambridge University Press, Cambridge (2002)
6. Bonikowski, Z., Bryniarski, E., Wybraniec-Skardowska, U.: Extensions and intentions in the rough set theory. Inf. Sci. **107**(1–4), 149–167 (1998). https://doi.org/10.1016/S0020-0255(97)10046-9
7. Celani, S.A.: Classical modal de morgan algebras. Stud. Logica. **98**, 251–266 (2011). https://doi.org/10.1007/s11225-011-9328-0
8. Davey, B.A., Priestley, H.A.: Introduction to Lattices and Order. Cambridge University Press, Cambridge (2002)
9. Ma, M., Chakraborty, M.K.: Covering-based rough sets and modal logics. Part I. Int. J. Approximate Reasoning **77**, 55–65 (2016). https://doi.org/10.1016/j.ijar.2016.06.002
10. Ma, M., Chakraborty, M.K.: Covering-based rough sets and modal logics. Part II. Int. J. Approximate Reasoning **95**, 113–123 (2018). https://doi.org/10.1016/j.ijar.2018.02.002
11. Ma, M., Chakraborty, M.K., Lin, Z.: Sequent calculi for varieties of topological quasi-boolean algebras. In: Nguyen, H.S., Ha, Q.-T., Li, T., Przybyła-Kasperek, M. (eds.) IJCRS 2018. LNCS (LNAI), vol. 11103, pp. 309–322. Springer, Cham (2018). https://doi.org/10.1007/978-3-319-99368-3_24
12. Małuszyński, J., Szałas, A., Vitória, A.: A four-valued logic for rough set-like approximate reasoning. In: Peters, J.F., Skowron, A., Düntsch, I., Grzymała-Busse, J., Orłowska, E., Polkowski, L. (eds.) Transactions on Rough Sets VI. LNCS, vol. 4374, pp. 176–190. Springer, Heidelberg (2007). https://doi.org/10.1007/978-3-540-71200-8_11
13. Małuszyński, J., Szałas, A., Vitória, A.: Paraconsistent logic programs with four-valued rough sets. In: Chan, C.-C., Grzymala-Busse, J.W., Ziarko, W.P. (eds.) RSCTC 2008. LNCS (LNAI), vol. 5306, pp. 41–51. Springer, Heidelberg (2008). https://doi.org/10.1007/978-3-540-88425-5_5
14. Pawlak, Z.: Rough sets. Int. J. Comput. Inf. Sci. **11**(5), 341–356 (1982). https://doi.org/10.1007/BF01001956
15. Pawlak, Z.: Rough Sets: Theoretical Aspects of Reasoning About Data, vol. 9. Springer Science & Business Media, Berlin (1991)
16. Samanta, P., Chakraborty, M.K.: Interface of rough set systems and modal logics: a survey. In: Peters, J.F., Skowron, A., Ślęzak, D., Nguyen, H.S., Bazan, J.G. (eds.) Transactions on Rough Sets XIX. LNCS, vol. 8988, pp. 114–137. Springer, Heidelberg (2015). https://doi.org/10.1007/978-3-662-47815-8_8
17. Sardar, M.R., Chakraborty, M.K.: Some implicative topological quasi-Boolean algebras and rough set models. Int. J. Approximate Reasoning **148**, 1–22 (2022). https://doi.org/10.1016/j.ijar.2022.05.008
18. Slowinski, R., Vanderpooten, D.: A generalized definition of rough approximations based on similarity. IEEE Trans. Knowl. Data Eng. **12**(2), 331–336 (2000). https://doi.org/10.1109/69.842271
19. Vitória, A., Małuszyński, J., Szałas, A.: Modeling and reasoning with paraconsistent rough sets. Fundamenta Informaticae **97**(4), 405–438 (2009). https://doi.org/10.3233/FI-2009-209
20. Vitória, A., Szałas, A., Małuszyński, J.: Four-valued extension of rough sets. In: Wang, G., Li, T., Grzymala-Busse, J.W., Miao, D., Skowron, A., Yao, Y. (eds.) RSKT 2008. LNCS (LNAI), vol. 5009, pp. 106–114. Springer, Heidelberg (2008). https://doi.org/10.1007/978-3-540-79721-0_19

21. Yao, Y.Y.: On generalizing pawlak approximation operators. In: Polkowski, L., Skowron, A. (eds.) RSCTC 1998. LNCS (LNAI), vol. 1424, pp. 298–307. Springer, Heidelberg (1998). https://doi.org/10.1007/3-540-69115-4_41
22. Zhu, W.: Generalized rough sets based on relations. Inf. Sci. **177**(22), 4997–5011 (2007). https://doi.org/10.1016/j.ins.2007.05.037
23. Zhu, W., Wang, F.Y.: Reduction and axiomization of covering generalized rough sets. Inf. Sci. **152**, 217–230 (2003). https://doi.org/10.1016/S0020-0255(03)00056-2

Applications

A Generalization of Bounded Commutative Rℓ-Monoids

Guangling Cao, Jinfan Xu, and Wenjuan Chen(✉)

University of Jinan, Jinan 250022, Shandong, People's Republic of China
{glcao,jinfanxu}@stu.ujn.edu.cn, wjchenmath@gmail.com

Abstract. In this paper, bounded commutative residuated quasi-ordered quasi-monoids (bounded commutative Rqℓ-monoids, for short) as a generalization of bounded commutative residuated lattice ordered monoids (bounded commutative Rℓ-monoids, for short) are introduced. First, the properties of bounded commutative Rqℓ-monoids are investigated and the relations among quasi-MV algebras, quasi-BL algebras and bounded commutative Rqℓ-monoids are discussed. Second, the filters and weak filters of bounded commutative Rqℓ-monoids are defined. The one-to-one correspondence between the set of filters and the set of filter congruences on a bounded commutative Rqℓ-monoid is given. Moreover, the relation between the set of weak filters and the set of congruences on a bounded commutative Rqℓ-monoid is showed. As an application of the study, the properties of the quotient algebra with respect to a weak filter are investigated. Finally, the properties of some special sets in a bounded commutative Rqℓ-monoid are showed.

Keywords: Bounded commutative Rqℓ-monoids · Quasi-BL algebras · Quasi-MV algebras · Weak filters

1 Introduction

Bounded commutative residuated lattice ordered monoids (bounded commutative Rℓ-monoids, for short) are the dual notion of bounded commutative dually residuated lattice ordered monoids (bounded commutative DRℓ-monoids, for short) which were given by Swamy in [17] as a generalization of abelian lattice ordered groups and Brouwerian algebras. Moreover, it is well known that both MV-algebras [2] which are the algebraic semantics of Lukasiewicz infinite valued logic and BL-algebras [7] which are the algebraic semantics of Hajek's basic fuzzy logic can be regarded as the special cases of bounded commutative Rℓ-monoids. Namely, a bounded commutative Rℓ-monoid **G** is an MV-algebra iff **G** satisfies the identity $\iota^{--} = \iota$, where $\iota^- = \iota \dashrightarrow 0$ (see [13,14]), a bounded commutative Rℓ-monoid **G** is a BL-algebra iff **G** satisfies the identity $(\iota \dashrightarrow \varrho) \sqcup (\varrho \dashrightarrow \iota) = 1$

Supported by Shandong Provincial Natural Science Foundation, China (No. ZR2020MA041).

Y. Chen and S. Zhang (Eds.): AILA 2022, CCIS 1657, pp. 183–198, 2022.
https://doi.org/10.1007/978-981-19-7510-3_14

(see [15]). Since bounded commutative Rℓ-monoids form larger class of algebras including some well-known logical algebras, lots of authors generalized and extended the previous investigation to bounded commutative Rℓ-monoids for studying the common properties and providing a more general algebraic foundations [5,11,12,16].

In the recent years, quantum computational logics arising from quantum computation have been received more and more attentions. In order to study the algebraic characterization of quantum computational logics, Ledda et al. introduced and investigated quasi-MV algebras in [6]. In 2020, Chen and Wang defined quasi-BL algebras [3] as a generalization of quasi-MV algebras similarly as BL-algebras generalized MV-algebras. Many more properties of quasi-MV algebras and quasi-BL algebras can be seen in [1,8–10]. The works on quasi-MV algebras and quasi-BL algebras were indicated that the study of quasi-$*$ algebras may play an important role in quantum computational logics. Moreover, introducing more quasi-$*$ algebras is benefit to improve and develop the study. Now, considering the relations among bounded commutative Rℓ-monoids, MV-algebras and BL-algebras, we want to introduce new structures which are generalization of bounded commutative Rℓ-monoids. The structure of the paper is as follows. In Sect. 2, some definitions and conclusions which will be used in what follows are recalled. In Sect. 3, bounded commutative Rqℓ-monoids as a generalization of bounded commutative Rℓ-monoids are defined and the basic properties of bounded commutative Rqℓ-monoids are investigated. Furthermore, the relation between bounded commutative Rqℓ-monoids and quasi-MV algebras is discussed. In Sect. 4, filters and weak filters of bounded commutative Rqℓ-monoids are defined and the one-to-one correspondence between the set of filters and the set of filter congruences on a bounded commutative Rqℓ-monoid is showed. Moreover, the relation between the set of weak filters and the set of congruences on a bounded commutative Rqℓ-monoid is discussed. As an application of the study, the properties of the quotient algebra with respect to a weak filter are investigated. Finally, the properties of some special sets in a bounded commutative Rqℓ-monoid are showed.

2 Preliminary

In this section, some definitions and conclusions of bounded commutative Rℓ-monoids, quasi-MV algebras and quasi-BL algebras which will be used in what follows are recalled.

Definition 1. *[12] Let* $\boldsymbol{G} = (G; \otimes, \sqcup, \sqcap, \dashrightarrow, 0, 1)$ *be an algebra of type (2,2,2,2,0,0). If it satisfies the following conditions for each* $\iota, \varrho, \zeta \in G$,

(RLM1) $(G; \otimes, 1)$ *is a commutative monoid,*
(RLM2) $(G; \sqcup, \sqcap, 0, 1)$ *is a bounded lattice,*
(RLM3) $\iota \otimes \varrho \leq \zeta \Leftrightarrow \iota \leq \varrho \dashrightarrow \zeta$,
(RLM4) $\iota \otimes (\iota \dashrightarrow \varrho) = \iota \sqcap \varrho$,

then $\mathbf{G} = (G; \otimes, \sqcup, \sqcap, \dashrightarrow, 0, 1)$ *is called a bounded commutative residuated lattice ordered monoid (bounded commutative Rℓ-monoid, for short).*

In [12], authors showed that a bounded commutative Rℓ-monoid $\mathbf{G}$ is an MV-algebra iff for each $\iota \in G$, $\iota \boxplus 0 = \iota$ where $\iota \boxplus 0 = (\iota' \otimes 0')'$. In [15], Rachunek showed that a bounded commutative Rℓ-monoid $\mathbf{G}$ is a BL-algebra iff $(\iota \dashrightarrow \varrho) \sqcup (\varrho \dashrightarrow \iota) = 1$ for each $\iota, \varrho \in G$.

In 2006, Ledda et al. introduced quasi-MV algebras as a generalization of MV-algebras.

Definition 2. *[6] Let* $(V; \boxplus, ', 0)$ *be an algebra of type (2,1,0). If it satisfies the following identities for each* $\iota, \varrho, \zeta \in V$,

(QM1) $\iota \boxplus (\varrho \boxplus \zeta) = (\iota \boxplus \varrho) \boxplus \zeta$,
(QM2) $0' = 1$,
(QM3) $\iota \boxplus 1 = 1$,
(QM4) $\iota'' = \iota$,
(QM5) $(\iota' \boxplus \varrho)' \boxplus \varrho = (\iota \boxplus \varrho')' \boxplus \iota$,
(QM6) $(\iota \boxplus 0)' = \iota' \boxplus 0$,
(QM7) $\iota \boxplus \varrho \boxplus 0 = \iota \boxplus \varrho$,
then $(V; \boxplus, ', 0)$ *is called a quasi-MV algebra.*

On each quasi-MV algebra $(V; \boxplus, ', 0)$, some operations were defined as follows: $\iota \sqcup \varrho = \iota \boxplus (\varrho' \boxplus \iota)'$, $\iota \sqcap \varrho = (\iota' \sqcup \varrho')'$, $\iota \otimes \varrho = (\iota' \boxplus \varrho')'$ and $\iota \dashrightarrow \varrho = \iota' \boxplus \varrho$. Also, a relation $\iota \leq \varrho$ was defined by $\iota \sqcup \varrho = \varrho \boxplus 0$, or equivalently, $\iota \sqcap \varrho = \iota \boxplus 0$.

Definition 3. *[4] Let* $(V; \sqcup, \sqcap)$ *be an algebra of type (2,2). If it satisfies the following identities for each* $\iota, \varrho, \zeta \in V$,

(QL1) $\iota \sqcup \varrho = \varrho \sqcup \iota$ *and* $\iota \sqcap \varrho = \varrho \sqcap \iota$,
(QL2) $\iota \sqcup (\varrho \sqcup \zeta) = (\iota \sqcup \varrho) \sqcup \zeta$ *and* $\iota \sqcap (\varrho \sqcap \zeta) = (\iota \sqcap \varrho) \sqcap \zeta$,
(QL3) $\iota \sqcup (\varrho \sqcap \iota) = \iota \sqcup \iota$ *and* $\iota \sqcap (\varrho \sqcup \iota) = \iota \sqcap \iota$,
(QL4) $\iota \sqcup \varrho = \iota \sqcup (\varrho \sqcup \varrho)$ *and* $\iota \sqcap \varrho = \iota \sqcap (\varrho \sqcap \varrho)$,
(QL5) $\iota \sqcup \iota = \iota \sqcap \iota$,
then $(V; \sqcup, \sqcap)$ *is called a quasi-lattice.*

On each quasi-lattice $(V; \sqcup, \sqcap)$, one can define a relation $\iota \leq \varrho$ by $\iota \sqcup \varrho = \varrho \sqcup \varrho$ and the relation $\leq$ is quasi-ordering.

Given a quasi-MV algebra $(V; \boxplus, ', 0)$, authors showed that $(V; \leq)$ is a quasi-ordered set and $(V; \sqcup, \sqcap, 0, 1)$ is a bounded quasi-lattice, i.e., $(V; \sqcup, \sqcap)$ is a quasi-lattice and it has the largest element 1 and the least element 0 (with respect to the quasi-ordering $\leq$) in [6].

Definition 4. *[3] Let* $(V; \otimes, 1)$ *be an algebra of type (2,0). If it satisfies the following identities for each* $\iota, \varrho, \zeta \in V$,

(QM1) $\iota \otimes (\varrho \otimes \zeta) = (\iota \otimes \varrho) \otimes \zeta$,
(QM2) $\iota \otimes 1 = 1 \otimes \iota$,
(QM3) $\iota \otimes \varrho \otimes 1 = \iota \otimes \varrho$,

(QM4) $1 \otimes 1 = 1$,
then $(V; \otimes, 1)$ *is called a quasi-monoid.*

Obviously, each quasi-monoid with $\iota \otimes 1 = \iota$ is a monoid. If $\iota \otimes \varrho = \varrho \otimes \iota$ for each $\iota, \varrho \in V$, then $(V; \otimes, 1)$ is a commutative quasi-monoid. Based on quasi-lattices and quasi-monoids, Chen and Wang gave the definition of quasi-BL algebras in [3].

Definition 5. *[3] Let* $(V; \otimes, \sqcup, \sqcap, \dashrightarrow, 0, 1)$ *be an algebra of type (2,2,2,2,0,0). If it satisfies the following conditions for each* $\iota, \varrho, \zeta \in V$,

(QB1) $(V; \otimes, 1)$ *is a commutative quasi-monoid,*
(QB2) $(V; \sqcup, \sqcap, 0, 1)$ *is a bounded quasi-lattice,*
(QB3) $\iota \sqcap \iota = \iota \otimes 1$ *and* $0 \sqcap 0 = 0$,
(QB4) $\iota \leq \varrho \dashrightarrow \zeta$ *iff* $\iota \otimes \varrho \leq \zeta$,
(QB5) $(\iota \dashrightarrow \varrho) \otimes 1 = \iota \dashrightarrow \varrho$,
(QB6) $\iota \sqcap \varrho = \iota \otimes (\iota \dashrightarrow \varrho)$,
(QB7) $(\iota \dashrightarrow \varrho) \sqcup (\varrho \dashrightarrow \iota) = 1$,
then $(V; \otimes, \sqcup, \sqcap, \dashrightarrow, 0, 1)$ *is called a quasi-BL algebra.*

On a quasi-BL algebra one can define the unary operation $' : V \to V$ by $\iota' \in V$ for each $\iota \in V$ and $\iota' \otimes 1 = (\iota \otimes 1)' = \iota \dashrightarrow 0$. In [3], authors showed that a quasi-BL algebra $(V; \otimes, \sqcup, \sqcap, \dashrightarrow, 0, 1)$ is a quasi-MV algebra iff $\iota'' = \iota$ for each $\iota \in V$.

3 Bounded Commutative Rqℓ-Monoids

In this section, we introduce bounded commutative Rqℓ-monoids as a generalization of bounded commutative Rℓ-monoids. The basic properties of a bounded commutative Rqℓ-monoid are discussed mainly.

Definition 6. *Let* $\boldsymbol{D} = (D; \otimes, \sqcup, \sqcap, \dashrightarrow, ', 0, 1)$ *be an algebra of type (2,2,2,2,1,0,0). If it satisfies the following conditions for each* $\iota, \varrho, \zeta \in D$,

(RQLM1) $(D; \otimes, 1)$ *is a commutative quasi-monoid,*
(RQLM2) $(D; \sqcup, \sqcap, 0, 1)$ *is a bounded quasi-lattice,*
(RQLM3) $\iota \otimes \varrho \leq \zeta \Leftrightarrow \iota \leq \varrho \dashrightarrow \zeta$,
(RQLM4) $\iota \otimes (\iota \dashrightarrow \varrho) = \iota \sqcap \varrho$,
(RQLM5) $\iota \dashrightarrow \varrho = (\iota \dashrightarrow \varrho) \otimes 1$,
(RQLM6) $\iota \sqcap \iota = \iota \otimes 1$ *and* $0 \sqcap 0 = 0$,
(RQLM7) $\iota' \otimes 1 = (\iota \otimes 1)' = \iota \dashrightarrow 0$,
then $\boldsymbol{D} = (D; \otimes, \sqcup, \sqcap, \dashrightarrow, ', 0, 1)$ *is called a bounded commutative residuated quasi-ordered quasi-monoid (bounded commutative Rqℓ-monoid, for short).*

In the following, a Rqℓ-monoid always means a bounded commutative Rqℓ-monoid. It is not difficult to see that each Rℓ-monoid is a Rqℓ-monoid. Conversely, let $\mathbf{D}$ be a Rqℓ-monoid. If $(D; \sqcup, \sqcap, 0, 1)$ is a bounded lattice, then $\iota \sqcap \iota = \iota$ and then we have $\iota \otimes 1 = \iota$ by (RQLM6), it follows that $(D; \otimes, 1)$ is a commutative monoid, so $\mathbf{D}$ is a Rℓ-monoid. On the other hand, if $(D; \otimes, 1)$ is a commutative monoid, then $\iota \otimes 1 = \iota$ and then we have $\iota \sqcap \iota = \iota$ by (RQLM6), it follows that $(D; \sqcup, \sqcap, 0, 1)$ is a bounded lattice, so $\mathbf{D}$ is also a Rℓ-monoid.

Proposition 1. *A Rqℓ-monoid $\mathbf{D}$ is a quasi-BL algebra iff $(\iota \dashrightarrow \varrho) \sqcup (\varrho \dashrightarrow \iota) = 1$ for each $\iota, \varrho \in D$.*

Proof. Follows from Definition 5 and Definition 6.

Example 1. Let $D = \{0, \rho, \sigma, 1\}$ and define the operations $\sqcup, \sqcap, \otimes, \dashrightarrow$ and $'$ as follows:

$\sqcup$	0	ρ	σ	1
0	0	σ	σ	1
ρ	σ	σ	σ	1
σ	σ	σ	σ	1
1	1	1	1	1

$\sqcap$	0	ρ	σ	1
0	0	0	0	0
ρ	0	σ	σ	σ
σ	0	σ	σ	σ
1	0	σ	σ	1

$\otimes$	0	ρ	σ	1
0	0	0	0	0
ρ	0	0	0	σ
σ	0	0	0	σ
1	0	σ	σ	1

$\dashrightarrow$	0	ρ	σ	1
0	1	1	1	1
ρ	σ	1	1	1
σ	σ	1	1	1
1	0	σ	σ	1

$'$	
0	1
ρ	σ
σ	σ
1	0

Then $\mathbf{D} = (D; \otimes, \sqcup, \sqcap, \dashrightarrow, ', 0, 1)$ is a Rqℓ-monoid. Since $\rho \sqcup \rho = \sigma \neq \rho$ and $\rho \otimes 1 = \sigma \neq \rho$, we have that $\mathbf{D}$ is not a Rℓ-monoid.

Example 2. Let $D = \{0, \rho, \sigma, \upsilon, 1\}$ and define the operations $\sqcup, \sqcap, \otimes, \dashrightarrow$ and $'$ as follows:

$\sqcup$	0	ρ	σ	υ	1
0	0	υ	υ	υ	1
ρ	σ	σ	σ	σ	1
σ	σ	σ	σ	σ	1
υ	σ	σ	σ	σ	1
1	1	1	1	1	1

$\sqcap$	0	ρ	σ	υ	1
0	0	0	0	0	0
ρ	0	σ	σ	σ	σ
σ	0	σ	σ	σ	σ
υ	0	σ	σ	σ	σ
1	0	σ	σ	σ	1

$\otimes$	0	ρ	σ	υ	1
0	0	0	0	0	0
ρ	0	0	0	0	σ
σ	0	0	0	0	σ
υ	0	0	0	0	σ
1	0	σ	σ	σ	1

$\dashrightarrow$	0	ρ	σ	υ	1
0	1	1	1	1	1
ρ	σ	1	1	1	1
σ	σ	1	1	1	1
υ	σ	1	1	1	1
1	0	σ	σ	σ	1

$'$	
0	1
ρ	σ
σ	σ
υ	σ
1	0

Then $\mathbf{D} = (D; \otimes, \sqcup, \sqcap, \dashrightarrow, ', 0, 1)$ is a Rqℓ-monoid.

Below we show some basic properties of a Rqℓ-monoid.

Proposition 2. *Let $\mathbf{D}$ be a Rqℓ-monoid. Then the following properties hold for each $\iota, \varrho, \zeta \in D$,*

(P1) $\iota \otimes 1 \leq \iota$ *and* $\iota \leq \iota \otimes 1$;
(P2) $\iota \sqcap \varrho \leq \iota \leq \iota \sqcup \varrho$ *and* $\iota \sqcap \varrho \leq \varrho \leq \iota \sqcup \varrho$;
(P3) $\iota \leq \varrho \dashrightarrow (\iota \otimes \varrho)$;
(P4) $\iota \leq \varrho \Rightarrow \iota \otimes \zeta \leq \varrho \otimes \zeta$;
(P5) $\iota \leq \varrho \Rightarrow \varrho \dashrightarrow \zeta \leq \iota \dashrightarrow \zeta$ *and* $\zeta \dashrightarrow \iota \leq \zeta \dashrightarrow \varrho$;
(P6) $\iota \leq \varrho \Rightarrow \iota \sqcap \zeta \leq \varrho \sqcap \zeta$ *and* $\iota \sqcup \zeta \leq \varrho \sqcup \zeta$;
(P7) $\iota \leq \varrho, \varsigma \leq \tau \Rightarrow \iota \otimes \varsigma \leq \varrho \otimes \tau$;
(P8) $\iota \leq \varrho, \varsigma \leq \tau \Rightarrow \iota \sqcap \varsigma \leq \varrho \sqcap \tau$ *and* $\iota \sqcup \varsigma \leq \varrho \sqcup \tau$;

(P9) $(\iota \otimes \varrho) \sqcap (\iota \otimes \varrho) = \iota \otimes \varrho$ *and* $(\iota \dashrightarrow \varrho) \sqcap (\iota \dashrightarrow \varrho) = \iota \dashrightarrow \varrho$;
(P10) $\iota \otimes \varrho \leq \iota, \varrho$ *and* $\iota \otimes \varrho \leq \iota \sqcap \varrho$;
(P11) $\varrho \leq \iota \dashrightarrow \varrho$;
(P12) $\iota \otimes (\varrho \dashrightarrow \zeta) \leq \varrho \dashrightarrow (\iota \otimes \zeta)$;
(P13) $\iota \dashrightarrow \varrho \leq (\varrho \dashrightarrow \zeta) \dashrightarrow (\iota \dashrightarrow \zeta)$;
(P14) $(\varrho \dashrightarrow \zeta) \otimes (\iota \dashrightarrow \varrho) \leq \iota \dashrightarrow \zeta$ *and* $(\zeta \dashrightarrow \iota) \otimes (\iota \dashrightarrow \varrho) \leq \zeta \dashrightarrow \varrho$;
(P15) $\iota \leq \varrho \Leftrightarrow \iota \dashrightarrow \varrho = 1$;
(P16) $\iota \leq \varrho \Rightarrow \iota \sqcup \varrho = ((\iota \dashrightarrow \varrho) \dashrightarrow \varrho) \sqcap ((\varrho \dashrightarrow \iota) \dashrightarrow \iota)$;
(P17) $\iota \dashrightarrow \iota = 1$;
(P18) $\iota \otimes 1 = 1 \dashrightarrow \iota$;
(P19) $\iota \dashrightarrow (\varrho \dashrightarrow \zeta) = (\iota \otimes \varrho) \dashrightarrow \zeta$;
(P20) $\iota \sqcap \varrho = (\iota \sqcap \varrho) \otimes 1 = (\iota \otimes 1) \sqcap \varrho = \iota \sqcap (\varrho \otimes 1)$;
(P21) $\iota \sqcup \varrho = (\iota \sqcup \varrho) \otimes 1 = (\iota \otimes 1) \sqcup \varrho = \iota \sqcup (\varrho \otimes 1)$;
(P22) $\iota \dashrightarrow \varrho = (\iota \dashrightarrow \varrho) \otimes 1 = (\iota \otimes 1) \dashrightarrow \varrho = \iota \dashrightarrow (\varrho \otimes 1)$;
(P23) $\iota \dashrightarrow \varrho = \iota \dashrightarrow (\iota \sqcap \varrho)$;
(P24) $\iota \otimes (\varrho \sqcup \zeta) = (\iota \otimes \varrho) \sqcup (\iota \otimes \zeta)$;
(P25) $(\iota \sqcup \varrho) \dashrightarrow \zeta = (\iota \dashrightarrow \zeta) \sqcap (\varrho \dashrightarrow \zeta)$;
(P26) $1' = 0$ *and* $0' = 1$;
(P27) $\iota' \otimes \iota = 0$;
(P28) $\iota \leq \iota' \dashrightarrow \varrho$;
(P29) $\iota \leq \iota''$;
(P30) $\iota \leq \varrho \Rightarrow \varrho' \leq \iota'$;
(P31) $\iota''' \otimes 1 = \iota' \otimes 1$;
(P32) $(\iota \otimes \varrho)' = \iota \dashrightarrow \varrho' = \varrho \dashrightarrow \iota'$;
(P33) $\iota \dashrightarrow \varrho \leq \varrho' \dashrightarrow \iota'$;
(P34) $\iota = \iota'' \Rightarrow \iota \dashrightarrow \varrho = \varrho' \dashrightarrow \iota'$;
(P35) $\iota' \sqcup \varrho' \leq (\iota \sqcap \varrho)'$;
(P36) $(\iota \sqcup \varrho)' = \iota' \sqcap \varrho'$;
(P37) $(\iota \dashrightarrow \varrho'')'' = \iota \dashrightarrow \varrho''$;
(P38) $\iota \otimes \varrho' \leq (\iota \dashrightarrow \varrho)'$;
(P39) $(\iota'' \dashrightarrow \iota)' = 0$;
(P40) $(\iota \dashrightarrow \varrho)'' = \iota'' \dashrightarrow \varrho''$;
(P41) $\iota'' \otimes \varrho'' \leq (\iota \otimes \varrho)''$;
(P42) $(\iota \sqcap \varrho)'' = \iota'' \sqcap \varrho''$;
(P43) $\iota \sqcup \varrho \leq \iota'' \sqcup \varrho'' \leq (\iota \sqcup \varrho)''$.

Proof.(1) By (QL4) and (RQLM6), we have $(\iota \otimes 1) \sqcap \iota = (\iota \otimes 1) \sqcap (\iota \sqcap \iota) = (\iota \otimes 1) \sqcap (\iota \otimes 1)$, so $\iota \otimes 1 \leq \iota$. Meanwhile, since $\iota \otimes 1 = \iota \sqcap \iota$, we have $\iota \sqcap (\iota \otimes 1) = \iota \sqcap (\iota \sqcap \iota) = \iota \sqcap \iota$ by (RQLM6) and (QL4), so $\iota \leq \iota \otimes 1$.

(2) Since $(\iota \sqcap \varrho) \sqcap \iota = (\iota \sqcap \iota) \sqcap \varrho = (\iota \sqcap \iota) \sqcap (\varrho \sqcap \varrho) = (\iota \sqcap \varrho) \sqcap (\iota \sqcap \varrho)$, we have $\iota \sqcap \varrho \leq \iota$. Meanwhile, since $(\iota \sqcup \varrho) \sqcup \iota = \iota \sqcup \varrho = (\iota \sqcup \varrho) \sqcup (\iota \sqcup \varrho)$, we have $\iota \leq \iota \sqcup \varrho$. Similarly, we have $\iota \sqcap \varrho \leq \varrho$ and $\varrho \leq \iota \sqcup \varrho$.

(3) Since $\iota \otimes \varrho \leq \iota \otimes \varrho$, we have $\iota \leq \varrho \dashrightarrow (\iota \otimes \varrho)$ by (RQLM3).

(4) If $\iota \leq \varrho$, then we have $\iota \leq \varrho \leq \zeta \dashrightarrow (\varrho \otimes \zeta)$ by (P3), so $\iota \otimes \zeta \leq \varrho \otimes \zeta$ by (RQLM3).

(5) If $\iota \leq \varrho$, then we have $(\varrho \dashrightarrow \zeta) \otimes \iota \leq (\varrho \dashrightarrow \zeta) \otimes \varrho = \varrho \sqcap \zeta \leq \zeta$ and $(\zeta \dashrightarrow \iota) \otimes \zeta = \zeta \sqcap \iota \leq \iota \leq \varrho$ by (P4) and (P2), so $\varrho \dashrightarrow \zeta \leq \iota \dashrightarrow \zeta$ and $\zeta \dashrightarrow \iota \leq \zeta \dashrightarrow \varrho$ by (RQLM3).

(6) If $\iota \leq \varrho$, then we have $\iota \sqcap \varrho = \iota \sqcap \iota$ and $\iota \sqcup \varrho = \varrho \sqcup \varrho$, hence $(\iota \sqcap \zeta) \sqcap (\varrho \sqcap \zeta) = (\iota \sqcap \varrho) \sqcap (\zeta \sqcap \zeta) = (\iota \sqcap \varrho) \sqcap \zeta = (\iota \sqcap \iota) \sqcap \zeta = (\iota \sqcap \zeta) \sqcap (\iota \sqcap \zeta)$ and $(\iota \sqcup \zeta) \sqcup (\varrho \sqcup \zeta) = (\iota \sqcup \varrho) \sqcup (\zeta \sqcup \zeta) = (\iota \sqcup \varrho) \sqcup \zeta = (\varrho \sqcup \varrho) \sqcup \zeta = (\varrho \sqcup \zeta) \sqcup (\varrho \sqcup \zeta)$, it turns out that $\iota \sqcap \zeta \leq \varrho \sqcap \zeta$ and $\iota \sqcup \zeta \leq \varrho \sqcup \zeta$.

(7) If $\iota \leq \varrho$ and $\varsigma \leq \tau$, then we have $\iota \otimes \varsigma \leq \varrho \otimes \varsigma$ and $\varrho \otimes \varsigma \leq \varrho \otimes \tau$ by (P4), so $\iota \otimes \varsigma \leq \varrho \otimes \tau$.

(8) If $\iota \leq \varrho$ and $\varsigma \leq \tau$, then we have $\iota \sqcap \varsigma \leq \varrho \sqcap \varsigma$ and $\varrho \sqcap \varsigma \leq \varrho \sqcap \tau$ by (P6), so $\iota \sqcap \varsigma \leq \varrho \sqcap \tau$. Similarly, we have $\iota \sqcup \varsigma \leq \varrho \sqcup \tau$.

(9) By (RQLM6) and (QM3), we have $(\iota \otimes \varrho) \sqcap (\iota \otimes \varrho) = (\iota \otimes \varrho) \otimes 1 = \iota \otimes \varrho$. Similarly, we have $(\iota \dashrightarrow \varrho) \sqcap (\iota \dashrightarrow \varrho) = \iota \dashrightarrow \varrho$.

(10) Since $\varrho \leq 1$, we have $\iota \otimes \varrho \leq \iota \otimes 1 \leq \iota$ by (P4) and (P1). Similarly, we have $\iota \otimes \varrho \leq \varrho$. Hence, $\iota \otimes \varrho = (\iota \otimes \varrho) \sqcap (\iota \otimes \varrho) \leq \iota \sqcap \varrho$ by (P9).

(11) By (P10), we have $\varrho \otimes \iota \leq \varrho$, so $\varrho \leq \iota \dashrightarrow \varrho$.

(12) By (P2) and (P4), we have $\iota \otimes (\varrho \dashrightarrow \zeta) \otimes \varrho = \iota \otimes (\varrho \sqcap \zeta) \leq \iota \otimes \zeta$, so $\iota \otimes (\varrho \dashrightarrow \zeta) \leq \varrho \dashrightarrow (\iota \otimes \zeta)$ by (RQLM3).

(13) Since $(\varrho \dashrightarrow \zeta) \otimes (\iota \dashrightarrow \varrho) \otimes \iota = (\varrho \dashrightarrow \zeta) \otimes (\iota \sqcap \varrho) \leq (\varrho \dashrightarrow \zeta) \otimes \varrho = \varrho \sqcap \zeta \leq \zeta$ by (RQLM4), (P4), and (P2), we have $(\varrho \dashrightarrow \zeta) \otimes (\iota \dashrightarrow \varrho) \leq \iota \dashrightarrow \zeta$ by (RQLM3), so $\iota \dashrightarrow \varrho \leq (\varrho \dashrightarrow \zeta) \dashrightarrow (\iota \dashrightarrow \zeta)$.

(14) Since $(\varrho \dashrightarrow \zeta) \otimes (\iota \dashrightarrow \varrho) \otimes \iota = (\varrho \dashrightarrow \zeta) \otimes (\iota \sqcap \varrho) \leq (\varrho \dashrightarrow \zeta) \otimes \varrho = \varrho \sqcap \zeta \leq \zeta$, we have $(\varrho \dashrightarrow \zeta) \otimes (\iota \dashrightarrow \varrho) \leq \iota \dashrightarrow \zeta$ by (RQLM3). Since $(\zeta \sqcap \iota) \otimes (\iota \dashrightarrow \varrho) \leq \iota \otimes (\iota \dashrightarrow \varrho) = \iota \sqcap \varrho \leq \varrho$, we have $(\zeta \sqcap \iota) \otimes (\iota \dashrightarrow \varrho) \leq \varrho$ iff $\zeta \otimes (\zeta \dashrightarrow \iota) \otimes (\iota \dashrightarrow \varrho) \leq \varrho$ iff $(\zeta \dashrightarrow \iota) \otimes (\iota \dashrightarrow \varrho) \leq (\zeta \dashrightarrow \varrho)$.

(15) If $\iota \leq \varrho$, then we have $1 \otimes \iota \leq \iota \leq \varrho$ by (P1), it follows that $1 \leq \iota \dashrightarrow \varrho$ by (RQLM3). Since $\iota \dashrightarrow \varrho \leq 1$, we have $\iota \dashrightarrow \varrho = 1$. Conversely, if $\iota \dashrightarrow \varrho = 1$, then we have $\iota \sqcap \varrho = \iota \otimes (\iota \dashrightarrow \varrho) = \iota \otimes 1 = \iota \sqcap \iota$ by (RQLM4) and (RQLM6), so $\iota \leq \varrho$.

(16) Since $\iota \sqcap \varrho = \iota \otimes (\iota \dashrightarrow \varrho) = \varrho \otimes (\varrho \dashrightarrow \iota) \leq \varrho$, we have $\iota \leq (\iota \dashrightarrow \varrho) \dashrightarrow \varrho$. Meanwhile, we have $\varrho \leq (\iota \dashrightarrow \varrho) \dashrightarrow \varrho$ by (P11), so $\iota \sqcup \varrho \leq (\iota \dashrightarrow \varrho) \dashrightarrow \varrho$. Similarly, we have $\iota \sqcup \varrho \leq (\varrho \dashrightarrow \iota) \dashrightarrow \iota$. Hence $\iota \sqcup \varrho \leq ((\iota \dashrightarrow \varrho) \dashrightarrow \varrho) \sqcap ((\varrho \dashrightarrow \iota) \dashrightarrow \iota)$. Denote $\varsigma = ((\iota \dashrightarrow \varrho) \dashrightarrow \varrho) \sqcap ((\varrho \dashrightarrow \iota) \dashrightarrow \iota)$. We have $\varsigma = \varsigma \otimes 1 = \varsigma \otimes (1 \sqcup (\varrho \dashrightarrow \iota)) = \varsigma \otimes ((\iota \dashrightarrow \varrho) \sqcup (\varrho \dashrightarrow \iota)) = (\varsigma \otimes (\iota \dashrightarrow \varrho)) \sqcup (\varsigma \otimes (\varrho \dashrightarrow \iota))$. Because $\varsigma \otimes (\iota \dashrightarrow \varrho) \leq ((\iota \dashrightarrow \varrho) \dashrightarrow \varrho) \otimes (\iota \dashrightarrow \varrho) = (\iota \dashrightarrow \varrho) \sqcap \varrho \leq \varrho$ and $\varsigma \otimes (\varrho \dashrightarrow \iota) \leq \iota$, we have $\varsigma \leq \iota \sqcup \varrho$ by (P6). Hence $\iota \sqcup \varrho = \varsigma = ((\iota \dashrightarrow \varrho) \dashrightarrow \varrho) \sqcap ((\varrho \dashrightarrow \iota) \dashrightarrow \iota)$.

(17) By (P1), we have $1 \otimes \iota \leq \iota$, so $1 \leq \iota \dashrightarrow \iota$ by (RQLM3). Meanwhile, we have $\iota \dashrightarrow \iota \leq 1$. So $\iota \dashrightarrow \iota = (\iota \dashrightarrow \iota) \otimes 1 = (\iota \dashrightarrow \iota) \sqcap (\iota \dashrightarrow \iota) = 1 \sqcap 1 = 1$.

(18) We have $1 \dashrightarrow \iota = (1 \dashrightarrow \iota) \otimes 1 = \iota \sqcap 1 = \iota \sqcap \iota = \iota \otimes 1$.

(19) We have $\varsigma \leq \iota \dashrightarrow (\varrho \dashrightarrow \zeta)$ iff $\varsigma \otimes \iota \leq \varrho \dashrightarrow \zeta$ iff $(\varsigma \otimes \iota) \otimes \varrho \leq \zeta$ iff $\varsigma \otimes (\iota \otimes \varrho) \leq \zeta$ iff $\varsigma \leq (\iota \otimes \varrho) \dashrightarrow \zeta$. Hence $\iota \dashrightarrow (\varrho \dashrightarrow \zeta) = (\iota \otimes \varrho) \dashrightarrow \zeta$.

(20) We have $\iota \sqcap \varrho = (\iota \dashrightarrow \varrho) \otimes \iota = (\iota \dashrightarrow \varrho) \otimes \iota \otimes 1 = (\iota \sqcap \varrho) \otimes 1$, $\iota \sqcap \varrho = \iota \sqcap (\varrho \sqcap \varrho) = \iota \sqcap (\varrho \otimes 1)$, and $\iota \sqcap \varrho = (\iota \sqcap \iota) \sqcap \varrho = (\iota \otimes 1) \sqcap \varrho$ by (RQLM4), (RQLM5), (QL4), and (RQLM6).

(21) We have $(\iota \sqcup \varrho) \otimes 1 = (\iota \sqcup \varrho) \sqcap (\iota \sqcup \varrho) = (\iota \sqcup \varrho) \sqcup (\iota \sqcup \varrho) = ((\iota \sqcup \iota) \sqcup \varrho) \sqcup \varrho = (\iota \sqcup \varrho) \sqcup \varrho = \iota \sqcup (\varrho \sqcup \varrho) = \iota \sqcup \varrho$ by (RQLM6), (QL5), and (QL4). Moreover, we have $\iota \sqcup \varrho = \iota \sqcup (\varrho \sqcup \varrho) = \iota \sqcup (\varrho \sqcap \varrho) = \iota \sqcup (\varrho \otimes 1)$ by (QL4), (QL5), and (RQLM6), and $\iota \sqcup \varrho = \varrho \sqcup \iota = \varrho \sqcup (\iota \sqcup \iota) = (\iota \sqcup \iota) \sqcup \varrho = (\iota \sqcap \iota) \sqcup \varrho = (\iota \otimes 1) \sqcup \varrho$.

(22) Since $\zeta \leq \iota \dashrightarrow (\varrho \otimes 1)$ iff $\zeta \otimes \iota \leq \varrho \otimes 1 \leq \varrho$ iff $\zeta \leq \iota \dashrightarrow \varrho$, we have $\iota \dashrightarrow (\varrho \otimes 1) = \iota \dashrightarrow \varrho$. Meanwhile, because $\zeta \leq (\iota \otimes 1) \dashrightarrow \varrho$ iff $\zeta \otimes (\iota \otimes 1) \leq \varrho$ iff $\zeta \otimes \iota \leq \varrho$ iff $\zeta \leq \iota \dashrightarrow \varrho$, we have $(\iota \otimes 1) \dashrightarrow \varrho = \iota \dashrightarrow \varrho$.

(23) Since $(\iota \dashrightarrow \varrho) \otimes \iota = \iota \sqcap \varrho$, we have $\iota \dashrightarrow \varrho \leq \iota \dashrightarrow (\iota \sqcap \varrho)$ by (RQLM3). Meanwhile, since $\iota \sqcap \varrho \leq \varrho$, we have $\iota \dashrightarrow (\iota \sqcap \varrho) \leq \iota \dashrightarrow \varrho$ by (P5). Hence $\iota \dashrightarrow \varrho = \iota \dashrightarrow (\iota \sqcap \varrho)$.

(24) By (P2), we have $\varrho \leq \varrho \sqcup \zeta$ and $\zeta \leq \varrho \sqcup \zeta$, it follows that $\iota \otimes \varrho \leq \iota \otimes (\varrho \sqcup \zeta)$ and $\iota \otimes \zeta \leq \iota \otimes (\varrho \sqcup \zeta)$ by (P4). So $(\iota \otimes \varrho) \sqcup (\iota \otimes \zeta) \leq (\iota \otimes (\varrho \sqcup \zeta)) \sqcup (\iota \otimes (\varrho \sqcup \zeta)) = \iota \otimes (\varrho \sqcup \zeta)$. On the other hand, since $\iota \otimes \varrho \leq (\iota \otimes \varrho) \sqcup (\iota \otimes \zeta)$ and $\iota \otimes \zeta \leq (\iota \otimes \varrho) \sqcup (\iota \otimes \zeta)$, we have $\varrho \leq \iota \dashrightarrow ((\iota \otimes \varrho) \sqcup (\iota \otimes \zeta))$ and $\zeta \leq \iota \dashrightarrow ((\iota \otimes \varrho) \sqcup (\iota \otimes \zeta))$ by (RQLM3), it follows that $\varrho \sqcup \zeta \leq \iota \dashrightarrow ((\iota \otimes \varrho) \sqcup (\iota \otimes \zeta))$, so $\iota \otimes (\varrho \sqcup \zeta) \leq (\iota \otimes \varrho) \sqcup (\iota \otimes \zeta)$ by (RQLM3). Hence $\iota \otimes (\varrho \sqcup \zeta) = (\iota \otimes \varrho) \sqcup (\iota \otimes \zeta)$.

(25) By (RQLM3) and (P24), we have $\varsigma \leq (\iota \sqcup \varrho) \dashrightarrow \zeta$ iff $(\iota \sqcup \varrho) \otimes \varsigma \leq \zeta$ iff $(\iota \otimes \varsigma) \sqcup (\varrho \otimes \varsigma) \leq \zeta$ iff $\iota \otimes \varsigma \leq \zeta$ and $\varrho \otimes \varsigma \leq \zeta$ iff $\varsigma \leq \iota \dashrightarrow \zeta$ and $\varsigma \leq \varrho \dashrightarrow \zeta$ iff $\varsigma \leq (\iota \dashrightarrow \zeta) \sqcap (\varrho \dashrightarrow \zeta)$. Hence $(\iota \sqcup \varrho) \dashrightarrow \zeta = (\iota \dashrightarrow \zeta) \sqcap (\varrho \dashrightarrow \zeta)$.

(26) We have $1' = (1 \otimes 1)' = 1 \dashrightarrow 0 = 0 \otimes 1 = 0 \sqcap 0 = 0$ by (RQLM7), (P18), and (RQLM6), and $0' = 0 \dashrightarrow 0 = 1$ by (P17).

(27) We have $\iota' \otimes \iota = 1 \otimes \iota' \otimes \iota = (\iota' \otimes 1) \otimes \iota = (\iota \dashrightarrow 0) \otimes \iota = \iota \sqcap 0 = 0 \sqcap 0 = 0$.

(28) By (P27), we have $\iota' \otimes \iota = 0 \leq \varrho$, so $\iota \leq \iota' \dashrightarrow \varrho$.

(29) By (P28), we have $\iota \leq \iota' \dashrightarrow 0 = (\iota')' \otimes 1 \leq \iota''$.

(30) Since $\iota \leq \varrho$, we have $\varrho \dashrightarrow 0 \leq \iota \dashrightarrow 0$ by (P6), it follows that $\varrho' \otimes 1 \leq \iota' \otimes 1$. Meanwhile, since $\varrho' \leq \varrho' \otimes 1$ and $\iota' \otimes 1 \leq \iota'$, we have $\varrho' \leq \iota'$.

(31) By (P29), we have $\iota' \leq \iota'''$. Meanwhile, since $\iota \leq \iota''$, we have $\iota''' \leq \iota'$ by (P30), so $\iota' \sqcap \iota' = \iota''' \sqcap \iota'''$ and then $\iota' \otimes 1 = \iota''' \otimes 1$.

(32) By (P22) and (P19), we have $\iota \dashrightarrow \varrho' = \iota \dashrightarrow (\varrho' \otimes 1) = \iota \dashrightarrow (\varrho \dashrightarrow 0) = (\iota \otimes \varrho) \dashrightarrow 0 = (\iota \otimes \varrho)'$ and $(\iota \otimes \varrho)' = (\varrho \otimes \iota)' = \varrho \dashrightarrow \iota'$.

(33) By (P13) and (P22), we have $\iota \dashrightarrow \varrho \leq (\varrho \dashrightarrow 0) \dashrightarrow (\iota \dashrightarrow 0) = (\varrho' \otimes 1) \dashrightarrow (\iota' \otimes 1) = \varrho' \dashrightarrow \iota'$.

(34) If $\iota'' = \iota$, then we have $\iota \dashrightarrow \varrho \leq \varrho' \dashrightarrow \iota' \leq \iota'' \dashrightarrow \varrho'' = \iota \dashrightarrow \varrho$ by (P33), so $\iota \dashrightarrow \varrho = \varrho' \dashrightarrow \iota'$.

(35) Since $\iota \sqcap \varrho \leq \iota$ and $\iota \sqcap \varrho \leq \varrho$, we have $\iota' \leq (\iota \sqcap \varrho)'$ and $\varrho' \leq (\iota \sqcap \varrho)'$ by (P30), so $\iota' \sqcup \varrho' \leq (\iota \sqcap \varrho)' \sqcup (\iota \sqcap \varrho)' = (\iota \sqcap \varrho)' \sqcap (\iota \sqcap \varrho)' = (\iota \sqcap \varrho)' \otimes 1 = ((\iota \sqcap \varrho) \otimes 1)' = (\iota \sqcap \varrho)'$.

(36) We have $(\iota \sqcup \varrho) \dashrightarrow 0 = (\iota \dashrightarrow 0) \sqcap (\varrho \dashrightarrow 0)$ by (P25), so $(\iota \sqcup \varrho)' = (\iota' \otimes 1) \sqcap (\varrho' \otimes 1) = \iota' \sqcap \varrho'$ by (P20).

(37) By (P32), we have $(\iota \dashrightarrow \varrho'')'' = (\iota \otimes \varrho')''' = (\iota \otimes \varrho')''' \otimes 1 = (\iota \otimes \varrho')' \otimes 1 = (\iota \otimes \varrho')' = \iota \dashrightarrow \varrho''$.

(38) By (RQLM4), (P4), and (P27), we have $(\iota \dashrightarrow \varrho) \otimes \iota \otimes \varrho' = (\iota \sqcap \varrho) \otimes \varrho' \leq \varrho \otimes \varrho' = 0$, so $\iota \otimes \varrho' \leq (\iota \dashrightarrow \varrho) \dashrightarrow 0 = (\iota \dashrightarrow \varrho)'$ by (RQLM3).

(39) Since $0 \le \iota$, we have $\iota'' \dashrightarrow 0 \le \iota'' \dashrightarrow \iota$ by (P5), it turns out that $\iota' \le \iota' \otimes 1 = \iota''' \otimes 1 = \iota'' \dashrightarrow 0 \le \iota'' \dashrightarrow \iota$ by (P1) and (P31), so $(\iota'' \dashrightarrow \iota)' \le \iota''$ and then $(\iota'' \dashrightarrow \iota)' = (\iota'' \dashrightarrow \iota)' \otimes 1 = (\iota'' \dashrightarrow \iota)' \sqcap (\iota'' \dashrightarrow \iota)' = (\iota'' \dashrightarrow \iota)' \sqcap \iota''$. By (RQLM4), (P32), (P15), and (P27), we have $(\iota'' \dashrightarrow \iota)' \sqcap \iota'' = \iota'' \otimes (\iota'' \dashrightarrow (\iota'' \dashrightarrow \iota)') = \iota'' \otimes ((\iota'' \dashrightarrow \iota) \otimes \iota'')' = \iota'' \otimes (\iota'' \sqcap \iota)' = \iota'' \otimes (\iota \otimes 1)' = \iota'' \otimes \iota' \otimes 1 = 0 \otimes 1 = 0$. Hence $(\iota'' \dashrightarrow \iota)' = 0$.

(40) By (P38), (P19), and (P22), we have $(\iota \dashrightarrow \varrho)'' = (\iota \dashrightarrow \varrho)' \dashrightarrow 0 \le (\iota \otimes \varrho') \dashrightarrow 0 = \iota \dashrightarrow (\varrho' \dashrightarrow 0) = \iota \dashrightarrow (\varrho'' \otimes 1) = \iota \dashrightarrow \varrho''$. Meanwhile, since $1 = 0' = (\varrho'' \dashrightarrow \varrho)'' \le ((\iota \sqcap \varrho'') \dashrightarrow \varrho)'' = (((\iota \dashrightarrow \varrho'') \otimes \iota) \dashrightarrow \varrho)'' = ((\iota \dashrightarrow \varrho'') \dashrightarrow (\iota \dashrightarrow \varrho))'' \le ((\iota \dashrightarrow \varrho'') \dashrightarrow (\iota \dashrightarrow \varrho)'')'' = (\iota \dashrightarrow \varrho'') \dashrightarrow (\iota \dashrightarrow \varrho)''$ by (P39), (P19), (P5), and (P37), and $(\iota \dashrightarrow \varrho'') \dashrightarrow (\iota \dashrightarrow \varrho)'' \le 1$, we have $(\iota \dashrightarrow \varrho'') \dashrightarrow (\iota \dashrightarrow \varrho)'' = 1$, so $\iota \dashrightarrow \varrho'' \le (\iota \dashrightarrow \varrho)''$ by (P15). Hence $\iota \dashrightarrow \varrho'' = (\iota \dashrightarrow \varrho)''$. By (P32), (P22), and (P31), we have $\iota'' \dashrightarrow \varrho'' = \varrho' \dashrightarrow \iota''' = \varrho' \dashrightarrow (\iota''' \otimes 1) = \varrho' \dashrightarrow (\iota' \otimes 1) = \varrho' \dashrightarrow \iota' = \iota \dashrightarrow \varrho'' = (\iota \dashrightarrow \varrho)''$.

(41) Since $\iota \otimes \varrho \le \iota$, we have $(\iota \otimes \varrho)'' \le \iota''$ by (P30), it turns out that $\iota'' \otimes \varrho'' = (\iota'' \otimes \varrho'') \otimes 1 = (\iota'' \otimes \varrho'') \sqcap (\iota'' \otimes \varrho'') \le \iota'' \sqcap (\iota'' \otimes \varrho'') = \iota'' \otimes (\iota'' \dashrightarrow (\iota'' \otimes \varrho'')) \le \iota'' \otimes (\iota'' \dashrightarrow (\iota'' \dashrightarrow \varrho')') = \iota'' \otimes (\iota'' \dashrightarrow (\iota'' \dashrightarrow \varrho''')') = \iota'' \otimes (\iota'' \dashrightarrow (\iota \dashrightarrow \varrho')''') = \iota'' \otimes (\iota'' \dashrightarrow (\iota \dashrightarrow \varrho')') = \iota'' \otimes (\iota'' \dashrightarrow (\iota \otimes \varrho)'') = (\iota \otimes \varrho)'' \sqcap \iota'' = (\iota \otimes \varrho)'' \sqcap (\iota \otimes \varrho)'' = (\iota \otimes \varrho)'' \otimes 1 = (\iota \otimes \varrho)''$ by (P38), (P22), (P40), (P32), (RQLM4) and (RQLM6).

(42) On the one hand, since $\iota \sqcap \varrho \le \iota$ and $\iota \sqcap \varrho \le \varrho$, we have $(\iota \sqcap \varrho)'' \le \iota''$ and $(\iota \sqcap \varrho)'' \le \varrho''$ by (P30), so $(\iota \sqcap \varrho)'' \le \iota'' \sqcap \varrho''$. On the other hand, we have $\iota'' \sqcap \varrho'' = (\iota'' \dashrightarrow \varrho'') \otimes \iota'' = (\iota \dashrightarrow \varrho)'' \otimes \iota'' \le (\iota \otimes (\iota \dashrightarrow \varrho))'' = (\iota \sqcap \varrho)''$ by (P40), (P41), and (RQLM4). Hence $(\iota \sqcap \varrho)'' = ((\iota \sqcap \varrho) \otimes 1)'' = (\iota \sqcap \varrho)'' \otimes 1 = (\iota \sqcap \varrho)'' \sqcap (\iota \sqcap \varrho)'' = (\iota'' \sqcap \varrho'') \sqcap (\iota'' \sqcap \varrho'') = \iota'' \sqcap \varrho''$.

(43) Since $\iota \le \iota''$ and $\varrho \le \varrho''$, we have $\iota \sqcup \varrho \le \iota'' \sqcup \varrho''$. Meanwhile, we have $\iota'' \sqcup \varrho'' \le (\iota' \sqcap \varrho')' = (\iota \sqcup \varrho)''$ by (P35) and (P36).

Given that **D** is a Rqℓ-monoid, the element $\iota \in D$ is called regular iff $\iota \otimes 1 = \iota$. We denote $R(D)$ the set of all regular elements of **D**, i.e., $R(D) = \{\iota \in D | \iota \otimes 1 = \iota\}$. It is easy to see that $0, 1 \in R(D)$ and then $R(D)$ is a non-empty subset of D. According to Definition 6 and Proposition 2, we know that if $\iota, \varrho \in D$, then $\iota \otimes \varrho$, $\iota \dashrightarrow \varrho$, $\iota \sqcap \varrho$ and $\iota \sqcup \varrho$ are regular elements. Hence $(R(D); \otimes, \sqcup, \sqcap, \dashrightarrow, ', 0, 1)$ is a Rℓ-monoid.

Let **D** be a Rqℓ-monoid. For $\iota, \varrho \in D$, we define $\iota \boxplus \varrho = (\iota' \otimes \varrho')'$.

Lemma 1. *Let **D** be a Rqℓ-monoid. Then we have for each $\iota, \varrho, \zeta \in D$,*

(1) $1 \boxplus 0 = 1$; *(2)* $\iota \boxplus \varrho = \varrho \boxplus \iota$;
(3) $\iota \boxplus (\varrho \boxplus \zeta) = (\iota \boxplus \varrho) \boxplus \zeta$; *(4)* $\iota \boxplus 1 = 1$;
(5) $(\iota \boxplus \varrho) \boxplus 0 = (\iota \boxplus \varrho) \otimes 1 = \iota \boxplus \varrho$; *(6)* $\iota \sqcup \varrho \le \iota'' \sqcup \varrho'' \le \iota \boxplus \varrho$;
(7) $\iota \boxplus 0 = \iota'' \sqcap \iota'' = \iota'' \otimes 1$; *(8)* $\iota \boxplus \iota' = 1$;
(9) $(\iota \boxplus 0)' = \iota' \boxplus 0$; *(10)* $(\iota \boxplus \varrho)'' = \iota'' \boxplus \varrho'' = \iota \boxplus \varrho$.

Proof.(1) We have $1 \boxplus 0 = (1' \otimes 0')' = (0 \otimes 1)' = (0 \sqcap 0)' = 0' = 1$.

(2) We have $\iota \boxplus \varrho = (\iota' \otimes \varrho')' = (\varrho' \otimes \iota')' = \varrho \boxplus \iota$.

(3) We have $\iota \boxplus (\varrho \boxplus \zeta) = \iota \boxplus (\varrho' \otimes \zeta')' = (\iota' \otimes (\varrho' \otimes \zeta')'')' = \iota' \dashrightarrow (\varrho' \otimes \zeta')''' = \iota' \dashrightarrow ((\varrho' \otimes \zeta')''' \otimes 1) = \iota' \dashrightarrow ((\varrho' \otimes \zeta')' \otimes 1) = \iota' \dashrightarrow (\varrho' \otimes \zeta')' = \iota' \dashrightarrow (\zeta' \dashrightarrow \varrho'') = (\iota' \otimes \zeta') \dashrightarrow \varrho'' = (\zeta' \otimes \iota') \dashrightarrow \varrho'' = \zeta' \dashrightarrow (\iota' \dashrightarrow \varrho'') = \zeta' \dashrightarrow (\iota' \otimes \varrho')' = \zeta' \dashrightarrow ((\iota' \otimes \varrho')' \otimes 1) = \zeta' \dashrightarrow ((\iota' \otimes \varrho')''' \otimes 1) = \zeta' \dashrightarrow (\iota' \otimes \varrho')''' = ((\iota' \otimes \varrho')'' \otimes \zeta')' = (\iota' \otimes \varrho')' \boxplus \zeta = (\iota \boxplus \varrho) \boxplus \zeta$.

(4) Since $\iota' \leq 1$, we have $1 = 1 \boxplus 0 = (0 \otimes 1)' \leq (0 \otimes \iota')' = 1 \boxplus \iota = \iota \boxplus 1 \leq 1$ by (1), (P4), (P30), and (2) so $\iota \boxplus 1 = 1$.

(5) By (P31), we have $(\iota \boxplus \varrho) \boxplus 0 = (\iota' \otimes \varrho')' \boxplus 0 = ((\iota' \otimes \varrho')'' \otimes 0')' = ((\iota' \otimes \varrho')'' \otimes 1)' = (\iota' \otimes \varrho')''' \otimes 1 = (\iota' \otimes \varrho')' \otimes 1 = (\iota' \otimes \varrho')' = \iota \boxplus \varrho$. Meanwhile, we have $(\iota \boxplus \varrho) \otimes 1 = (\iota' \otimes \varrho')' \otimes 1 = (\iota' \otimes \varrho')' = \iota \boxplus \varrho$ by (P18).

(6) Since $\iota' \otimes \varrho' \leq \iota' \otimes 1 \leq \iota'$ and $\iota' \otimes \varrho' \leq 1 \otimes \varrho' \leq \varrho'$ by (P1) and (P4), we have $\iota'' \leq (\iota' \otimes \varrho')'$ and $\varrho'' \leq (\iota' \otimes \varrho')'$ by (P30), so $\iota \sqcup \varrho \leq \iota'' \sqcup \varrho'' \leq \iota \boxplus \varrho$ by (P43).

(7) We have $\iota \boxplus 0 = (\iota' \otimes 0')' = (\iota' \otimes 1)' = \iota'' \otimes 1 \leq \iota''$. Meanwhile, since $\iota'' \leq \iota'' \sqcup 0 = \iota'' \sqcup 0'' \leq \iota \boxplus 0$ by (5), we have $(\iota \boxplus 0) \sqcap (\iota \boxplus 0) = \iota'' \sqcap \iota''$. Because $(\iota \boxplus 0) \sqcap (\iota \boxplus 0) = (\iota \boxplus 0) \otimes ((\iota \boxplus 0) \dashrightarrow (\iota \boxplus 0)) = (\iota \boxplus 0) \otimes 1 = \iota \boxplus 0$ by (5). Hence $\iota \boxplus 0 = \iota'' \sqcap \iota'' = \iota'' \otimes 1$.

(8) By (P27), we have $\iota \boxplus \iota' = (\iota' \otimes \iota'')' = 0' = 1$.

(9) We have $(\iota \boxplus 0)' = (\iota'' \otimes 1)' = (\iota'' \otimes 0')' = \iota' \boxplus 0$ by (7).

(10) We have $(\iota \boxplus \varrho)'' \sqcap (\iota \boxplus \varrho)'' = \iota \boxplus \varrho \boxplus 0 = \iota \boxplus \varrho$ by (7) and (5). Meanwhile, we have $\iota'' \boxplus \varrho'' = (\iota''' \otimes \varrho''')' = ((\iota''' \otimes 1) \otimes (\varrho''' \otimes 1))' = (\iota' \otimes \varrho')' = \iota \boxplus \varrho$ by (P31). Since $\iota \boxplus \varrho = (\iota \boxplus \varrho) \otimes 1$, we have $(\iota \boxplus \varrho)' = ((\iota \boxplus \varrho) \otimes 1)' = (\iota \boxplus \varrho)' \otimes 1$ by (5) and (RQLM7), it follows that $(\iota \boxplus \varrho)'' = ((\iota \boxplus \varrho)' \otimes 1)' = (\iota \boxplus \varrho)'' \otimes 1$, so $(\iota \boxplus \varrho)'' \sqcap (\iota \boxplus \varrho)'' = (\iota \boxplus \varrho)'' \otimes 1 = (\iota \boxplus \varrho)''$. Hence $(\iota \boxplus \varrho)'' = \iota'' \boxplus \varrho'' = \iota \boxplus \varrho$.

Proposition 3. *A Rqℓ-monoid* **D** *is a quasi-MV algebra iff* $\iota'' = \iota$ *for each* $\iota \in D$.

Proof. If a Rqℓ-monoid **D** is a quasi-MV algebra, then $\iota'' = \iota$ is obvious. Conversely, if **D** is a Rqℓ-monoid and $\iota'' = \iota$ for each $\iota \in D$, then $\iota \boxplus (\varrho' \boxplus \iota)' = \iota'' \boxplus (\varrho' \boxplus \iota)' = (\iota' \otimes (\varrho' \boxplus \iota))' = (\iota' \otimes (\varrho' \boxplus \iota''))' = (\iota' \otimes (\varrho \otimes \iota')')' = (\iota' \otimes (\varrho \dashrightarrow \iota))' = (\iota' \otimes (\iota' \dashrightarrow \varrho'))' = (\iota' \sqcap \varrho')'$ and $\varrho \boxplus (\iota' \boxplus \varrho)' = \varrho'' \boxplus (\iota' \boxplus \varrho)' = (\varrho' \otimes (\iota' \boxplus \varrho))' = (\varrho' \otimes (\iota' \boxplus \varrho''))' = (\varrho' \otimes (\iota \otimes \varrho')')' = (\varrho' \otimes (\iota \dashrightarrow \varrho))' = (\varrho' \otimes (\varrho' \dashrightarrow \iota'))' = (\varrho' \sqcap \iota')'$ by (P32) and (P34). Since $\iota' \sqcap \varrho' = \varrho' \sqcap \iota'$, we have $\varrho \boxplus (\iota' \boxplus \varrho)' = \iota \boxplus (\varrho' \boxplus \iota)'$. The rest can be obtained by Lemma 1, so **D** is a quasi-MV algebra.

4 Filters and Filter Congruences

In this section, the notions of filters and weak filters in a Rqℓ-monoid are introduced. We discuss the relation between the set of filters and the set of filter congruences on a Rqℓ-monoid and then generalize the result to weak filters.

Definition 7. *Let* **D** *be a Rqℓ-monoid. If a non-empty subset* H *of* D *satisfies the following conditions,*

(F1) $\iota, \varrho \in H \Rightarrow \iota \otimes \varrho \in H$, *(F2)* $\iota \in H$ *and* $\varrho \in D$ *with* $\iota \leq \varrho \Rightarrow \varrho \in H$, *then* H *is called a filter of* **D**.

Definition 8. *Let* $\mathbf{D}$ *be a Rqℓ-monoid. If a non-empty subset* H *of* D *satisfies the following conditions,*

(F1) $\iota, \varrho \in H \Rightarrow \iota \otimes \varrho \in H$, *(WF2)* $\iota \in H$ *and* $\varrho \in D$ *with* $\iota \leq \varrho \Rightarrow \varrho \otimes 1 \in H$, *then* H *is called a weak filter of* $\mathbf{D}$.

Lemma 2. *Let* $\mathbf{D}$ *be a Rqℓ-monoid and* H *be a filter of* $\mathbf{D}$. *Then* H *is a weak filter of* $\mathbf{D}$.

Proof. For each $\iota \in H$ and $\varrho \in D$ with $\iota \leq \varrho$, then we have $\varrho \in H$. Moreover, since $\iota \leq 1$, we also have $1 \in H$, so $\varrho \otimes 1 \in H$. Hence H is a weak filter of $\mathbf{D}$.

Proposition 4. *Let* $\mathbf{D}$ *be a Rqℓ-monoid and* H *be a weak filter of* $\mathbf{D}$. *Then* $\iota \boxplus \varrho \in H$ *for each* $\iota \in H$ *and* $\varrho \in D$.

Proof. Let $\iota \in H$ and $\varrho \in D$. Then $\iota \leq \iota \sqcup \varrho \leq \iota \boxplus \varrho$ by (P2) and Lemma 1(6). Since H is a weak filter of $\mathbf{D}$, we have $\iota \boxplus \varrho = (\iota \boxplus \varrho) \otimes 1 \in H$ by (WF2) and Lemma 1(5).

Definition 9. *Let* $\mathbf{D}$ *be a Rqℓ-monoid and* θ *be a congruence on* $\mathbf{D}$. *If for each* $\iota, \varrho \in D$, $\langle \iota \otimes 1, \varrho \otimes 1 \rangle \in \theta$ *implies* $\langle \iota, \varrho \rangle \in \theta$, *then* θ *is called a filter congruence on* $\mathbf{D}$.

Given that $\mathbf{D}$ is a Rqℓ-monoid. If H is a filter of $\mathbf{D}$ and θ is a filter congruence on $\mathbf{D}$, then we define the relation $\varphi(H) \sqsubseteq D \times D$ as follows: $\langle \iota, \varrho \rangle \in \varphi(H)$ iff $(\iota \dashrightarrow \varrho) \sqcap (\varrho \dashrightarrow \iota) \in H$. The set $\psi(\theta) \sqsubseteq D$ is defined by $\psi(\theta) = \{\iota \in D | \langle \iota, 1 \rangle \in \theta\}$. It turns out that we have the following results.

Lemma 3. *Let* $\mathbf{D}$ *be a Rqℓ-monoid and* H *be a filter of* $\mathbf{D}$. *Then* $\varphi(H)$ *is a filter congruence on* $\mathbf{D}$.

Proof. Firstly, we show that $\varphi(H)$ is an equivalence relation on $\mathbf{D}$. For each $\iota \in D$, since $\iota \dashrightarrow \iota = 1 \in H$, we have $(\iota \dashrightarrow \iota) \sqcap (\iota \dashrightarrow \iota) = 1 \sqcap 1 = 1 \in H$ and then $\langle \iota, \iota \rangle \in \varphi(H)$. If $\langle \iota, \varrho \rangle \in \varphi(H)$, since the operation $\sqcap$ is commutative, we have that $\langle \varrho, \iota \rangle \in \varphi(H)$ is obvious. Let $\langle \iota, \varrho \rangle \in \varphi(H)$ and $\langle \varrho, \zeta \rangle \in \varphi(H)$. Then $(\iota \dashrightarrow \varrho) \sqcap (\varrho \dashrightarrow \iota) \in H$ and $(\varrho \dashrightarrow \zeta) \sqcap (\zeta \dashrightarrow \varrho) \in H$, it follows that $\iota \dashrightarrow \varrho, \varrho \dashrightarrow \iota \in H$ and $\varrho \dashrightarrow \zeta, \zeta \dashrightarrow \varrho \in H$. Since $(\varrho \dashrightarrow \zeta) \otimes (\iota \dashrightarrow \varrho) \leq \iota \dashrightarrow \zeta$ and H is a filter, we have $\iota \dashrightarrow \zeta \in H$. Similarly, we have $\zeta \dashrightarrow \iota \in H$. Since $(\iota \dashrightarrow \zeta) \otimes (\zeta \dashrightarrow \iota) \leq (\iota \dashrightarrow \zeta) \sqcap (\zeta \dashrightarrow \iota)$ and $(\iota \dashrightarrow \zeta) \otimes (\zeta \dashrightarrow \iota) \in H$, we have $(\iota \dashrightarrow \zeta) \sqcap (\zeta \dashrightarrow \iota) \in H$. Hence $\langle \iota, \zeta \rangle \in \varphi(H)$.

Secondly, we show that $\varphi(H)$ keeps the operations. Let $\langle \iota, \varrho \rangle \in \varphi(H)$ and $\langle \varsigma, \tau \rangle \in \varphi(H)$. Then $(\iota \dashrightarrow \varrho) \sqcap (\varrho \dashrightarrow \iota) \in H$ and $(\varsigma \dashrightarrow \tau) \sqcap (\tau \dashrightarrow \varsigma) \in H$, it follows that $\iota \dashrightarrow \varrho, \varrho \dashrightarrow \iota \in H$ and $\varsigma \dashrightarrow \tau, \tau \dashrightarrow \varsigma \in H$. (1) Since $\iota \otimes (\iota \dashrightarrow \varrho) \leq \iota$ and $\tau \otimes (\tau \dashrightarrow \varsigma) \leq \varsigma$, we have $(\iota \dashrightarrow \varrho) \otimes (\tau \dashrightarrow \varsigma) \otimes (\iota \otimes \tau) = \iota \otimes (\iota \dashrightarrow \varrho) \otimes \tau \otimes (\tau \dashrightarrow \varsigma) \leq \iota \otimes \varsigma$ by (P7), it turns out that $(\iota \dashrightarrow \varrho) \otimes (\tau \dashrightarrow \varsigma) \leq (\iota \otimes \tau) \dashrightarrow (\iota \otimes \varsigma)$, so $(\iota \otimes \tau) \dashrightarrow (\iota \otimes \varsigma) \in H$. Similarly, we have $(\iota \otimes \varsigma) \dashrightarrow (\iota \otimes \tau) \in H$. So $\langle \iota \otimes \varsigma, \iota \otimes \tau \rangle \in \varphi(H)$. Analogously, we obtain $\langle \iota \otimes \tau, \varrho \otimes \tau \rangle \in \varphi(H)$. Hence $\langle \iota \otimes \varsigma, \varrho \otimes \tau \rangle \in \varphi(H)$. (2) Since $(\iota \dashrightarrow \varrho) \otimes (\varsigma \dashrightarrow \iota) \leq \varsigma \dashrightarrow \varrho$ and $(\varrho \dashrightarrow \iota) \otimes (\varsigma \dashrightarrow \varrho) \leq \varsigma \dashrightarrow \iota$ by (P14), we have $\iota \dashrightarrow \varrho \leq (\varsigma \dashrightarrow \iota) \dashrightarrow (\varsigma \dashrightarrow \varrho)$

and $\varrho \dashrightarrow \iota \leq (\varsigma \dashrightarrow \varrho) \dashrightarrow (\varsigma \dashrightarrow \iota)$, it follows that $(\varsigma \dashrightarrow \iota) \dashrightarrow (\varsigma \dashrightarrow \varrho) \in H$ and $(\varsigma \dashrightarrow \varrho) \dashrightarrow (\varsigma \dashrightarrow \iota) \in H$, so $\langle \varsigma \dashrightarrow \iota, \varsigma \dashrightarrow \varrho \rangle \in \varphi(H)$. Similarly, we have $\langle \varsigma \dashrightarrow \varrho, \tau \dashrightarrow \varrho \rangle \in \varphi(H)$. Thus $\langle \varsigma \dashrightarrow \iota, \tau \dashrightarrow \varrho \rangle \in \varphi(H)$. (3) By (1) and (2), we have $\langle \iota \otimes (\iota \dashrightarrow \varsigma), \varrho \otimes (\varrho \dashrightarrow \tau) \rangle \in \varphi(H)$, so $\langle \iota \sqcap \varsigma, \varrho \sqcap \tau \rangle \in \varphi(H)$. (4) Since $\iota \otimes (\iota \dashrightarrow \varrho) \otimes (\varsigma \dashrightarrow \tau) \leq \iota \otimes (\iota \dashrightarrow \varrho) = \iota \sqcap \varrho \leq \varrho \sqcup \tau$ and $\varsigma \otimes (\iota \dashrightarrow \varrho) \otimes (\varsigma \dashrightarrow \tau) \leq \varsigma \otimes (\varsigma \dashrightarrow \tau) = \varsigma \sqcap \tau \leq \varrho \sqcup \tau$, we have $(\iota \dashrightarrow \varrho) \otimes (\varsigma \dashrightarrow \tau) \leq \iota \dashrightarrow (\varrho \sqcup \tau)$ and $(\iota \dashrightarrow \varrho) \otimes (\varsigma \dashrightarrow \tau) \leq \varsigma \dashrightarrow (\varrho \sqcup \tau)$, so $(\iota \dashrightarrow \varrho) \otimes (\varsigma \dashrightarrow \tau) \leq (\iota \dashrightarrow (\varrho \sqcup \tau)) \sqcap (\varsigma \dashrightarrow (\varrho \sqcup \tau)) = (\iota \sqcup \varsigma) \dashrightarrow (\varrho \sqcup \tau)$ by (P25). Since $\iota \dashrightarrow \varrho, \varsigma \dashrightarrow \tau \in H$, we have $(\iota \dashrightarrow \varrho) \otimes (\varsigma \dashrightarrow \tau) \in H$, so $(\iota \sqcup \varsigma) \dashrightarrow (\varrho \sqcup \tau) \in H$. Similarly, we have $(\varrho \sqcup \tau) \dashrightarrow (\iota \sqcup \varsigma) \in H$. Hence $\langle \iota \sqcup \varsigma, \varrho \sqcup \tau \rangle \in \varphi(H)$. (5) Since $\iota \dashrightarrow \varrho \leq \varrho' \dashrightarrow \iota'$ and $\varrho \dashrightarrow \iota \leq \iota' \dashrightarrow \varrho'$, we have $\iota' \dashrightarrow \varrho' \in H$ and $\varrho' \dashrightarrow \iota' \in H$, so $\langle \iota', \varrho' \rangle \in \varphi(H)$. Hence $\varphi(H)$ is a congruence on **D**.

Finally, suppose that $\langle \iota \otimes 1, \varrho \otimes 1 \rangle \in \varphi(H)$. Then $((\iota \otimes 1) \dashrightarrow (\varrho \otimes 1)) \sqcap ((\varrho \otimes 1) \dashrightarrow (\iota \otimes 1) \in H)$, we have $(\iota \otimes 1) \dashrightarrow (\varrho \otimes 1) \in H$ and $(\varrho \otimes 1) \dashrightarrow (\iota \otimes 1) \in H$, it follows that $\iota \dashrightarrow \varrho \in H$ and $\varrho \dashrightarrow \iota \in H$ by (P22), so $(\iota \dashrightarrow \varrho) \sqcap (\varrho \dashrightarrow \iota) \in H$ and then $\langle \iota, \varrho \rangle \in \varphi(H)$. Hence $\varphi(H)$ is a filter congruence on **D**.

Lemma 4. *Let* **D** *be a Rqℓ-monoid and* θ *be a filter congruence on* **D**. *Then* $\psi(\theta)$ *is a filter of* **D**.

Proof. It is easy to see that $1 \in \psi(\theta)$. Suppose that $\iota, \varrho \in \psi(\theta)$. Then $\langle \iota, 1 \rangle \in \theta$ and $\langle \varrho, 1 \rangle \in \theta$. Since θ is a congruence on **D**, we have $\langle \iota \otimes \varrho, 1 \rangle = \langle \iota \otimes \varrho, 1 \otimes 1 \rangle \in \theta$, so $\iota \otimes \varrho \in \psi(\theta)$. Let $\iota \in \psi(\theta)$ and $\varrho \in D$ with $\iota \leq \varrho$. Then we have $\langle \iota, 1 \rangle \in \theta$ and $\iota \sqcup \varrho = \varrho \sqcup \varrho = \varrho \sqcap \varrho = \varrho \otimes 1$. Since θ is a congruence on **D**, we have $\langle \varrho, \varrho \rangle \in \theta$, it follows that $\langle \iota \dashrightarrow \varrho, 1 \dashrightarrow \varrho \rangle \in \theta$, so $\langle (\iota \dashrightarrow \varrho) \dashrightarrow \varrho, (1 \dashrightarrow \varrho) \dashrightarrow \varrho \rangle \in \theta$. Since $(1 \dashrightarrow \varrho) \dashrightarrow \varrho = (\varrho \otimes 1) \dashrightarrow \varrho = \varrho \dashrightarrow \varrho = 1$, we get $\langle (\iota \dashrightarrow \varrho) \dashrightarrow \varrho, 1 \rangle \in \theta$. Similarly, we have $\langle (\varrho \dashrightarrow \iota) \dashrightarrow \iota, (\varrho \dashrightarrow 1) \dashrightarrow \iota \rangle \in \theta$. Because $\varrho \dashrightarrow 1 = 1$, we have $\langle (\varrho \dashrightarrow 1) \dashrightarrow \iota, 1 \dashrightarrow \iota \rangle \in \theta$ and then $\langle (\varrho \dashrightarrow \iota) \dashrightarrow \iota, 1 \dashrightarrow \iota \rangle \in \theta$. Since $\iota \otimes 1 = 1 \dashrightarrow \iota = (1 \dashrightarrow \iota) \otimes 1 = (1 \dashrightarrow \iota) \sqcap (1 \dashrightarrow \iota) = 1 \sqcap (1 \dashrightarrow \iota)$ and $\iota \sqcup \varrho = ((\iota \dashrightarrow \varrho) \dashrightarrow \varrho) \sqcap ((\varrho \dashrightarrow \iota) \dashrightarrow \iota)$ by (P16), we have $\langle \varrho \otimes 1, \iota \otimes 1 \rangle = \langle \iota \sqcup \varrho, \iota \otimes 1 \rangle = \langle \iota \sqcup \varrho, 1 \dashrightarrow \iota \rangle = \langle ((\iota \dashrightarrow \varrho) \dashrightarrow \varrho) \sqcap ((\varrho \dashrightarrow \iota) \dashrightarrow \iota), 1 \sqcap (1 \dashrightarrow \iota) \rangle \in \theta$. Note that $\langle \iota \otimes 1, 1 \otimes 1 \rangle \in \theta$, we have $\langle \varrho \otimes 1, 1 \otimes 1 \rangle \in \theta$. Because θ is a filter congruence on **D**, we have $\langle \varrho, 1 \rangle \in \theta$, so $\varrho \in \psi(\theta)$. Hence $\psi(\theta)$ is a filter of **D**.

Then, we get the relation between the set of filters of **D** and the set of filter congruences on **D**.

Theorem 1. *Let* **D** *be a Rqℓ-monoid,* H *be a filter of* **D** *and* θ *be a filter congruence on* **D**. *Then* $\psi(\varphi(H)) = H$ *and* $\varphi(\psi(\theta)) = \theta$, *so there exists a one-to-one correspondence between the set of filters of* **D** *and the set of filter congruences on* **D**.

Proof. By Lemma 3 and Lemma 4, we have $\varphi(H)$ is a filter congruence on **D**, $\psi(\theta)$ is a filter of **D**. We have $\psi(\varphi(H)) = \{\iota \in D | \langle \iota, 1 \rangle \in \varphi(H)\} = \{\iota \in D | 1 \dashrightarrow \iota = \iota \otimes 1 \in H\} = H$ for $\iota \otimes 1 \in H \Leftrightarrow \iota \in H$. For each $\langle \iota, \varrho \rangle \in \varphi(\psi(\theta))$, we have

$\iota \dashrightarrow \varrho \in \psi(\theta)$ and $\varrho \dashrightarrow \iota \in \psi(\theta)$, i.e., $\langle \iota \dashrightarrow \varrho, 1\rangle \in \theta$ and $\langle \varrho \dashrightarrow \iota, 1\rangle \in \theta$. Since $\langle \iota, \iota\rangle \in \theta$, we have $\langle \iota \sqcap \varrho, \iota \otimes 1\rangle = \langle \iota \otimes (\iota \dashrightarrow \varrho), \iota \otimes 1\rangle \in \theta$ and $\langle \iota \sqcap \varrho, \varrho \otimes 1\rangle = \langle \varrho \otimes (\varrho \dashrightarrow \iota), \varrho \otimes 1\rangle \in \theta$, so $\langle \iota \otimes 1, \varrho \otimes 1\rangle \in \theta$. Since θ is a filter congruence on **D**, we have $\langle \iota, \varrho\rangle \in \theta$. Conversely, if $\langle \iota, \varrho\rangle \in \theta$, then $\langle \varrho, \iota\rangle \in \theta$, we have $\langle \iota \dashrightarrow \varrho, 1\rangle = \langle \iota \dashrightarrow \varrho, \varrho \dashrightarrow \varrho\rangle \in \theta$ and $\langle \varrho \dashrightarrow \iota, 1\rangle = \langle \varrho \dashrightarrow \iota, \iota \dashrightarrow \iota\rangle \in \theta$, it turns out that $\iota \dashrightarrow \varrho, \varrho \dashrightarrow \iota \in \psi(\theta)$, so $\langle \iota, \varrho\rangle \in \varphi(\psi(\theta))$. Hence $\varphi(\psi(\theta)) = \theta$.

Below suppose that H is a weak filter of a Rqℓ-monoid **D** and θ is a congruence on **D**. We define the relation $\xi(H) \sqsubseteq D \times D$ as follows: $\langle \iota, \varrho\rangle \in \xi(H)$ iff $(\iota \dashrightarrow \varrho) \sqcap (\varrho \dashrightarrow \iota) \in H$. The set $\eta(\theta) \sqsubseteq D$ is defined by $\eta(\theta) = \{\iota \in D | \langle \iota, 1\rangle \in \theta\}$. Similarly to the proofs of Lemma 3, Lemma 4 and Theorem 1, we have the following results.

Proposition 5. *Let **D** be a Rqℓ-monoid, H be a weak filter of **D** and θ be a congruence on **D**. Then (1) $\xi(H)$ is a congruence on **D**; (2) $\eta(\theta)$ is a weak filter of **D**; (3) $\eta(\xi(H)) \sqsubseteq H$ and $\xi(\eta(\theta)) \sqsubseteq \theta$.*

Let **D** be a Rqℓ-monoid and H be a weak filter of **D**. Then $D/H = \{\iota/H | \iota \in D\}$ where $\iota/H = \{\varrho \in D | \iota \dashrightarrow \varrho \in H \text{ and } \varrho \dashrightarrow \iota \in H\}$ is a quotient set with respect to H. Some operations on D/H are defined as follows:

(1) $(\iota/H) \sqcup (\varrho/H) = (\iota \sqcup \varrho)/H$;
(2) $(\iota/H) \sqcap (\varrho/H) = (\iota \sqcap \varrho)/H$;
(3) $(\iota/H) \otimes (\varrho/H) = (\iota \otimes \varrho)/H$;
(4) $(\iota/H) \dashrightarrow (\varrho/H) = (\iota \dashrightarrow \varrho)/H$;
(5) $(\iota/H)' = \iota'/H$.

Then, it is direct to see that the algebraic structure of D/H is inherited from the algebra **D**. So $\mathbf{D}/H = \{D/H; \sqcup, \sqcap, \otimes, \dashrightarrow, ', 0/H, 1/H\}$ is a Rqℓ-monoid.

Proposition 6. *Let **D** be a Rqℓ-monoid and H be a weak filter of **D**. Then $\mathbf{D}/H = \{D/H; \sqcup, \sqcap, \otimes, \dashrightarrow, ', 0/H, 1/H\}$ is a Rℓ-monoid.*

Proof. We only need show that $(\iota/H) \otimes (1/H) = \iota/H$ for each $\iota/H \in D/H$. For each $\zeta \in (\iota/H) \otimes (1/H) = (\iota \otimes 1)/H$, we have $\zeta \dashrightarrow (\iota \otimes 1) \in H$ and $(\iota \otimes 1) \dashrightarrow \zeta \in H$, it turns out that $\zeta \dashrightarrow \iota = \zeta \dashrightarrow (\iota \otimes 1) \in H$ and $\iota \dashrightarrow \zeta = (\iota \otimes 1) \dashrightarrow \zeta \in H$ by (P22), so $\zeta \in \iota/H$ and then $(\iota/H) \otimes (1/H) \sqsubseteq \iota/H$. Similarly, we can show that $\iota/H \sqsubseteq (\iota \otimes 1)/H$. Hence $(\iota/H) \otimes (1/H) = (\iota \otimes 1)/H = \iota/H$ and then $\mathbf{D}/H$ is a Rℓ-monoid.

Definition 10. *A Rqℓ-monoid **D** is normal, if **D** satisfies the identity $(\iota \otimes \varrho)'' = \iota'' \otimes \varrho''$ for each $\iota, \varrho \in D$.*

Let **D** be a Rqℓ-monoid. We denote $E(D) = \{\iota \in D | \iota'' = 1\}$ and elements in $E(D)$ is called dense.

Proposition 7. *Let **D** be a normal Rqℓ-monoid. Then $E(D)$ is a weak filter of **D**.*

Proof. Let $\iota, \varrho \in E(D)$. Then we have $\iota'' = 1$ and $\varrho'' = 1$, it follows that $1 = 1 \otimes 1 = \iota'' \otimes \varrho'' \leq (\iota \otimes \varrho)''$. Since $(\iota \otimes \varrho)'' \leq 1$, we have $(\iota \otimes \varrho)'' = 1$, so $\iota \otimes \varrho \in E(D)$. Let $\iota \in E(D)$ and $\varrho \in D$ with $\iota \leq \varrho$. Then we have $\iota'' = 1$ and $\iota'' \leq \varrho''$, so $1 \leq \varrho''$. Since $\varrho'' \leq 1$, we have $\varrho'' \otimes 1 = 1$, so $(\varrho \otimes 1)'' = \varrho'' \otimes 1 = 1$ and then $\varrho \otimes 1 \in E(D)$. Hence $E(D)$ is a weak filter of $\mathbf{D}$.

Proposition 8. *Let* $\mathbf{D}$ *be a normal Rqℓ-monoid. Then the quotient Rℓ-monoid* $\mathbf{D}/E(D)$ *is an MV-algebra.*

Proof. According to Proposition 3, we only need show $\iota/E(D) = (\iota/E(D))''$ for each $\iota/E(D) \in D/E(D)$. For each $\iota \in D$, since $\iota \leq \iota''$, we have $\iota \dashrightarrow \iota'' = 1$ by (P15), so $(\iota \dashrightarrow \iota'')'' = 1'' = 1$. Meanwhile, we have $(\iota'' \dashrightarrow \iota)'' = 0' = 1$ by (P39). So $\iota \dashrightarrow \iota'' \in E(D)$ and $\iota'' \dashrightarrow \iota \in E(D)$, it turns out that $\iota \in \iota''/E(D)$ and $\iota'' \in \iota/E(D)$, we have $\iota/E(D) = \iota''/E(D) = (\iota/E(D))''$. Hence $\mathbf{D}/E(D)$ is an MV-algebra.

Let $\mathbf{D}$ be a Rqℓ-monoid and denote $M(D) = \{\iota \in D | \iota'' = \iota\}$. Then $M(D)$ is a non-empty subset of D. We define the algebra $\mathbf{M(D)} = (M(D); \otimes, \sqcup_{M(D)}, \sqcap, \dashrightarrow, ', 0, 1)$, where $\iota \sqcup_{M(D)} \varrho = (\iota \sqcup \varrho)''$ for each $\iota, \varrho \in M(D)$ and the remaining operations are the restrictions of the original ones in Rqℓ-monoid $\mathbf{D}$ on $M(D)$.

Theorem 2. *Let* $\mathbf{D}$ *be a normal Rqℓ-monoid. Then* $\mathbf{M(D)}$ *is a quasi-MV subalgebra of* $\mathbf{D}$.

Proof. Since $\mathbf{D}$ is a normal Rqℓ-monoid, we have $(\iota \otimes \varrho)'' = \iota'' \otimes \varrho'' = \iota \otimes \varrho$. Moreover, we have $(\iota \sqcup_{M(D)} \varrho)'' = (\iota \sqcup \varrho)'''' = (\iota \sqcup \varrho)'' = \iota \sqcup_{M(D)} \varrho$, $(\iota \sqcap \varrho)'' = \iota'' \sqcap \varrho'' = \iota \sqcap \varrho$, $(\iota \dashrightarrow \varrho)'' = \iota'' \dashrightarrow \varrho'' = \iota \dashrightarrow \varrho$, and $(\iota')'' = (\iota'')' = \iota'$. So $\mathbf{M(D)} = (M(D); \otimes, \sqcup_{M(D)}, \sqcap, \dashrightarrow, ', 0, 1)$ is a subalgebra of $\mathbf{D}$. Furthermore, $\iota'' = \iota$ for each $\iota \in M(D)$. Hence $\mathbf{M(D)} = (M(D); \otimes, \sqcup_{M(D)}, \sqcap, \dashrightarrow, ', 0, 1)$ is a quasi-MV algebra by Proposition 3.

Let $\mathbf{D}$ be a Rqℓ-monoid. We denote $I(D) = \{\iota \in D | \iota \otimes \iota = \iota\}$ and $I(D)$ is a non-empty subset of D. Moreover, if $\iota \in I(D)$, then we have $\iota \in R(D)$.

Lemma 5. *Let* $\mathbf{D}$ *be a Rqℓ-monoid and* $\iota \in I(D)$. *For each* $\varrho \in D$, *then* $\iota \sqcap \varrho = \iota \otimes \varrho$.

Proof. By (P10), we have $\iota \otimes \varrho \leq \iota \sqcap \varrho$. On the other hand, since $\iota \sqcap \varrho \leq \varrho$, we have $\iota \sqcap \varrho = \iota \otimes (\iota \dashrightarrow \varrho) = \iota \otimes \iota \otimes (\iota \dashrightarrow \varrho) = \iota \otimes (\iota \sqcap \varrho) \leq \iota \otimes \varrho$ by (P4). Hence $\iota \sqcap \varrho = \iota \otimes \varrho$.

Theorem 3. *Let* $\mathbf{D}$ *be a Rqℓ-monoid. Then* $\mathbf{I(D)}$ *is a subalgebra of the reduct* $(D; \otimes, \sqcup, \sqcap, 0, 1)$ *of* $\mathbf{D}$.

Proof. For each $\iota, \varrho \in I(D)$, then we have $(\iota \otimes \varrho) \otimes (\iota \otimes \varrho) = (\iota \otimes \iota) \otimes (\varrho \otimes \varrho) = \iota \otimes \varrho$, $(\iota \sqcup \varrho) \otimes (\iota \sqcup \varrho) = (\iota \otimes \iota) \sqcup (\iota \otimes \varrho) \sqcup (\varrho \otimes \varrho) = \iota \sqcup \varrho \sqcup (\iota \otimes \varrho) = \iota \sqcup \varrho$, and $(\iota \sqcap \varrho) \otimes (\iota \sqcap \varrho) = (\iota \otimes \varrho) \otimes (\iota \otimes \varrho) = \iota \otimes \varrho = \iota \sqcap \varrho$ by Lemma 5. Hence $\mathbf{I(D)}$ is a subalgebra of the reduct $(D; \otimes, \sqcup, \sqcap, 0, 1)$ of $\mathbf{D}$.

5 Conclusion

In this paper, we have introduced and investigated bounded commutative Rqℓ-monoids. Bounded commutative Rqℓ-monoids not only generalize bounded commutative Rℓ-monoids but also generalize quasi-MV algebras and quasi-BL algebras. Hence they may play an important role in many-valued logics and quantum computational logics. In the future, we will continue to study the algebraic and topological structures of bounded commutative Rqℓ-monoids.

Acknowledgements. The authors express their sincere gratitude to the reviewers for their comments.

References

1. Bou, F., Paoli, F., Ledda, A., Freytes, H.: On some properties of quasi-MV algebras and $\sqrt{\prime}$ quasi-MV algebras. Part II. Soft Comput. **12**(4), 341–352 (2008). https://doi.org/10.1007/s00500-007-0185-8
2. Chang, C.C.: A new proof of the completeness of Lukasiewicz axioms. Trans. Am. Math. Soc. **93**(1), 74–80 (1959)
3. Chen, W.J., Wang, H.K.: Filters and ideals in the generalization of pseudo-BL algebras. Soft. Comput. **24**(2), 795–812 (2020). https://doi.org/10.1007/s00500-019-04528-9
4. Chajda, I.: Lattices in quasiordered sets. Acta Univ. Palack. Olomuc. Fac. Rerum Natur. Math. **31**(1), 6–12 (1992)
5. Dvurecenskij, A., Rachunek, J.: Bounded commutative residuated ℓ-monoids with general comparability and state. Soft. Comput. **10**(3), 212–218 (2006). https://doi.org/10.1007/s00500-005-0473-0
6. Ledda, A., Konig, M., Paoli, F., Giuntini, R.: MV algebras and quantum computation. Stud. Logica. **82**(2), 245–270 (2006). https://doi.org/10.1007/s11225-006-7202-2
7. Hajek, P.: Metamathematics of Fuzzy Logic. Kluwer Academic Publishers, Dordrecht (1998)
8. Jipsen, P., Ledda, A., Paoli, F.: On some properties of quasi-MV algebras and $\sqrt{\prime}$ quasi-MV algebras. Part IV. Rep. Math. Logic **48**, 3–36 (2013)
9. Kowalski, T., Paoli, F.: On some properties of quasi-MV algebras and $\sqrt{\prime}$ quasi-MV algebras. Part III. Rep. Math. Logic **45**, 161–199 (2010)
10. Paoli, F., Ledda, A., Giuntini, R., Freytes, H.: On some properties of quasi-MV algebras and $\sqrt{\prime}$ quasi-MV algebras. Part I. Rep. Math. Logic **44**, 31–63 (2009)
11. Rachunek, J., Salounova, D.: Modal operators on bounded residuated ℓ-monoids. Math. Bohem. **133**(3), 299–311 (2008)
12. Rachunek, J., Svrcek, F.: Interior and closure operators on bounded commutative residuated ℓ-monoids. Discussiones Math. Gen. Algebra Appl. **28**(1), 11–27 (2008)
13. Rachunek, J.: DRℓ-semigroups and MV-algebras. Czechoslov. Math. J. **48**(123), 365–372 (1998)
14. Rachunek, J.: MV-algebras are categorically equivalent to a class of $\mathrm{DR}\ell_{1(i)}$-semigroups. Math. Bohem. **123**(4), 437–441 (1998)
15. Rachunek, J.: A duality between algebras of basic logic and bounded representable DRℓ-monoids. Math. Bohem. **126**(3), 561–569 (2001)

16. Rachunek, J., Salounova, D.: Local bounded commutative residuated ℓ-monoids. Czechoslov. Math. J. **57**(132), 395–406 (2007). https://doi.org/10.1007/s10587-007-0068-2
17. Swamy, K.L.N., Subba Rao, B.V.: Isometries in dually residuated lattice ordered semigroups. Math. Semin. Notes Kobe Univ **8**, 369–380 (1980)

On the Weak Dominance Relation Between Conjunctors

Lizhu Zhang, Qigao Bo, and Gang Li[(✉)]

School of Mathematics and Statistics, Qilu University of Technology (Shandong Academy of Sciences), Jinan 250353, China
sduligang@163.com

Abstract. The concept of weak dominance relation as a generalization of the dominance relation between binary fuzzy connectives was introduced in literature. In this paper we mainly discuss the weak dominance relation on the class of conjunctors. First, we provide some properties of the weak dominance relation and show that the weak dominance relation still holds under duality and isomorphism. Secondly, we deal with the weak dominance relation between conjunctors. Finally, we present some results about the summand-wise nature of the weak dominance relations between ordinal sum conjunctors.

Keywords: Weak dominance · Conjunctor · Ordinal sum · Quasi-copula · Triangular norm

1 Introduction

Fuzzy connectives play the important role in fuzzy logic [1]. Various classes of fuzzy connectives have been studied, including t-norms and t-conorms, copulas, quasi-copulas, conjunctors and disjunctors, fuzzy negations and fuzzy implications, etc. The properties and related functional equations were discussed for different fuzzy connectives in literature. As a kind of binary relation between binary fuzzy connectives, the dominant relation first appears in the probability metric space [3,4] and is closely related to the construction of Cartesian product of probability metric space. Later, the dominance relation was applied to the construction of fuzzy equivalence relations [5,7,8], fuzzy orderings [6] and the open problem about the transitivity [11–13], which involve t-norms, t-conorms. With the expansion of application fields of the dominance relation, more general classes of fuzzy connectives were studied [9,10,14–17]. Moreover, the dominance relation between binary fuzzy connectives is related to Minkowski inequality, convex function and bisymmetry equation.

Some generalization of dominance relation was proposed in literature. The graded version of dominance relation was discussed in [2]. In [19] the authors introduce the weak dominance relation between t-norms, which can also be regarded as another generalization of the dominance relation. On the other hand,

Y. Chen and S. Zhang (Eds.): AILA 2022, CCIS 1657, pp. 199–210, 2022.
https://doi.org/10.1007/978-981-19-7510-3_15

weak dominance can also be regarded as an inequality generalization of modularity equation [20] which is related to some associative equations and widely used in fuzzy theory. It is noted that only some primary results about the weak dominance between strict t-norms are given in [19]. Whether the weak dominance relation of various classes of binary fuzzy connectives (Triangular norms, Copulas, Weaker operators) has similar results needs further study. Therefore, more attention should be paid to the weak dominance relation of more general fuzzy connectives.

The aim of the present contribution is to provide results on the weak dominance relation between conjunctors. We present some basic properties about the weak dominance relation, and discuss the weak dominance relation between the classes of conjunctors. Then, we mainly deal with the weak dominance relation between two ordinal sum conjunctors. Some sufficient and necessary conditions are provided.

The structure of the paper is as follows. Firstly, we recall some basic definitions of conjunctors which will be used in the sequel and the notion of weak domination concerning two conjunctors. In Sect. 3, we show the weak domination between conjunctors of their dual or isomorphism. In Sect. 4, we present the necessary conditions of conjunctors weakly dominating continuous Archimedean t-norm. In Sect. 5, we provide some results about the weak dominance relations between ordinal sum conjunctors. Finally, we will close the contribution with a short summary.

2 Preliminaries

We recall here definitions of some conjunctors which will be used in the sequel.

Definition 1. *A binary operation* $C : [0,1]^2 \to [0,1]$ *is called a conjunctor if it is increasing in each place and has a neutral element* 1 *satisfying* $C(x,1) = C(1,x) = x$ *for any* $x \in [0,1]$.

A conjunctor which is associative and commutative is called a t-norm.

Definition 2 [18]. *A t-norm is a two place function* $T : [0,1]^2 \to [0,1]$, *such that for all* $x, y, z \in [0,1]$ *the following conditions are satisfied:*

$(T1)$ $T(x,y) = T(y,x)$.
$(T2)$ $T(T(x,y),z) = T(x,T(y,z))$.
$(T3)$ $T(x,y) \le T(y,z)$ *whenever* $y \le z$.
$(T4)$ $T(x,1) = x$.

The four basic t-norms T_M, T_P, T_L, and T_D are usually discussed in literature. They are defined by, respectively:

$$T_M(x,y) = \min(x,y),$$
$$T_P(x,y) = x \cdot y,$$
$$T_L(x,y) = \max(x+y-1,0),$$
$$T_D(x,y) = \begin{cases} 0 & (x,y) \in [0,1[^2, \\ \min(x,y) & \textit{otherwise}. \end{cases}$$

Remark 1. If a conjunctor $C : [0,1]^2 \to [0,1]$ satisfies commutativity and associativity, then C is a t-norm. For any conjunctor C it holds that $T_D \le C \le T_M$.

Definition 3 [18].

1. *A t-norm T is called Archimedean if for each $(x, y) \in]0,1[^2$ there is an $n \in N$ with $x_T^{(n)} < y$.*
2. *A t-norm T is called strict if it is continuous and strictly monotonic.*
3. *A t-norm T is called nilpotent if it is continuous and if each $a \in]0,1[$ is a nilpotent element of T.*

T_L and T_P are Archimedean and each strict t-norm is isomorphic to T_P, each nilpotent t-norm is isomorphic to T_L.

Definition 4. *A unary operation $n : [0,1] \to [0,1]$ is called a negation if it is decreasing and compatible with the classical logic, i.e., $n(0) = 1$ and $n(1) = 0$. A negation is strict if it is strictly decreasing and continuous. A negation is strong if it is involutive, i.e., $n(n(x)) = x$ for all $x \in [0,1]$.*

Definition 5. *Let C be a conjunctor and n be a strong negation on $[0,1]$. A disjunctor C^* defined by for all $x, y \in [0,1]$,*

$$C^*(x,y) = n(C(n(x), n(y)))$$

is call the n-dual of C.

Definition 6. *Let C be a conjunctor and $\varphi : [0,1] \to [0,1]$ be an increasing bijection. A conjunctor C_φ defined by for all $x, y \in [0,1]$,*

$$C_\varphi(x,y) = \varphi^{-1}(C(\varphi(x), \varphi(y)))$$

is said to be $\varphi-$ isomorphic to C.

The conjunctors C and C_φ are isomorphic to each other.

Definition 7 [21]. *A conjunctor $C : [0,1]^2 \to [0,1]$ is called a quasi-copula if it satisfies 1-Lipschitz property: for any $x_1, x_2, y_1, y_2 \in [0,1]$ it holds that*

$$|C(x_1,y_1) - C(x_2,y_2)| \le |x_1 - x_2| + |y_1 - y_2|$$

Definition 8. *If $(C_i)_{i \in I}$ is a family of conjunctors and $(]a_i, b_i[)_{i \in I}$ is a family of non-empty, pairwise disjoint open subintervals of $[0,1]$. Then the ordinal sum $C = (\langle a_i, b_i, C_i \rangle)_{i \in I} : [0,1]^2 \to [0,1]$ is a conjunctor defined by*

$$C(x,y) = \begin{cases} a_i + (b_i - a_i) C_i\left(\frac{x-a_i}{b_i-a_i}, \frac{y-a_i}{b_i-a_i}\right) & (x,y) \in [a_i, b_i]^2, \\ \min(x,y) & \text{otherwise.} \end{cases}$$

We refer to the triplets $\langle a_i, b_i, C_i\rangle$ as the summands of the ordinal sum, to the intervals $[a_i, b_i]$ as its summand carriers and to the conjunctors C_i as its summand operations.

Definition 9. *A conjunctor C that has no ordinal sum representation different from $(\langle 0, 1, C\rangle)$ is called ordinally irreducible.*

Now we recall the definition of (weak) dominance about two conjunctors.

Definition 10. *Consider two conjunctors C_1 and C_2. We say that C_1 dominates C_2, and denoted by $C_1 \gg C_2$, if for all $x_1, x_2, y_1, y_2 \in [0, 1]$ it holds that*

$$C_2(C_1(x_1, y_2), C_1(x_2, y_1)) \leq C_1(C_2(x_1, x_2), C_2(y_1, y_2)). \tag{1}$$

Definition 11. *Consider two conjunctors C_1 and C_2. We say that C_1 weakly dominates C_2, and denoted by $C_1 >> C_2$, if for all $x_1, x_2, y_1 \in [0, 1]$ it holds that*

$$C_2(x_1, C_1(x_2, y_1)) \leq C_1(C_2(x_1, x_2), y_1). \tag{2}$$

On the one hand, the weak dominance relation can be viewed as a generalization of the dominance relation. It was applied to the construction of fuzzy equivalence relations, fuzzy orderings and the open problem about the transitivity. On the other hand, weak dominance can also be treated as an inequality generalization of modularity equation. It plays an important role in fuzzy sets and fuzzy logic theory.

Remark 2. (i) Let C_1 and C_2 be two conjunctors. If $C_1 \gg C_2$, then $C_1 >> C_2$. Indeed, due to the fact that 1 is the common neutral element of all conjunctors, by setting $y_2 = 1$ in (1), we have the conclusion. Therefore, the weak dominance relation can be regarded as a generalization of the dominance relation.

(ii) Since 1 is the common neutral element of all conjunctors, the weak dominance relation between two conjunctors implies their comparability: $C_1 >> C_2$ implies $C_1 \geq C_2$ by setting $x_2 = 1$ in (1). Obviously, the converse does not hold.

(iii) According to the monotonicity of conjunctors, it is clear that T_M weakly dominates any conjunctor C. Conversely, since weak dominance implies comparability, T_M is the only conjunctor weakly dominating T_M. Moreover, it is easily verified that the weakest conjunctor T_D is weakly dominated by any conjunctor C.

3 Duality and Isomorphism on the Weak Dominance

In this section, we discuss the weak dominance relation in the sense of duality and isomorphism.

Theorem 1. *Let C_1 and C_2 be conjunctors on $[0, 1]$, C_1^* and C_2^* be n-dual to them respectively. If C_1 and C_2 are commutative, then C_1 weakly dominates C_2 if and only if C_2^* weakly dominates C_1^*.*

Proof. Suppose that conjunctor C_1 weakly dominates C_2, i.e., for all $x_1, x_2, y_1 \in [0,1]$ it holds that

$$C_2(x_1, C_1(x_2, y_1)) \leq C_1(C_2(x_1, x_2), y_1).$$

Let $x_1, x_2, y_1 \in [0,1]$, then, by using the assumptions of C_1, C_2, C_1^* and C_2^*, we obtain

$$\begin{aligned}
C_1^*(x_1, C_2^*(x_2, y_1)) &= n(C_1(n(x_1), n(C_2^*(x_2, y_1)))) \\
&= n(C_1(n(x_1), n(n(C_2(n(x_2), n(y_1)))))) \\
&= n(C_1(n(x_1), C_2(n(x_2), n(y_1)))) \\
&= n(C_1(C_2(n(x_2), n(y_1)), n(x_1))) \\
&\leq n(C_2(n(y_1), C_1(n(x_2), n(x_1)))) \\
&= n(C_2(n(y_1), n(C_1^*(x_2, x_1)))) \\
&= C_2^*(y_1, C_1^*(x_2, x_1)) \\
&= C_2^*(C_1^*(x_1, x_2), y_1).
\end{aligned}$$

Thus, C_2^* weakly dominates C_1^*. The converse implication is obvious since $(C_1^*)^* = C_1$ and $(C_2^*)^* = C_2$.

Theorem 2. *Let C_1 and C_2 be conjunctors on $[0,1]$. φ is an increasing bijection and $(C_1)_\varphi, (C_2)_\varphi$ are their $\varphi-$isomorphism. Then C_1 weakly dominates C_2 if and only if $(C_1)_\varphi$ weakly dominates $(C_2)_\varphi$.*

Proof. Suppose that conjunctor C_1 weakly dominates C_2, i.e., for all $x_1, x_2, y_1 \in [0,1]$ it holds that

$$C_2(x_1, C_1(x_2, y_1)) \leq C_1(C_2(x_1, x_2), y_1)$$

Let $x_1, x_2, y_1 \in [0,1]$, φ be increasing. We obtain

$$\begin{aligned}
(C_2)_\varphi(x_1, (C_1)_\varphi(x_2, y_1)) &= \varphi^{-1}(C_2(\varphi(x_1), \varphi((C_1)_\varphi(x_2, y_1)))) \\
&= \varphi^{-1}(C_2(\varphi(x_1), \varphi(\varphi^{-1}(C_1(\varphi(x_2), \varphi(y_1)))))) \\
&= \varphi^{-1}(C_2(\varphi(x_1), C_1(\varphi(x_2), \varphi(y_1)))) \\
&\leq \varphi^{-1}(C_1(C_2(\varphi(x_1), \varphi(x_2)), \varphi(y_1))) \\
&= \varphi^{-1}(C_1(\varphi(\varphi^{-1}(C_2(\varphi(x_1), \varphi(x_2))))), \varphi(y_1)) \\
&= \varphi^{-1}(C_1(\varphi((C_2)_\varphi(x_1, x_2)), \varphi(y_1))) \\
&= (C_1)_\varphi((C_2)_\varphi(x_1, x_2), y_1).
\end{aligned}$$

Thus, $(C_1)_\varphi$ weakly dominates $(C_2)_\varphi$.

4 Weak Dominance Between Conjunctor and T-Norm

In this section, we discuss the weak dominance between a conjunctor and some t-norms.

Proposition 1. *Let C be a commutative conjunctor. If C weakly dominates T_L, then it is a quasi-copula.*

Proof. Suppose that a conjunctor C weakly dominates T_L, i.e. for all $x_1, x_2, y_1 \in [0,1]$ it holds that

$$T_L(x_1, C(x_2, y_1)) \leq C(T_L(x_1, x_2), y_1). \tag{3}$$

Then we need to prove that C satisfies the 1-Lipschitz property, i.e.,

$$C(a,b) - C(a-\varepsilon, b-\delta) \leq \varepsilon + \delta$$

whenever $a, b \in [0,1]$, $0 \leq \varepsilon \leq a$, $0 \leq \delta \leq b$. Next, we set $x_1 = 1-\varepsilon$, $x_2 = a$, $y_1 = b$ in (3), we have

$$\begin{aligned} C(a-\varepsilon, b) &= C(T_L(1-\varepsilon, a), b) \\ &\geq T_L(1-\varepsilon, C(a,b)) \\ &= \max(C(a,b)-\varepsilon, 0) \\ &= C(a,b) - \varepsilon. \end{aligned}$$

Analogously, by putting $x_1 = 1-\delta$, $x_2 = b$, $y_1 = a$ in (3) and according to the commutativity of C, we obtain

$$C(a, b-\delta) \geq C(a,b) - \delta.$$

Therefore,

$$C(a-\varepsilon, b-\delta) \geq C(a-\varepsilon, b) - \delta \geq C(a,b) - \varepsilon - \delta.$$

As a consequence, C satisfies 1-Lipschitz property and C is a quasi-copula.

Example 1. Let conjunctor $C = T_M$. It is obvious that C weakly dominates T_L and C satisfies 1-Lipschitz property.

The 1-Lipschitz property is a necessary condition for a conjunctor to weak dominate T_L. It is note that the same condition applies for a conjunctor to weakly dominate T_P.

Proposition 2. *Let C be a commutative conjunctor. If C weakly dominates T_P, then it is a quasi-copula.*

Proof. Suppose that a conjunctor C weakly dominates T_P, i.e. for all $x_1, x_2, y_1 \in [0,1]$ it holds that

$$x_1 C(x_2, y_1)) \leq C(x_1 x_2, y_1). \tag{4}$$

Then we have to prove that C satisfies the 1-Lipschitz property, according to the increasingness of C, we need to show that

$$C(a,b) - C(a-\varepsilon, b-\delta) \leq \varepsilon + \delta$$

whenever $a, b \in [0,1]$, $0 \leq \varepsilon \leq a$, $0 \leq \delta \leq b$. We divide the proof into two cases:

1. If $a = 0$ (resp. $b = 0$), it holds that $\varepsilon = 0$ (resp. $\delta = 0$), and the inequality is trivially fulfilled.
2. If $a > 0, b > 0$, we set $x_1 = 1 - \frac{\varepsilon}{a}$, $x_2 = a$, $y_1 = b$ in (4),

$$C(a - \varepsilon, b) \geq (1 - \frac{\varepsilon}{a})C(a, b).$$

Since $C \leq T_M$, for arbitrary $0 \leq a \leq 1$, $0 \leq b \leq 1$ and $0 \leq \varepsilon \leq a$, it holds that

$$\begin{aligned} C(a,b) - C(a-\varepsilon,b) &\leq C(a,b)(1-(1-\frac{\varepsilon}{a})) \\ &= \frac{\varepsilon}{a}C(a,b) \leq \varepsilon. \end{aligned}$$

Analogously, by putting $x_1 = 1 - \frac{\delta}{b}$, $x_2 = b$, $y_1 = a$ in (4) and according to the commutativity of C, we obtain

$$C(b - \delta, a) \geq (1 - \frac{\delta}{b})C(a, b).$$

Since $C \leq T_M$, we conclude that

$$\begin{aligned} C(a,b) - C(a,b-\delta) &\leq C(a,b)(1-(1-\frac{\delta}{b})) \\ &= \frac{\delta}{b}C(a,b) \leq \delta. \end{aligned}$$

As a consequence,

$$\begin{aligned} C(a,b) - C(a-\varepsilon,b-\delta) &= C(a,b) - C(a,b-\delta) + C(a,b-\delta) - C(a-\varepsilon,b-\delta) \\ &\leq \varepsilon + \delta \end{aligned}$$

Hence, C satisfies 1-Lipschitz property and C is a quasi-copula.

5 Weak Dominance Between Ordinal Sum Conjunctors

In this section, we discuss the weak dominance relation between two ordinal sum conjunctors.

Proposition 3. *Consider two ordinal sum conjunctors* $C_1 = (\langle a_i, b_i, C_{1,i}\rangle)_{i\in I}$ *and* $C_2 = (\langle a_i, b_i, C_{2,i}\rangle)_{i\in I}$. *Then* C_1 *weakly dominates* C_2 *if and only if* $C_{1,i}$ *weakly dominates* $C_{2,i}$*, for all* $i \in I$.

Proof. Suppose that conjunctor C_1 weakly dominates C_2, i.e., for all $x_1, x_2, y_1 \in [0, 1]$ it holds that

$$C_2(x_1, C_1(x_2, y_1)) \leq C_1(C_2(x_1, x_2), y_1). \tag{5}$$

Choose arbitrary $x_1, x_2, y_1 \in [0, 1]$ and some $i \in I$. Since $\varphi_i : [a_i, b_i] \to [0, 1]$, $x \to \frac{x-a_i}{b_i-a_i}$ is an increasing bijection, there exist unique $x_1', x_2', y_1' \in [a_i, b_i]$ such that $\varphi_i(x_1') = x_1$, $\varphi_i(x_2') = x_2$ and $\varphi_i(y_1') = y_1$. According to the ordinal sum structure of C_1 and C_2, for $x_1', x_2', y_1' \in [a_i, b_i]$, Eq. (5) can be equivalently expressed as

$$\varphi_i^{-1} \circ C_{2,i}(\varphi_i(x_1'), C_{1,i}(\varphi_i(x_2'), \varphi_i(y_1'))) \leq \varphi_i^{-1} \circ C_{1,i}(C_{2,i}(\varphi_i(x_1'), \varphi_i(x_2')), \varphi_i(y_1')).$$

The above inequality is equivalent to

$$\varphi_i^{-1} \circ C_{2,i}(x_1, C_{1,i}(x_2, y_1)) \leq \varphi_i^{-1} \circ C_{1,i}(C_{2,i}(x_1, x_2), y_1).$$

Applying φ_i to both sides of the above inequality yields $C_{2,i} << C_{1,i}$.

Conversely, suppose that for all $i \in I$ it holds that $C_{2,i} << C_{1,i}$. Obviously, Eq. (5) is satisfied for all $x_1, x_2, y_1 \in [a_i, b_i]$ due to the isomorphism property. Next, for any $x_1, x_2, y_1 \in [0, 1]$, we can distinguish the following cases.

1. $x_2 = \min(x_1, x_2, y_1)$ and $y_1, x_2 \in [a_i, b_i]$ for some $i \in I$. Let $x_1^* = \min(x_1, b_i) \in [a_i, b_i]$. Then

$$\begin{aligned} C_2(x_1, C_1(x_2, y_1)) &= C_2(x_1^*, C_1(x_2, y_1)) \\ &\leq C_1(C_2(x_1^*, x_2), y_1) \\ &\leq C_1(C_2(x_1, x_2), y_1). \end{aligned}$$

2. $x_2 = \min(x_1, x_2, y_1)$ and $x_2 \in [a_i, b_i]$ for some $i \in I$ and $y_1 \notin [a_i, b_i]$ for any $i \in I$. Then $y_1 > b_i$,

$$C_1(x_2, y_1) = \min(x_2, y_1) = x_2,$$

and

$$\begin{aligned} C_2(x_1, C_1(x_2, y_1)) &= C_2(x_1, x_2) \\ &= \min(C_2(x_1, x_2), y_1) \\ &= C_1(C_2(x_1, x_2), y_1). \end{aligned}$$

3. $x_2 = \min(x_1, x_2, y_1)$ and $x_2 \notin [a_i, b_i]$ for any $i \in I$. Then

$$\begin{aligned} C_2(x_1, x_2) &= \min(x_1, x_2) \\ &= x_2 C_1(x_2, y_1) \\ &= \min(x_2, y_1) \\ &= x_2, \end{aligned}$$

and

$$\begin{aligned} C_2(x_1, C_1(x_2, y_1)) &= C_2(x_1, \min(x_2, y_1)) \\ &= C_2(x_1, x_2) \\ &= x_2 \\ &= C_1(x_2, y_1) \\ &= C_1(\min(x_1, x_2), y_1) \\ &= C_1(C_2(x_1, x_2), y_1). \end{aligned}$$

4. $x_1 = \min(x_1, x_2, y_1)$ and $x_1, x_2 \in [a_i, b_i]$ for some $i \in I$. Let $y_1^* = \min(x_1, b_i) \in [a_i, b_i]$. Then

$$\begin{aligned}C_2(x_1, C_1(x_2, y_1)) &= C_2(x_1, C_1(x_2, y_1^*))\\ &\leq C_1(C_2(x_1, x_2), y_1^*)\\ &\leq C_1(C_2(x_1, x_2), y_1).\end{aligned}$$

5. $x_1 = \min(x_1, x_2, y_1)$ and $x_1 \in [a_i, b_i]$ for some $i \in I$ and $x_2 \notin [a_i, b_i]$ for any $i \in I$. Then $C_2(x_1, x_2) = \min(x_1, x_2) = x_1$.
   If $y_1 \in [a_i, b_i]$, then

$$\begin{aligned}C_2(x_1, C_1(x_2, y_1)) &= C_2(x_1, C_1(b_i, y_1))\\ &\leq C_1(C_2(x_1, b_i), y_1))\\ &= C_1(C_2(x_1, x_2), y_1).\end{aligned}$$

   If $y_1 > b_i$, then

$$\begin{aligned}C_2(x_1, C_1(x_2, y_1)) &\leq x_1\\ &= C_1(x_1, y_1)\\ &= C_1(C_2(x_1, x_2), y_1).\end{aligned}$$

6. $x_1 = \min(x_1, x_2, y_1)$ and $x_1 \notin [a_i, b_i]$ for any $i \in I$. Then

$$\begin{aligned}C_2(x_1, C_1(x_2, y_1)) &= \min(x_1, C_1(x_2, y_1))\\ &\leq x_1\\ &= C_1(x_1, y_1)\\ &= C_1(\min(x_1, x_2), y_1)\\ &= C_1(C_2(x_1, x_2), y_1).\end{aligned}$$

7. $y_1 = \min(x_1, x_2, y_1)$. The proof is similar to that of cases 1, 2 and 3.

   This completes the proof that C_1 weakly dominates C_2.

Example 2. Consider two ordinal sum conjunctors $C_1 = (\langle 0, \frac{1}{2}, T_3\rangle, \langle \frac{1}{2}, 1, T_2\rangle)$ and $C_2 = (\langle 0, \frac{1}{2}, T_2\rangle, \langle \frac{1}{2}, 1, T_1\rangle)$ where T_α is defined by

$$T_\alpha(x, y) = \left[\max(x^{-\alpha} + y^{-\alpha} - 1, 0)\right]^{-1/\alpha}, \alpha = 1, 2, 3.$$

By the computation, we can demonstrate that $C_1 >> C_2$. Moreover, according to Theorem 4.2.10 in [19] and Remark 2.(i), $T_3 >> T_2$, $T_2 >> T_1$.

Proposition 4. *If a conjunctor C_1 weakly dominates a conjunctor C_2, then $C_1 = T_M$ whenever $C_2 = T_M$.*

Proof. Suppose that conjunctor C_1 weakly dominates C_2, i.e., for all $x_1, x_2, y_1 \in [0, 1]$ it holds that

$$C_2(x_1, C_1(x_2, y_1)) \leq C_1(C_2(x_1, x_2), y_1).$$

By putting $x_1 = 1$, we have

$$T_M(x_1, C_1(1, y_1)) \leq C_1(T_M(x_1, 1), y_1),$$

i.e., $T_M(x_1, y_1) \leq C_1(x_1, y_1)$.

Since for any conjunctor $C \leq T_M$, we obtain $C_1(x_1, y_1) \leq T_M(x_1, y_1)$. So $C_1 = T_M$.

Corollary 1. *Let C_1 and C_2 be two conjunctors. If C_1 weakly dominates C_2, then $\mathrm{Im}(C_2) \subseteq \mathrm{Im}(C_1)$, where $\mathrm{Im}(C_i)$ is the set of idempotent elements of $C_i, i = 1, 2$.*

Corollary 2. *Let $C_1 = (\langle a_i, b_i, C_{1,i}\rangle)_{i\in I}$ and $C_2 = (\langle a_j, b_j, C_{2,j}\rangle)_{j\in J}$ be two ordinal sum conjunctors with ordinally irreducible summand operations only. If C_1 weakly dominates C_2 then for any $i \in I$, there exists $j \in J$, such that*

$$[a_{1,i}, b_{1,i}] \subseteq [a_{2,j}, b_{2,j}]. \tag{6}$$

Hence, for each $j \in J$ let us consider the following subset of I:

$$I_j = \{i \in I | [a_{1,i}, b_{1,i}] \subseteq [a_{2,j}, b_{2,j}]\}. \tag{7}$$

Based on Proposition 4, we have the following result about the weak dominance relation between two ordinal sum conjunctors with different summand carriers.

Proposition 5. *Let $C_1 = (\langle a_{1i}, b_{1i}, C_{1,i}\rangle)_{i\in I}$ and $C_2 = (\langle a_{2j}, b_{2j}, C_{2,j}\rangle)_{j\in J}$ be two ordinal sum conjunctors with ordinally irreducible summand operations only. Then C_1 weakly dominates C_2 if and only if*

(i) $\cup_{j\in J} I_j = I$,
(ii) $C_1^j >> C_{2,j}$ *for all* $j \in J$ *with*

$$C_1^j = (\langle \varphi_j(a_{1,i}), \varphi_j(b_{1,i}), C_{1,i}\rangle)_{i\in I_j} \tag{8}$$

and $\varphi_j : [a_{2,j}, b_{2,j}] \to [0, 1]$, $\varphi_j(x) = \frac{x - a_{2,j}}{b_{2,j} - a_{2,j}}$.

Proof. Under condition (i) we can prove that C_1 can be formulated as an ordinal sum based on the summand carriers of C_2 in the following way

$$C_1 = (\langle a_{2,j}, b_{2,j}, C_1^j\rangle)_{j\in J} \tag{9}$$

where C_1^j is defined by Eq. (8). For any $x, y \in [0, 1]^2$, we divide the proof into two cases:

1. If x or $y \notin [a_{1i}, b_{1i}]$ for arbitrary $i \notin I$, then due to the definition of ordinal sum conjunctor, we have

$$C_1(x, y) = T_M(x, y).$$

According to Eq. (9), we have

$$C_1(x, y) = T_M(x, y).$$

2. If $x, y \in [a_{1i}, b_{1i}]$, for some $i \in I$, then by condition (i), there exists $j \in J$, such that $i \in I_j$, according to the definition of ordinal sum conjunctor, we have

$$C_1(x,y) = a_{1i} + (b_{1i} - a_{1i})C_{1,i}(\frac{x - a_{1i}}{b_{1i} - a_{1i}}, \frac{y - a_{1i}}{b_{1i} - a_{1i}}).$$

According to Eq. (9), we have

$$C_1(x,y) = a_{2j} + (b_{2j} - a_{2j})C_1^j(\frac{x - a_{2j}}{b_{2j} - a_{2j}}, \frac{y - a_{2j}}{b_{2j} - a_{2j}})$$

and $C_1^j(\frac{x-a_{2j}}{b_{2j}-a_{2j}}, \frac{y-a_{2j}}{b_{2j}-a_{2j}})$

$$= \frac{a_{1i} - a_{2j}}{b_{2j} - a_{2j}} + \left(\frac{b_{1i} - a_{2j}}{b_{2j} - a_{2j}} - \frac{a_{1i} - a_{2j}}{b_{2j} - a_{2j}}\right) C_{1,i}\left(\frac{\frac{x-a_{2j}}{b_{2j}-a_{2j}} - \frac{a_{1i}-a_{2j}}{b_{2j}-a_{2j}}}{\frac{b_{1i}-a_{2j}}{b_{2j}-a_{2j}} - \frac{a_{1i}-a_{2j}}{b_{2j}-a_{2j}}}, \frac{\frac{y-a_{2j}}{b_{2j}-a_{2j}} - \frac{a_{1i}-a_{2j}}{b_{2j}-a_{2j}}}{\frac{b_{1i}-a_{2j}}{b_{2j}-a_{2j}} - \frac{a_{1i}-a_{2j}}{b_{2j}-a_{2j}}}\right)$$

$$= \frac{a_{1i} - a_{2j}}{b_{2j} - a_{2j}} + \left(\frac{b_{1i} - a_{2j}}{b_{2j} - a_{2j}} - \frac{a_{1i} - a_{2j}}{b_{2j} - a_{2j}}\right) C_{1,i}(\frac{x - a_{1i}}{b_{1i} - a_{1i}}, \frac{y - a_{1i}}{b_{1i} - a_{1i}}).$$

Hence,

$$C_1(x,y) = a_{1i} + (b_{1i} - a_{1i})C_{1,i}(\frac{x - a_{1i}}{b_{1i} - a_{1i}}, \frac{y - a_{1i}}{b_{1i} - a_{1i}}).$$

Based on Corollary 2 and Proposition 4, we can immediately conclude the proposition.

Example 3. Consider two conjunctors $C_1 = (\langle 0, \frac{1}{4}, T_3\rangle, \langle \frac{1}{4}, \frac{1}{2}, T_2\rangle, \langle \frac{1}{2}, 1, T_4\rangle)$ and $C_2 = (\langle 0, \frac{1}{2}, T_2\rangle, \langle \frac{1}{2}, 1, T_1\rangle)$ where T_α is defined by

$$T_\alpha(x,y) = \left[\max(x^{-\alpha} + y^{-\alpha} - 1, 0)\right]^{-1/\alpha}, \alpha = 1,2,3,4.$$

By the computation, we can demonstrate that $C_1 >> C_2$. By the notation in Eq. (7) and Proposition 5, we have $I_1 = \{1,2\}, I_2 = \{3\}$ and $I = \{1,2,3\}$. Furthermore, $C_1^1 = (\langle 0, \frac{1}{2}, T_3\rangle, \langle \frac{1}{2}, 1, T_2\rangle), C_1^2 = (\langle \frac{1}{2}, 1, T_4\rangle)$ and $C_1^1 >> T_2, C_1^2 >> T_1$ according to Theorem 4.2.10 in [19] and Remark 2.(i).

6 Conclusion

In this paper, we deal with the weak dominance relation between the classes of conjunctors: the class of triangular norms and the class of quasi-copulas. Furthermore, we also provide the sufficient and necessary conditions of weak dominance relation between two ordinal sum conjunctors.

Acknowledgements. This work is supported by National Nature Science Foundation of China under Grant 61977040 and Natural Science Foundation of Shandong Province under Grant ZR2019MF055.

References

1. Nguyen, H.T., Walker, E.A.: A First Course in Fuzzy Logic, 3 edn. DBLP (2005)
2. Běhounek, L., et al.: Graded dominance and related graded properties of fuzzy conncectives. Fuzzy Sets Syst. **262**, 78–101 (2015)
3. Tardiff, R.M.: Topologies for probabilistic metric spaces. Pacific J. Math. **65**(1), 233–251 (1976)
4. Schweizer, B., Sklar, A.: Probabilistic Metric Spaces. North-Holland Series in Probability and Applied Mathematics, North-Holland Publishing Co., New York (1983)
5. Saminger, S., Mesiar, R., Bodenhofer, U.: Domination of aggregation operators and preservation of transitivity. Internat. J. Uncertain. Fuzziness Knowl.-Based Syst. **10**(1), 11–35 (2002)
6. Bodenhofer, U.: A similarity-based generalization of fuzzy orderings. Universitätsverlag Rudolf Trauner Linz, Austria (1999)
7. De Baets, B., Mesiar, R.: T-partitions. Fuzzy Sets Syst. **97**(2), 211–223 (1998)
8. Dlłaz, S., Montes, S., De Baets, B.: Transitivity bounds in additive fuzzy preference structures. IEEE Trans. Fuzzy Syst. **15**(2), 275–286 (2007)
9. Saminger, S., De Baets, B., De Meyer, H.: On the dominance relation between ordinal sums of conjunctors. Kybernetika **42**, 337–350 (2006)
10. Saminger, S.: The dominance relation in some families of continuous Archimedean t-norms and copulas. Fuzzy Sets Syst. **160**, 2017–2031 (2009)
11. Sarkoci, P.: Dominance is not transitive on continuous triangular norms. Aequationes Math. **75**(3), 201–207 (2008). https://doi.org/10.1007/s00010-007-2915-5
12. Petrílk, M.: Dominance on strict triangular norms and Mulholland inequality. Fuzzy Sets Syst. **335**, 3–17 (2018)
13. Petrlłk, M.: Dominance on continuous Archimedean triangular norms and generalized Mulholland inequality. Fuzzy Sets Syst. **403**, 88–100 (2021)
14. Drewniak, J., Drygaś, P., Dudziak, U.: Relation of domination, Abstracts FSTA, pp. 43–44 (2004)
15. Drewniak, J., Król, A.: On the problem of domination between triangular norms and conorms. J. Elect. Eng. **56**(12), 59–61 (2005)
16. Bentkowska, U., Drewniak, J., Drygaś, P., Król, A., Rak, E.: Dominance of binary operations on posets. In: Atanassov, K.T., et al. (eds.) IWIFSGN 2016. AISC, vol. 559, pp. 143–152. Springer, Cham (2018). https://doi.org/10.1007/978-3-319-65545-1_14
17. Sarkoci, P.: Conjunctors dominating classes of t-conorms. In: International Conference on fuzzy Sets Theory and its Applications, FSTA (2006)
18. Klement, E.P., Mesiar, R., Pap, E.: Triangular Norms. Springer, Heidelberg (2013)
19. Alsina, C., Schweizer, B., Frank, M.J.: Associative Functions: Triangular Norms and Copulas. World Scientific, Singapore (2006)
20. Su, Y., Riera, J.V., Ruiz-Aguilera, D., Torrens, J.: The modularity condition for uninorms revisted. Fuzzy Sets Syst. **357**, 27–46 (2019)
21. Saminger, S., De Baets, B., De Meyer, H.: On the dominance relation between ordinal sums of conjunctors. Kybernetika **42**(3), 337–350 (2006)

New Modification to Toulmin Model as an Analytical Framework for Argumentative Essays

Donghong Liu(✉)

Southeast University, Nanjing 211189, China
101013190@seu.edu.cn

Abstract. The Toulmin model has provided a clear and flexible set of categories for conducting research on both oral and written argumentation. All kinds of modifications of the model contribute much to the application of Toulmin model to various genres. However, there is still a lack of an appropriate Toulmin framework for argumentative essays because of the deficiencies in those modified models. In this paper a synthesis of modifications to Toulmin model is proposed for analyzing argumentative essays not only to update the Toulmin model, but also to broaden the scope of its application. It reveals justification depth by displaying the hierarchical relationship in arguments, merges Backing into Warrant considering the nature of actual writing. The new model takes in the merits of the previous modified models, clarifies the vagueness of Warrant and avoid the deficiencies of the previous models so that it can explain argumentative essays more efficiently.

Keywords: Toulmin · Argumentation · Syllogism · Warrant · Data

1 Introduction

Data science and artificial intelligence (AI) based on logic have made great progress. In the meantime problems such as weak accountability and inadequate inference, arise in the new generation of AI that is centered on big data and machine learning. To solve the problems, interdisciplinary research is necessary. For example, AI can solve the problems in inference by reference to logic theories. "The combination of formal argumentation with existing big data and machine learning technologies can be expected to break through existing technical bottle necks to some extent" [1]. We contend that the combination is not restricted to formal argumentation. Informal argumentation (e.g. Toulmin theory) combined with AI and data science will also play a critical role in the exploration of argumentative essays.

Toulmin [2] put forward his initial three-component model of argumentation in 1958 and revised it by adding another three triad. The famous six-component model, Toulmin model, is quite influential and applied to many research areas such as court debate, rhetoric, philosophy, medicine, science, first and second language argumentation instruction. Various modifications of Toulmin model emerge in the process of the wide

Y. Chen and S. Zhang (Eds.): AILA 2022, CCIS 1657, pp. 211–224, 2022.
https://doi.org/10.1007/978-981-19-7510-3_16

application, such as Crammond's [3], Qin and Karabacak [4], Voss [5] and Jackson and Schneider [6]. However, deficiencies of these modified versions and discrepancies of views on certain Toulmin components, for example Warrant, are more and more obvious and unavoidable. Solutions have to be found.

This paper attempts discussing several influential modified models, analyzing their strong points as well as the weaknesses, and putting forward a new model as an analytical framework especially for argumentative essays.

2 Toulmin Model of Argumentation

Toulmin [2] invented a variation of categorical syllogism to deal with the deficiencies of syllogistic reasoning. He contended that the syllogism failed to represent the very nature of argument because of its arbitrary restriction to a three-part structure. In fact, most arguments have a more complex structure than the syllogism [2].

Toulmin used formal logical demonstration as his point of departure. His initial model has only three basic components: data, claim and warrant. An argument begins with an accepted data, moves through a warrant and finally reaches a claim. According to Toulmin [2], Data (D) provides evidence for Claim (C). Warrant (W) certifies the claim as true and bridges the gap between Data and Claim. Claim is a conclusion and the final proposition in an argument. The two elements answer the following corresponding questions:

Data (D)—"What have you got to go on?".

Warrant (W)—"How do you get there?".

The argumentation sequence can be "Data, therefore Claim" or "Claim because Data". The initial model is in fact a variation of categorical syllogism. Toulmin then compensated the inadequacies of the three-component model with a second triad of components: Backing, Rebuttal, and Qualifier (see Fig. 1). Backing (B) supports the warrant and enables it convincing enough. Rebuttal (R) acknowledges certain conditions under which the claim does not hold water. Qualifier (Q) expresses the degree of certainty in the claim. Rebuttal and Qualifier in Toulmin model of argumentation anticipate the challenging questions raised by people of different opinions. Backing anticipates a challenge to the legitimacy of the warrant.

For example, "Li Ming is a Chinese and since a Chinese can be taken almost certainly not to be a Christian, Li Ming is not a Christian" [7]. A three-component argument goes in this way: "Li Ming is a Chinese" is the datum; "A Chinese can be taken almost certainly not to be a Christian" is the warrant, and "Li Ming is not a Christian" is the claim inferred via the warrant that licenses the inference. A six-component argument goes like this: "Li Ming is a Chinese; and a Chinese can be taken almost certainly not to be a Christian because Chinese children are educated to be atheists when they go to school; therefore, Li Ming is almost certainly not a Christian unless he has chosen by himself to believe in God" [7]. The six-component model illustrates the example in Fig. 2.

The Toulmin model is a great advance in argumentation. It is developed from traditional logic but differs from it. First, the claim in the Toulmin model is open-ended

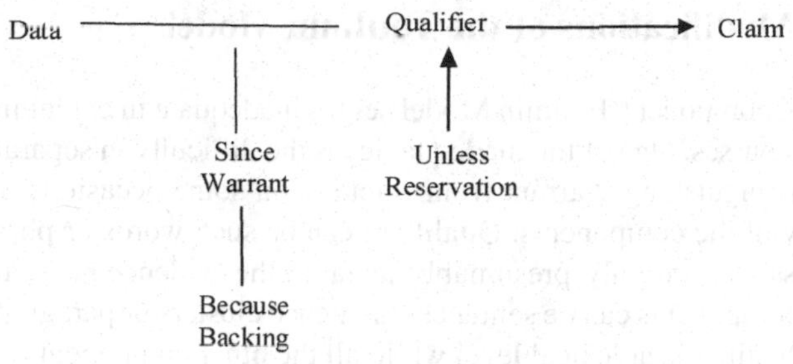

Fig. 1. Toulmin model

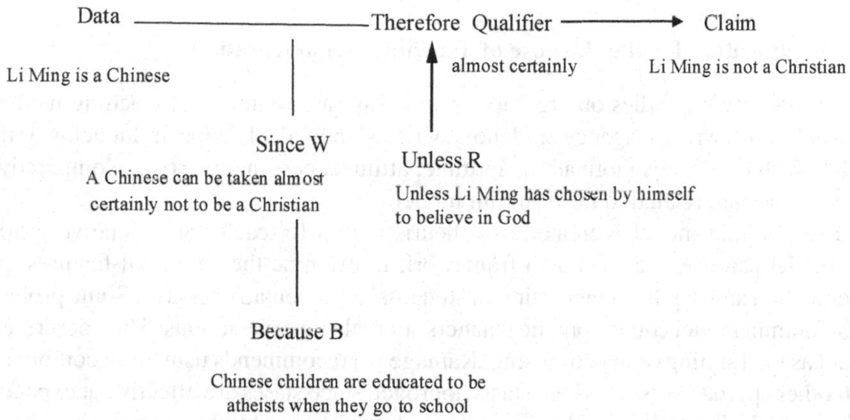

Fig. 2. Example of Li Ming in Toulmin model

and can be challenged while traditional logic produces uncontroversial claims by means of syllogism or enthymeme. Second, the Toulmin model conditions a claim by qualifiers and rebuttal instead of universal propositions that are preferred in traditional logic. Third, while traditional logic is weak in material proof, the Toulmin model puts weight on elements such as facts, evidence and statistics as the data.

The Toulmin model is considered as the most influential work of Toulmin, particularly in the field of rhetoric, logic and computer science. Just like Rhetorical Structure Theory (RST) that is fairly accepted and studied linguists, rhetoricians and computer scientists, the Toulmin model makes the process of argumentation more reasonable and transparent than the traditional syllogism.

3 Various Modifications of the Toulmin Model

The original six-component Toulmin Model seems inadequate in explaining and analyzing modern discourses. One of the inadequacies is the difficulty in separating Data from Warrant and distinguishing Warrant from Backing on some occasions. Another one is the discrepancy of the components. Qualifiers can be such words or phrases: probably, possible, impossible, certainly, presumably, as far as the evidence goes, and necessarily. The other five components can be sentences, sentence clusters or paragraphs. To put it in another way, Qualifier is at lexical level while all the other components are at discourse level. The Toulmin model has been modified by many researchers so as to facilitate the application for many purposes.

3.1 Modification for the Purpose of Teaching Argumentation

A vast majority of studies on argumentative essays are centered on teaching methods, feedback types, writers' agency or identity, writers' individual factors influencing writing quality such as anxiety, motivation, aptitude, attitude, personality, style. Comparatively fewer studies are related to the Toulmin model.

The Toulmin model is treated as a heuristic tool to teach argumentative writing. The model can even be used as a framework to examine the structural features, goal specification and depth of elaboration in students' argumentative essays. Some problems with Toulmin model come from the teachers' unrealistic expectations. They describe the model as too limiting or flat confusing. Ramage [8] recommends using it in combination with other approaches such as the stasis approach since stases are effective at expanding claims and the Toulmin model is efficient in sharpening and tightening claims.

Some of the Toulmin components have been modified by the textbooks for composition or resources for writing. In order to make these components more transparent to students, some components have been given different names. For example, "Purdue Owl" (Purdue University Online Writing Lab) uses different names for Claim (conclusion, opinion), Data (ground, evidence, reasons), Warrant (link, assumption) and Reservation (Rebuttal). The Warrant and Backing are hard to understand by the students. In order to facilitate argumentation instruction, Qin and Karabacak [4] even change the components by eliminating Warrant, Backing and Qualifier, but splitting Rebuttal into Counterargument claim, Counterargument data, Rebuttal claim and Rebuttal data. Their model consists of six components as is shown in Fig. 3. This model is adopted to teach argumentation or analyze the written work of argumentation, for example Abdollahzadeh et al. [9] employ this model to analyze Iran postgraduates' argumentation, exploring the relationship between the frequencies of the six components and the writing quality.

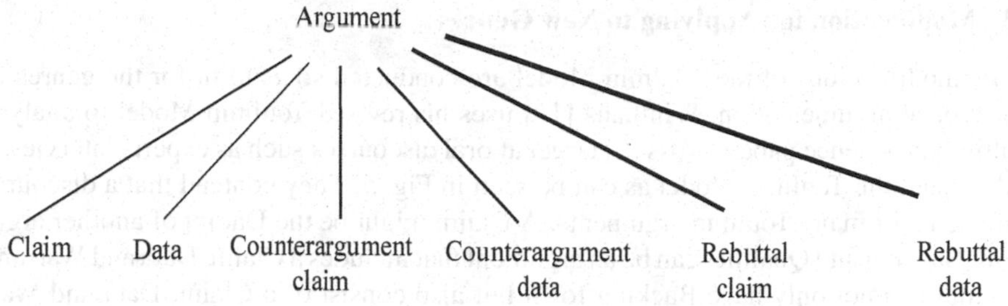

Fig. 3. Modified Toulmin model for teaching argumentative writing

3.2 Modification for the Purpose of Finer Analysis and In-Depth Study

New elements have been added and redefined. For one thing, to reveal the complicated nature of the modern argumentative essays written by all kinds of writers, and for another to make it more convenient to conduct comprehensive analyses. For instance, Crammond [3] expands the qualifier to include not only modality operators but also 'constraints', divides the component of backing into 'Warrant backing' and 'Data backing' and also recognizes possible 'Alternative solution' as well as "Countered rebuttal" and "Reservation". He classifies the components as necessary and optional. According to Crammond [3], "a Claim and the Data offered in support of this Claim are considered to be elements that are required or necessary for an argument structure. Warrants, along with the remaining substructures … are classified as being optional or elaborative". Figure 4 illustrates his idea. Even more and more recent empirical studies are being conducted within Crammond's [3] framework of the modified Toulmin model. The modified model of Crammond [3] is used in many studies such as Liu and Wan [10], Cheng and Chen [11].

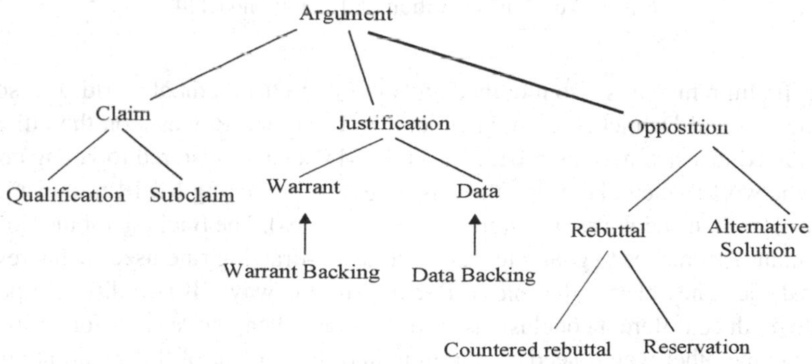

Fig. 4. Crammond's modification of Toulmin model

3.3 Modification for Applying to New Genres

Some modifications of the Toulmin Model are conducted so as to fit for the genres of non-typical argumentation. Whithaus [12] uses his revised Toulmin Model to analyze multimodal science papers. Voss [5] target at oral discourses such as experts interviews, and expand the Toulmin Model as can be seen in Fig. 5. They contend that a discourse is made up of many Toulmin arguments. A Claim might be the Datum of another argument. Backing and Qualifier can be an argument that includes a Claim, Data and Warrant. Rebuttal can not only have Backing for it but also consist of a Claim, Data and Warrant. However, this kind of modification of Toulmin Model is limited to certain type of discourse. Thus, it has narrow scope of application.

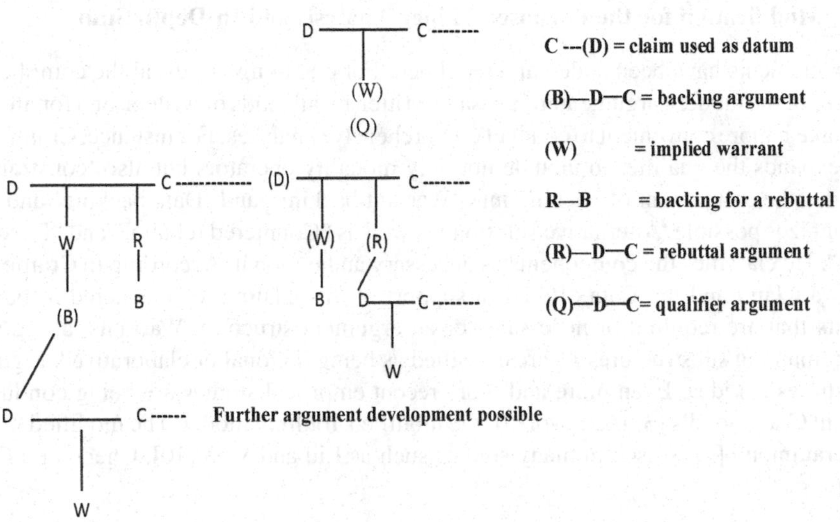

Fig. 5. Voss's modification of Toulmin model [4]

The Toulmin model is also modified and applied to the medical world. Jackson and Schneider [6] and Demicheli et al. [13] analyze Cochrane Reviews on the efficacy of the combined measles-mumps-rubella vaccine (MMR) administered to young children around the world. They claim that there is no evidence linking MMR to autism. They propose more than one Warrant (different inference rules). The Backing for the Cochrane rule contains assurances to guarantee the conclusion-drawing rule used by the reviewer is dependable. They clarify the role of Backing in this way: "If two different possible warrants produce different conclusions from the same data, the reasons for trusting one rather than the other would need exploration, and the content of this exploration is the backing for each of the alternatives' warrants" [6].

Jackson and Schneider [6] stress the field-specificity of Warrant. They argue that the purpose of a Warrant device is to offer a convincing conclusion to people who know the workings of the device well and have confidence in it. Cochrane Reviews have a well-defined context with a primary readership composed of medical experts. Warranting devices demand consideration of context, including not only the composition of the community where they emerge but also the state of working within that community. Therefore, in their model material, procedural and institutional assurances are included. Figure 6 presents a proposed analysis of the argument.

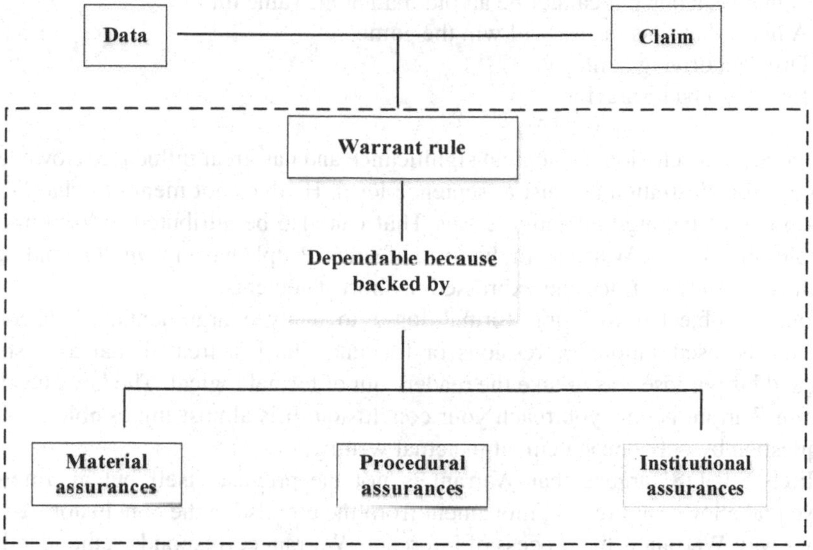

Fig. 6. The revised Toulmin model by Jackson and Schneider [5]

4 The Problem of Warrant

According to Toulmin [2], Warrant can be different kinds of propositions such as rules, principles, and licenses for inference. It is a bridge linking Data to Claim. The role of Warrant is to make sure that the procedure from Data to Claim is sound and valid. However, Toulmin's description and expression of Warrant is still general and vague. That causes many scholars to understand and presume in their own ways.

Some people consider Warrant as the major premise in syllogism, such as Warren [14], Deng [15], Jin and Zhao [16]. From this perspective, the example of Li Ming can be formulated in the following syllogism pattern:

Major premise: A Chinese can be taken almost certainly not to be a Christian.
Minor premise: Li Ming is a Chinese.
Conclusion: Li Ming is not a Christian.

However, Freeman [17] takes a different view. He contends that Warrant is determined by human intuition. He mentions four kinds of intuition: a priori intuition, empirical intuition, institutional intuition and evaluative intuition. The four kinds of intuition lead to four kinds of Warrants: a priori Warrant, empirical Warrant, institutional Warrant and evaluative Warrant. Take the following statements as examples. (A) is a priori

Warrant since it involves common sense knowledge. (B) is empirical Warrant since it depends on previous experience. (C) is institutional Warrant on account of the legal rules and (D) is evaluative Warrant because of its moral nature.

(A) A male is a boy but cannot be an old man at the same time.
(B) A horse that runs fastest will win the game.
(C) Drunken driving is illegal.
(D) Lying is a bad behavior.

Freeman's discussion has logical significance and has great influence. However, his examples for illustration are just at sentence level. He does not mention what Warrant is like in an actual argumentative essay. That can also be attributed to Toulmin's [2] examples that display Warrants in the form of rules. People have misunderstanding that Warrants can only be rules and expressed in short statements.

Toulmin objected to using formal logic to analyze argumentative discourses. Although he used simple expressions or formulas, he just treated that as a starting point and his purpose was to take the readers out of formal logical. The key question of Toulmin Warrant is how you reach your conclusion. It is almost impossible to answer this question by only one statement in actual writing.

Hitchcock [18] argues that Warrant is not the premise itself but an inference-license that allows an inferring movement from the premise to the conclusion. To put it another way, Warrant is the process of inference. Warrant corresponds semantically to a generalized formula:

"If P1, P2, P3 … Pn, then C"

Hitchcock [18] sticks to Toulmin's [1] consideration of warrant, i.e. inference license. Thus, his view is closest to Toulmin's initial description of Warrant, which answers the question—How do you get there? This view is supported by Jin [19] who applies Toulmin model to describe and display the inference process of Mo Zi, a great ancient Chinese thinker and philosopher. Liu [20] investigates the modern Chinese argumentation and supports the view as well.

5 Synthesized Model as a New Analytical Framework

In this section we put forward a new model for argumentative essays by synthesizing the modified models of Voss [5], Jackson and Schneider [6], Crammond [3] and Qin and Karabacak [4]. We take Crammond's [3] and Voss's [5] view of complicated hierarchical structure inside Toulmin model. The new model relies more heavily on Crammond [3], Qin and Karabacak [4] but avoid their inadequacies so as to fit for argumentative essays.

Despite the contribution of Crammond [3], Qin and Karabacak [4], their analytical frameworks have some inadequacies or infeasibilities. The first inadequacy is the exclusion of Warrant in Qin and Karabacak [4] which alters the Toulmin model. The important element Warrant makes up the defining features of an argument structure, together with Claim and Data. And according to Connor [21] and Ferris [22], Warrant plays a critical role in predicting the writing quality of an essay. Second, Qin and Karabacak [4] confuse

Data with Warrant. They define Data as "facts, statistics, anecdotes, research studies, expert opinions, definitions, analogies, and logical explanations" (pp. 449). However, according to Toulmin [2], Warrant is a bridge linking Data and Claim. Analogies and logical explanations can help the readers understand the inference process. Thus, they are more like Warrant rather than Data.

An infeasibility is that Crammond [3] divides Toulmin's [2] Rebuttal into "Alternative solution" and "Reservation." We speculate this would be too subtle to be applied to some of the argumentative essays. Qin and Karabacak [4] did not adopt this distinction in their framework probably out of the same reason. Another infeasibility involves the qualifier. In the original Toulmin model, Qualifier can be a word or a phrase. It differs greatly from the other elements that can be a sentence or sentence clusters at sentence level or at discourse level. It performs the function of moderating the tone in the conclusion. Considering the discrepancy with the other elements, Qualifier would be excluded in our present analytical framework.

As for Warrant in Toulmin model, we adopt Hitchcock's [18] view of Warrant and treat it as a process of inference. Only in this way can the Toulmin model, especially the important component—Warrant be applied to actual argumentative essays that are influenced greatly by context and culture. It is impossible to find a universal principle to explain or describe all kinds of warrant. This kind of argumentation in written discourse does not belong to formal logic but the argumentation in broad sense proposed by Ju and He [23]. That is to say, this kind of argumentation involves the knowledge in certain contexts and should conform to the pragmatic features in those contexts. In this sense, Warrant in a piece of discourse may be several sentences and even paragraphs, which are called "segmented discourse" by Asher [24]. A segmented discourse can be identified by means of logic relations, semantic structures, subject matters and rhetorical structures. It is the same case with the identification of Warrant in Toulmin model.

Moreover, we support Jackson and Schneider's [6] view that Warrant can be of different kinds and can be employed at the same time to connect Data with Claim. Moreover, we roughly agree with Freeman [14] in terms of his classification of Warrant. There are four kinds of Warrants: (A) a priori Warrant, (B) empirical Warrant, (C) institutional Warrant and (D) evaluative Warrant. However, we disagree with Freeman in the identification of Warrant. We hold the view that Warrant is a segmented discourse consisting of a sentence, or sentence clusters, or even paragraphs. The role of Warrant in an argumentative essay is to explain, assume, or comment and so on for the purpose of guiding the readers to get the writer's viewpoint correctly and accurately from the given data.

With all of the above mentioned in mind, we propose a new model for argumentative essays, which displays hierarchical relationship in logic and complexity in thought. The main argument structure consists of the claim, justification and opposition. Justification might include more than one argument that is basically made up of Subclaim, Data, and Warrant. The component warrant is kept in our analytical framework so that the basic Toulmin model is intact. The component Backing is merged into Warrant, considering that Warrant has more significance than Backing [22] and that it is rather too difficult to distinguish the two elements [6]. Merging the two components is also conducive to

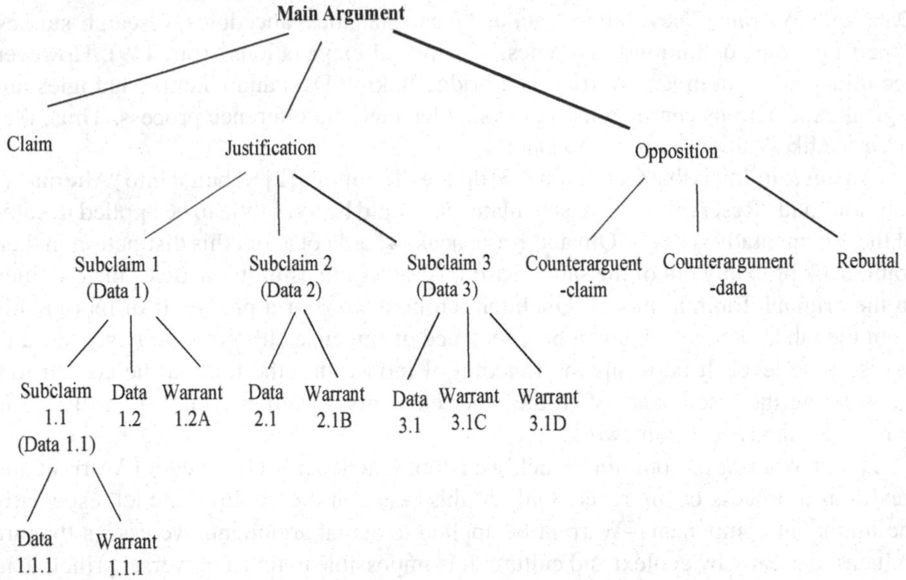

Fig. 7. Synthesized model for argumentative essays

coding in quantitative research. Warrant can be of various kinds. There can be more than one kind of Warrant to connect Data and Claim.

The claim may be sufficiently justified; thus, justification has certain depth. As an argument structure is usually the starting point for a further argument, the arguments in an argumentative essay are hierarchically arranged in the form of argument chains [3]. To put it in another way, an argument chain is composed of embedded argument structures and therefore the argument depth means the longest argument chain identified in an essay. In Fig. 7 Subclaim 1, Subclaim 2 and Subclaim 3 may be Data 1, Data 2 and Data 3 if there is no further argument structure. All of the three Data support Claim. However, if there are further argument structures, the three Data just serve the function of subclaim and are in turn supported by their own data and warrants. There can be still further division. For example, Data 1.1 can be a subclaim of a still further argument structure and it can be labeled as Subclaim 1.1.1 that may be further supported by Data 1.1.1 and Warrant 1.1.1.

Warrant has four kinds. The writer may choose (A) a priori warrant to link Data 1.2 to Subclaim 1, use (B) empirical warrant to connect Data 2.1 to Subclaim 2. He can even use more than one warrant, for example, use both (C) institutional warrant and (D) evaluative warrant to bridge the gap between Data 3.1 and Subclaim 3.

Counterargument-claims and counterargument-data usually go with rebuttals in the sound and effective argumentation. The three oppositional elements are generally viewed as obvious evidence of reader consideration and therefore the use of these components can in a sense strengthen the persuasiveness of an argumentative essay [3]. Thus, our model includes opposition, which is composed of counterargument-claims and counterargument-data and rebuttal.

Our new analytical framework is applied to an argumentative essay written by a volunteer student in our project to test its feasibility and effect. The labeled essay is in Appendix.

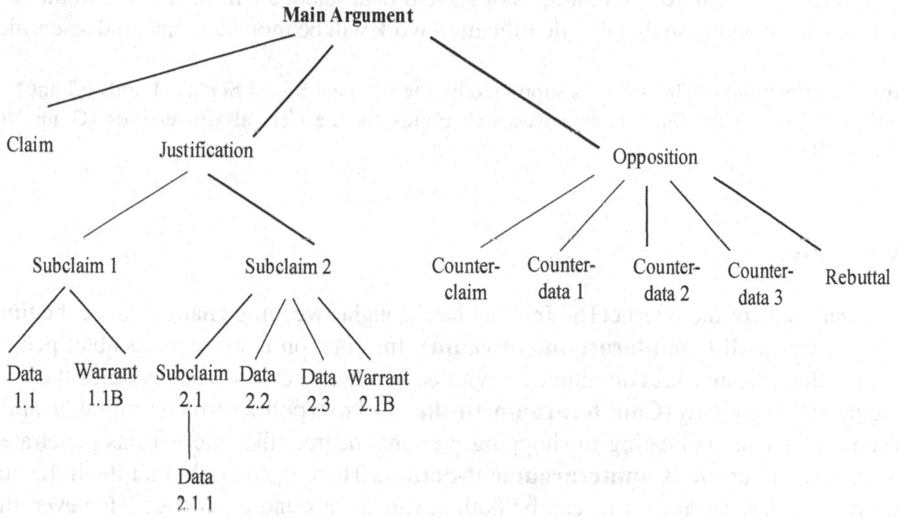

Fig. 8. Toulmin structure of the example

This essay has Claim, Justification and Opposition. Justification consists of two Subclaims at the same level. Subclaim 1 is supported by Data 1.1, and Warrant 1.1B (empirical Warrant) links them. Subclaim 2 has two levels consisting of Subclaim 2.1, Data 2.2 and Data 2.3. Subclaim 2.1 is further supported by Data 2.1.1. Warrant 2.3 B links Data 2.3 to Subclaim 2. Oppositions includes Counterargument-claim, three Counterargument-data and one Rebuttal. Figure 8 illustrates the structure of the essay.

6 Conclusion

Toulmin model, whether original or modified by other people, has provided a clear and flexible set of categories for conducting research on both oral and written argumentation. It can be taken as a method for describing, analyzing and evaluating not only typical argumentative discourse but also new genres. Like the other structural modifications, the synthesized model is based on the original Toulmin model. It is supposed not only to update Toulmin model, but also broaden the scope of its application. The merits of the previous modified models are taken in and synthesized, for example the view of complicated hierarchical structure of Crammond [3] and Voss [5], the counterarguments components of Qin and Karabacak [4], the views of warrant put forward by Freeman [17], Hitchcock's [18], Jackson and Schneider [6]. However, this paper also points out the demerits of the previous modifications. The synthesized model tries to avoid their deficiencies so as to explain argumentative essays more efficiently.

One of the limitations of this study is the lack of examination or empirical study to test this model just because we are still working at presuming and perfecting the model at present stage. Another limitation is the exclusion of Qualifier. In our future study we may have a reconsideration of Qualifier and try to do more empirical studies to test the model. Furthermore, technologies of AI and data science will be applied to the text analysis and labeling so that the identification work will be more accurate and scientific.

Acknowledgements. This study is supported by the National Social Science Funds (Grant No. 21FYYB016) and the Fundamental Research Funds for the Central Universities (Grant No. 2242022R10038).

Appendix

The Dark Side of the InternetThe Internet has brought sweeping changes since the time when it was used **(Counterargument-claim)**: Information is more accessible; people from different countries communicate with each other; e-commerce becomes the new engine of the economy **(Counterargument-data1)**. From politics to economy, education to entertainment, and eating to shopping, to some degree, the Internet has penetrated every area of our life **(Counterargument-data 2)**. There is no doubt that the influence the Internet has exerted on us can be both advantageous and disastrous. However, the bright side of the Internet is so frequently stressed that the dark side tend to be neglected. From my perspective, it's necessary to disclose its dark side so as to make better use of it **(Claim)**.

First and foremost, the Internet unites its users but at the same time alienates them (**Subclaim1**). On one hand, the Internet facilitates the exchange of different ideas and the communication on a global scale **(Counterargument-data 3)**. On the other hand, those who are burying themselves in a virtual world often ignore people around them (**Rebuttal**). Nowadays, it is a common phenomenon that some participants in a gathering, no matter what kind of gathering it is, always look down at their cellphones. The precious time that is supposed to be spent in talking with loved ones is wasted in surfing the Internet (**Data1.1**). Consequently, conflicts between lovers and even between family members gradually arise, and it's no longer easy for one to maintain a harmonious relationship. Obviously, the Internet shortens the physical distance from one user to another, but it lengthens the psychological distance from one heart to another. **(Warrant 1.1B)**.

Secondly, the Internet, extolled as "information superhighway", has caused a series of problems due to its huge and patchy information (**Subclaim2**). From one side, too much information can weaken our capability of thinking and solving problems (**Subclaim 2.1**). When netizens look through the information online, they are inclined to get a general idea without meditation **(Data 2.1.1)**. Besides, when more and more people rely on the Internet to find answers, their ability of solving problems declines **(Data2.2)**. From another, improper information will mislead teenagers and corrupt their mind **(Data2.3)**. Although the Internet is a powerful tool of spreading knowledge, it is also the hotbed of inappropriate information and false ideas. As few teenagers possess the ability of judging what is good and evil, they are very likely to become the victims of the wrong information **(Warrant 2.3B)**.

References

1. Liao, B.: On the cross over study of the new generation of artificial intelligence and logic. Soc. Sci. China **205**(3), 37–54 (2022)
2. Toulmin, S.: The Uses of Argument, Updated Cambridge University Press, Cambridge (2003)
3. Crammond, J.G.: The uses and complexity of argument structures in expert and student persuasive writing. Writ. Commun. **15**, 230–268 (1998). https://doi.org/10.1177/0741088398015002004
4. Qin, J., Karabacak, E.: The analysis of Toulmin elements in Chinese EFL university argumentative writing. System **38**, 444–456 (2010)
5. Voss, J.F.: Toulmin's model and the solving of ill-structured problems. Argumentation **19**, 321–329 (2005). https://doi.org/10.1007/s10503-005-4419-6
6. Jackson, S., Schneider, J.: Cochrane review as a "warranting device" for reasoning about health. Argumentation **32**(2), 241–272 (2017). https://doi.org/10.1007/s10503-017-9440-z
7. Liu, D., Lloyd, K.: Rhetoric and Composition Studies. Central China Normal University, Wuhan (2020)
8. Ramage, J.: Argument in Composition. Anderson. Parlor Press and the WAC Clearinghouse, New York (2007)
9. Abdollahzadeh, E., Amini Farsani, M., Beikmohammadi, M.: Argumentative writing behavior of graduate EFL learners. Argumentation **31**(4), 641–661 (2017). https://doi.org/10.1007/s10503-016-9415-5
10. Liu, D., Wan, F.: What makes proficient writers' essays more persuasive?—a toulmin perspective. Int. J. TESOL Stud. **1**, 1–13 (2020). https://doi.org/10.46451/ijts.2020.06.01
11. Cheng, F., Chen, Y.: Taiwanese argumentation skills: contrastive rhetoric perspective. Taiwan Int. ESP J. **1**(1), 23–50 (2009)
12. Whithaus, C.: Claim-Evidence structures in environmental science writing: modifying Toulmin's model to account for multimodal arguments. Tech. Commun. Q. **21**, 105–128 (2012). https://doi.org/10.1080/10572252.2012.641431
13. Demicheli, V., Rivetti, A., Debalini, M.G., Pietrantonj, D.P.: Vaccines for measles mumps and rubella in children. Cochrane Database Syst. Rev. **2**, CD004407 (2012). https://doi.org/10.1002/14651858.CD004407.pub3
14. Warren, J.E.: Taming the warrant in Toulmin's model of argument. Engl. J. **99**(6), 41–46 (2010)
15. Deng, J.: Does logic matter in assessment arguments?—On the rational logic of arguments and the building of a progressive argument. Foreign Lang. **35**(4), 70–79 (2012)
16. Jin, L., Zhao, J.: Logical analysis of the analogy based on Toulmin model. Fujian Forum (Humanity Soc. Sci. version) **1**, 81–86 (2016)
17. Freeman, J.B.: Systemizing Toulmin's warrants: an epistemic approach. Argumentation **19**, 331–346 (2006). https://doi.org/10.1007/s10503-005-4420-0
18. Hitchcock, D.: Good reasoning on the Toulmin model. Argumentation **19**, 373–391 (2005). https://doi.org/10.1007/s10503-005-4422-y
19. Jin, R.: The theory of Tuilei and the justification for the characteristics of ancient Chinese logic. Soc. Sci. **4**, 127–136 (2014)
20. Liu, D.: Studies on Rhetoricity of Chinese and English Argumentative Genre. Guangming Daily Press, Beijing (2021)
21. Connor, U.: Linguistic/rhetorical measures for international persuasive student writing. Res. Teach. Engl. **24**(1), 67–87 (1990)
22. Ferris, D.R.: Rhetorical strategies in student persuasive writing: differences between native and non-native English speakers. Res. Teach. Engl. **28**, 45–62 (1994)

23. Ju, S., He, Y.: A study of ancient Chinese logic based on universal argumentation: anthems and poems in Spring and Autumn period. Philos. Study **1**, 103–110 (2014)
24. Asher, N.: Reference to Abstract Objects in Discourse. Kluwer Academic Publishers, Dordrecht (1993)

Value-Based Preference Aggregation Argument Framework and Its Application

Yuhui An and Jianying Cui(✉)

Institute of Logic and Cognition and Department of Philosophy, Sun Yat-sen University, Guangzhou 510275, China
anyh@mail2.sysu.edu.cn, cuijiany@mail.sysu.edu.cn

Abstract. In this paper, a value-based preference graph G of a decision-making problem is presented by using a method similar to Borda-counting to quantify the preferences of agents holding different values for the alternatives in the decision problem, and then a Value-based Preference Aggregation Argument Framework ($VPAAF$, an extended VAF theory) generated fromelimiate G is introduced. In $VPAAF$, a defeating relation between arguments is redefined by an aggregate function, which is used to aggregate individual preferences into group preferences. The defeating relation helps us almost eliminate the odd-cycles (or even-cycles) in the $VPAAF$, allowing obtain an extension containing only a single set of arguments. The argument framework not only achieves the goal of exploring some decision-making problems, which include resolving voting paradox to a certain extent and expanding the application field of VAF, but also explicitly reveals the value factors implied in the decision results, which makes it possible to do retrospective reasoning of the results by virtue of arguments.

Keywords: Value · Preference aggregation · Value-based preference aggregation argument framework

1 Introduction

Classical decision theory focuses on using the mathematical model of the problem in which each alternative is evaluated numerically according to some relevant criteria, looking for principles for comparing different alternatives, and then making clear what rational decision makers are. However, the principles defined for comparing alternatives are usually expressed by analytical expressions that summarize the entire decision-making process, which makes it difficult for ones who are not familiar with abstract decision-making methods to understand why one proposed alternative is better than another. One needs a method to help

The paper is supported by National Office for Philosophy and Social Science, P.R. China, for the project "Deep-Expansion and Application of Pragmatic Logic" (NO. 19ZDA042) and Philosophy and Social Science Youth Projects of Guangdong Province (No. GD19CZX03).

Y. Chen and S. Zhang (Eds.): AILA 2022, CCIS 1657, pp. 225–238, 2022.
https://doi.org/10.1007/978-981-19-7510-3_17

us better understand the basis of the evaluation and document the reasons for and against each of the alternatives. As a result, researchers in the field of artificial intelligence tend to use preference statement to describe, explain and reason preferences, and even to make decisions. Propositional logic, preference logic and temporal logic are often used for preference representation and reasoning. This is also known as a cognitively-oriented approach. However, in practical decision-making problems, agents usually need to make choices under incomplete and uncertain information, which results in the exposure of many disadvantages of the above approach based on classical monotonous logic. The facts that argumentation helps people identify one or more alternatives, or to explain and justify the choices adopted are the main reason why abstract argumentation theory, one of non-monotonic theories, has been a new research hotspot and attracted more and more attention in the study of decision theory.

Specifically, with the help of the following example, we show the process of resolving a decision-making problem in Value-based Preference Aggregation Argument Frameworks $VPAAF$.

Example 1. ([6]). Trevor and Katie need to travel to Paris for a conference.

Trevor believes that it is better to take a plane than a train, while Katie's view is the opposite. It is obvious that Trevor and Katie cannot agree on the choice of transportation, which requires a further in-depth analysis of the reasons for their choice preference. It means that the two decision-makers' ranking of alternatives will correspond to an argument that is consistent with the value of the decision maker and supports this ranking. In fact, according to the definition of $VPAAF$, for Trevor, there exists an argument A"we should travel to Paris by plane because it is fastest"; for Katie, there exists an argument B "we should travel to Paris by train because it is the most comfortable".

Thus, a decision-making problem is transformed into an argument evaluation problem in Value-based Argumentation Frameworks VAF. Bench-Capon focused on how to integrate value elements into the AF theory in [6], which enables us to understand the reasons behind the decision makers' choices while does not provide a comparison method of the value strengths attached to the argument. Therefore, we can not be sure of the defeat relation between the generated arguments when facing actual problems, and then are unable to eliminate the possible attack cycles that may appear in the corresponding directed graph. In Example 1, there is an attack even-cycle between A and B, which will cause neither plane nor train to be skeptically acceptable, and at the same time, both are credulously acceptable, which makes decision-makers still unable to make a choice.

We first propose a value-based preference graph G for decision-making problems by using a method similar to Broda-counting [16] to quantify the preferences of agents holding different values for alternatives in the decision-making problem. For example, Trevor(p_1) holds a value of speed(s), so he can be assigned to alternatives (plane, train) $(f_s(l), f_s(t))$ by a quantization function f and we have $f_s^{p_1}(l) > f_s^{p_1}(t)$ because the plane is faster than the train; Secondly, we provide a Value-based Preference Aggregation Argument Framework corresponding to value-based preference graph G, an extended preference graph in [10]. In $VPAAF$, the defeat relation is redefined by a value aggregation function F, which aims to aggregate individual preferences into group preferences. In this example, if A attacks B and $F(Con(A)) > F(Con(B))$, where $Con(A) = l, Con(B) = t$ and $F(l) = f_s^{p_1}(l) + f_c^{p_2}(l)$, $F(t) = f_s^{p_1}(t) + f_c^{p_2}(t)$, then we can say that A defeats B. Finally, we prove that the newly defined defeat relation helped us to almost eliminate attack cycles of the argumentation framework. Furthermore, under the four classic semantics in [13] characterized by this defeat relation, we can obtain an extension containing only a single set of arguments, and make the extension both skeptically and credulously acceptable, that is, we can obtain the only decision result. In fact, $VPAAF$ is similar to Claim-augmented argumentation frameworks $CAFs$ in [19] and [20], which are minimal structure on arguments, i.e. they have a conclusion.

The rest of the paper is organized as follows: In Sect. 2 we review the background on argumentation framework and the value-based argumentation framework; Sect. 3 then shows the Value-based Preference Aggregation Argumentation Framework ($VPAAF$) and discusses its related properties; The application research of $VPAAF$ in theory–the Voting paradox resolution is further explored in Sect. 4. Finally, We conclude with some discussion and consider future work in Sect. 5.

2 Background

2.1 Abstract Argumentation Theory

Many researchers have done lots of pioneering work based on argumentation framework (*AF*) [13], involving making and explaining decisions [1], argumentation in legal reasoning [7], the reasoning of inconsistent knowledge [8], argumentation-based negotiation [11], argumentation in multi-agent systems [14] and so on. The role of arguments is only determined by their relation with other arguments, without considering the internal structure of the argument.

Definition 1 ([13]). *An argumentative framework is a pair*

$$AF = (AR, attacks)$$

Where

- AR is a set of arguments;
- $attacks$ is a binary relation on AR, i.e. $attacks \subseteq AR \times AR$.

$attacks(A, B)$ means argument A attacks argument B. An argument set S attacks argument B if B is attacked by an element in S [13].

Definition 2 ([13]). *A set S of arguments is said to be conflict-free if there are no arguments A and B in S such that A attacks B.*

Definition 3 ([13]). *An argument $A \in AR$ is acceptable with respect to set of argument S iff for each argument $B \in AR$: if B attacks A then B is attacked by S.*

Definition 4 ([13]). *A conflict-free set of arguments S is admissible iff each argument in S is acceptable with respect to S.*

Definition 5 ([13]). *A preferred extension of an Argumentation Framework AF is a maximal (with respect to set inclusion) admissible set of AF.*

Definition 6 ([13]). *A conflict-free set of arguments S is called a stable extension iff S attacks each argument which does not belong to S.*

In general, a standard AF always has a preferred extension and its preferred extensions is usually not unique. [13] In a dispute, if two parties adopt different positions, then this dispute can't be solved by AF, such as Example 1.

2.2 Value-Based Argument Framework

When people make decisions in the face of conflicting arguments, the reliability of arguments is often not the only consideration. Many researchers believe that one should study imperfect decision-relevant information from the perspective of different preferences of values advocated by conflicting arguments: arguments can be defended not only by defeating the attacker, but also by ranking their value higher than that of the attacker. [5] This is also the main insight in [15] for resolving the problem of moral and legal differences. As [9] points out: in politics, many policies are often justified on the basis of the values they promote; in law, the reasons behind legal decisions can be reflected through the values of the public; in morality, the individual and collective ethical views play an important role in reasoning and action [3]. In Example 1, they still disagree on the choice of transportation, but they can't deny each other's opinions that are both acceptable. The concept of argument in AF is too abstract for us to model this debate. *We may need relate arguments to values, and to allow these values to be ranked to reflect the preferences of agents.* [5] Under this consideration the Value-based Argumentation Framework is established and introduces an element that can be used to make rational choices from multiple reasonable choices, that is value.

Definition 7 ([5]). *A Value-based Argumentation Framework (VAF) is a 5-tuple:*

$$VAF = (AR, attacks, V, val, valpref)$$

Where

- AR and *attacks* are the same as defined in Definition 1;
- V is a non-empty set of values;
- *val* is a function which maps from elements of AR to elements of V;

- *valpref* is a preference relation (transitive, irreflexive, and asymmetric) on $V \times V$.

An argument A relates to value v if accepting A promotes or defends v: the value in question is given by $val(A)$. $\forall A \in AR, val(A) \in V$ [5].

VAF allows one to distinguish the attack from the defeat relation between arguments. In Example 1, Trevor and Katie both attacked each other's arguments, but did not defeat them.

Definition 8 ([5]). *An argument $A \in AR$ defeats an argument $B \in AR$ iff both $attacks(A, B)$ and not $valpref(val(B), val(A))$*

If both arguments relate to the same value, or if there is no preference between values, then the attack is successful. An arguments set S defeats argument B if B is defeated by an element in S. It is Obvious that if V contains a single value, then the VAF is a standard AF [5].

Since the concept of value is introduced into VAF and a distinction is made between attack and defeat, the concepts of Conflict-free and Acceptable need to be redefined.

Definition 9 ([5]). *An argument $A \in AR$ is acceptable with respect to a set of argument $S(acceptable(A, S))$ if $(\forall B)(B \in AR \;\&\; defeats(B, A)) \longrightarrow (\exists C)(C \in AR) \;\&\; defeats(C, B)))$.*

Definition 10 ([5]). *A set S of arguments is conflict-free if $(\forall A)(\forall B)((A \in S \;\&\; B \in S) \longrightarrow (\neg attacks(A, B) \vee valpref(Val(B), Val(A))))$.*

Definition 11 ([13]). *A conflict-free set of arguments S is admissible if $\forall A(A \in S) \longrightarrow (acceptable(A, S))$.*

Definition 12 ([13]). *A set of arguments S in an value-based argumentation framework VAF is a preferred extension if it is a maximal (with respect to set inclusion) admissible set of AR.*

Definition 13 ([13]). *A conflict-free set of arguments S is called a stable extension iff S attacks each argument which does not belong to S.*

AF is often described as a directed graph where vertices represent arguments and directed edges from the attacker to the attacked represent attack relations. [13] In VAF, vertices of different colors are often used to represent different values, and the cycles are called monochromatic if they contain arguments related to a single value; they are called dichromatic cycles if they contain arguments related to exactly two values, and they are called polychromatic cycles if they contain two or more values. [5] We also use the same notion in the following section.

The agent's acceptance of the argument is reflected by the concepts of credulous acceptance and skeptical acceptance.

Definition 14 ([4]). *Given an argumentation framework AF, an extension E based on the semantics S is denoted as $\mathcal{E}_S(AF)$: an argument A is skeptically accepted if $\forall E \in \mathcal{E}_S(AF), A \in E$; an argument A is credulously accepted if $\exists E \in \mathcal{E}_S(AF), A \in E$.*

If the preferred extension of a dispute is unique, then the ideal situation is that every agent accepts the same arguments. In this case, the dispute can be resolvable. If the preferred extension is not unique, then we wish to reach a consensus among the parties, which are manifested as skeptically acceptable arguments. However, if an argument does not fit this consensus, an agent may wish to show that it is at least defensible. [5] To realize this situation, the argument must be credulously acceptable. Generally, it is not easy for a standard AF to determine such things.

3 Value-Based Preference Aggregation Argumentation Framework

In order to enable conflicting individuals to finally reach a consensus on an issue, we need to aggregate individual preferences. Here, we use a location-based method – Borda-counting [16] to aggregate individual preferences into group preferences. The main idea of the Borda-counting is to assign a point value to the alternatives according to their position in the individual preference order. For example, we have $x_1 > x_2 > x_3$, then we can assign 3 to x_1, assign 2 to x_2, and assign 1 to x_3. In this paper, a variant of Borda-counting, namely the preference graph [10] method based on graph theory, is used for case the study.

Therefore, in order to quantify the preference of agents holding different values for the alternatives in the decision problem, we define a value-based preference graph G as follows:

Definition 15. *Value-based preference graph $G = (X, R, V, P, f_v^p, val)$is a 6-tuple, where finite set $X = x_1, x_2, \ldots, x_n$ is a set of alternatives; the directed edge $R = (x_i, x_j)(i \in n, j \in n$ and $i \neq j)$ is the oppositional relation[1] between x_i and x_j; V is a set of values; P is a set of agents; $f_v^p : X \longrightarrow N^+$ is a quantization function that each agent assigns a positive integer to each $x_i \in X$ based on value v; $val : P \longrightarrow V$ is a function that assigns a value $v \in V$ to each agent $p \in P$.*

In order to achieve the purpose of individual preference aggregation, we require different agents to assign alternatives based on different value preferences under the same assignment standard, that is:

[1] The oppositional relation here means that any two alternatives are in conflict with each other, that is, any two alternatives can't coexist, and an agent can only choose one of the oppositional alternatives.

Condition 1. *For any $x_1, x_2, \ldots, x_n \in X$ and for any $p_1, p_2, p_3 \ldots p_n \in P$, $f_{v_1}^{p_1}(x_1) + f_{v_1}^{p_1}(x_2) + \ldots + f_{v_1}^{p_1}(x_n) = f_{v_2}^{p_2}(x_1) + f_{v_2}^{p_2}(x_2) + \ldots + f_{v_2}^{p_2}(x_n) = \ldots = f_{v_n}^{p_n}(x_1) + f_{v_n}^{p_n}(x_2) + \ldots + f_{v_n}^{p_n}(x_n) = N$, where we call N the total value*[2].

We will aggregate individual preferences into group preferences by following Aggregation function F:

Definition 16. *Aggregation function F:*

$$F(x_1) = f_{v_1}^{p_1}(x_1) + f_{v_2}^{p_2}(x_1) + \ldots + f_{v_n}^{p_n}(x_1)$$

Therefore, in Example 1, $F(t) = f_c^{p_1}(t) + f_s^{p_2}(t)$, $F(l) = f_c^{p_1}(l) + f_s^{p_2}(l)$ and we have $f_c^{p_1}(t) + f_c^{p_2}(l) = f_s^{p_2}(t) + f_s^{p_2}(l) = N$.

The rationality of this aggregation method has been proved in [18]. In fact, the aggregate function F form extends the concept of joint audience proposed by Trevor Bench-Capon in [6], which requires comprehensive consideration of the choices made by all agents based on different values. The joint audience is to achieve the purpose of Trevor and Katie traveling together in Paris and the aggregation function F is to consider the different choices made by different agents holding different values. To a certain extent, it also follows the majority principle.

Assuming $f_c^{p_2}(t) + f_c^{p_2}(l) = f_s^{p_1}(t) + f_s^{p_1}(l) - N = 10$, we can assign (9,1) or (8, 2) or (7,3) or (6,4) to $(f_c^{p_2}(t), f_c^{p_2}(l))$, similarly, $(f_s^{p_1}(l), f_s^{p_1}(t))$ can also be assign to (9,1) or (8, 2) or (7,3) or (6,4), from the Definition 15, we have $F(l) = f_s^{p_1}(l) + f_c^{p_2}(l)$, $F(t) = f_s^{p_1}(t) + f_c^{p_2}(t)$ as follows:

Table 1. Trevor and Katie's preference aggregation table

$(F(t), F(l))$ \ $(f_c^{p_2}(t), f_c^{p_2}(l))$ / $(f_s^{p_1}(l), f_s^{p_1}(t))$	(9,1)	(8,2)	(7,3)	(6,4)
(9,1)	(10,10)	(9,11)	(8,12)	(7,13)
(8,2)	(11,9)	(10,10)	(9,11)	(8,12)
(7,3)	(12,8)	(11,9)	(10,10)	(9,11)
(6,4)	(13,7)	(12,8)	(11,9)	(10,10)

Definition 17. *Given a value-based preference graph $G = (X, R, V, P, f_v^p, val)$, we can generate a value-based preference aggregation argument framework (VPAAF), which is a 7-tuple:*

$$VPAAF = (AR, attacks, Con, V, P, f_v^p, val')$$

[2] If the total value is not equal, preference aggregation cannot be carried out under the condition of fairness. In other words, if the total value based on which one agent assigns the alternative is higher than that of other agents, it will have a full advantage in the result of the operation of the aggregation function, so it is unfair to other agents.

- if $\exists x_i \in X$, $\exists p \in P$ such that $\forall x_j \in X(j \neq i)$, $f_v^p(x_i) > f_v^p(x_j)$, then exists an argument $A \in AR$ with conclusion x_i for p, denoted as $Con(A) = x_i$;
- if $A, B \in AR$ and $(A, B) \in attacks$ iff there exists $Con(A) = x_1$ and $Con(B) = x_2$ such that $(x_1, x_2) \in R$;
- V, $P$and f_v^p are the same in G and $VPAAF$;
- $val' : AR \longrightarrow V$ is a function that assigns a value $v \in V$ to each argument $A \in AR$. $val(P) = val'(AR)$.

Definition 18. *Given a value-based preference aggregation argument framework (VPAAF) and an aggregate function F, we can define the notion of defeat as follows:*

Argument $A \in AR$ *defeats argument* $B \in AR$ *iff* $attacks(A, B)$ *and* $F(Con(A)) > F(Con(B))$.

The important concepts of Acceptable and Conflict-free are define in $VPAAF$ as follows.

Definition 19. *An argument* $A \in AR$ *is acceptable with respect to a set of argument* $S(acceptable(A, S))$ *if* $(\forall B)(B \in AR \;\&\; defeats(B, A)) \longrightarrow (\exists C)(C \in AR) \;\&\; defeats(C, B)))$.

Definition 20. *A set S of arguments is conflict-free if* $(\forall A)(\forall B)((A \in S \;\&\; B \in S) \longrightarrow \neg defeats(A, B))$.

In $VPAAF$, the definitions of admissible, preferred extension and stable extension are the same as in AF.

Continuation to the Example 1:

Trevor: Choose plane instead of train based on value speed, that is, $f_s^{p_1}(l) > f_s^{p_1}(t)$.
Katie: Choose train instead of plane based on value comfort, that is, $f_c^{p_2}(t) > f_c^{p_2}(l)$;

According to the Definition 17, for Trevor, there exists an argument A with conclusion t, i.e. $Con(A) = t$; For Katie, there exists an argument B with conclusion l, i.e. $Con(B) = l$; and $(t, l) \in R$, so $(A, B) \in attacks$.

As shown in Table 1, we can get 16 sets of $(F(t), F(l))$, so when $F(t) > F(l)$, according to the Definitions 18, 5 and Definition 6, we can get the preferred extension and grounded extension of the arguments set $\{A, B\}$ are both $\{A\}$, so Trevor and Kaite will choose the train to travel; Similarly, when $F(l) > F(t)$, the preferred extension and grounded extension of the set $\{A, B\}$ are both $\{B\}$, so Trevor and Kaite will choose the plane to travel; but we still find that except for $F(t) > F(l)$ and $F(l) > F(t)$, there is still a situation of $F(l) = F(t)$ in Table 1. Our solution to this is to increase N to re-assign the value. The rationality of the scheme is obtained by the property of the following probability function P'.

Proposition 1. *Let* P' *be the probability of two alternatives aggregate function having equal values, and* N *be the total value, then*

$$P' = \begin{cases} \frac{2}{N-1} & when\ N\ is\ odd\ number\ \&\ N \geq 3) \\ \frac{2}{N-2} & when\ N\ is\ even\ number\ \&\ N > 3) \end{cases} \tag{1}$$

and $\lim_{N \to \infty} P' = 0$.

Proof. Let N be the total value, because agents needs to have a clear preference for the alternatives, then when N is an even number, $\frac{N}{2} - 1$ different sum formulas will be generated.

And there are $\frac{N}{2} - 1$ cases in which the value of the aggregate function is equal in the $\frac{N}{2} - 1$ different sum formulas, therefore $P' = \frac{\frac{N}{2}-1}{(\frac{N}{2}-1)^2} = \frac{2}{N-2}$;

Therefore, $\lim_{N \to \infty} P' = \lim_{N \to \infty} \frac{2}{N-2} = 0$;

Similarly, when N is an odd number, $P' = \frac{\frac{N-1}{2}}{(\frac{N-1}{2})^2} = \frac{2}{N-1}$;

So, we have $\lim_{N \to \infty} P' = \lim_{N \to \infty} \frac{2}{N-1} = 0$.

So, as N increases, the probability of the two alternatives aggregate functions having equal values will be closer and closer to 0.

Theorem 1. *In $VPAAF$, a unique extension set will be generated based on four basic semantics (preferred semantics, stable semantic, grounded semantics, and complete semantics)*

This is because, with the continuous increase of N, the probability of equal aggregate function values will closer and closer to 0. Therefore, according to the Definition 18, any two mutually attacking arguments can only choose one, and the cycle will be destroyed. In this case, a unique extension set will be generated based on these basic semantics.

Bench-Capon has done a thorough analysis and discussion on the dichromatic cycle in [5]. However, in real life, conflicting cases between different decisions made based on two or more values are also common. For example, the Voting paradox to be discussed in the next section is a typical one.

4 Case Study: The Voting Paradox

In modern society, voting is a common method for collective decision-making or collective selection, it is an important method to realize democracy. However, in order to achieve democracy, "majority principle" will lead to a difficult problem, which is the voting paradox. [2] The research of voting paradox has been attracting many scholars to explore it continuously. [12,17] We believe that an important reason for this paradox is the different values based on which voters make their choices. Through the work in the previous section, we can reasonably solve the paradox to some extent.

Assuming that there are three voters p_1, p_2, and p_3, each voter has three candidates x, y, and z. The order of preference of each voter is shown in the following table (Table 2):

Table 2. List of voters' preference order

Voter	The order of preference for candidates
p_1	$x > y > z$
p_2	$y > z > x$
p_3	$z > x > y$

$>$ means better than, for example, p_1 thinks $x > y$ means that in the two candidates of x and y, p_1 believes that x is better than y.

Obviously, there is a cycle $x > y > z > x$ derived from majority principle, and that is the voting paradox.

We believe that an important reason for voting paradox is the different in values based on which voters make their choices and provide a solution to the paradox based on $VPAAF$. For clarity, we put the voting paradox in a specific case.

Example 2. Student Union President Election

Suppose that the Department of Philosophy of Sun Yat-Sen University wants to elect the new president of the student union among three candidates x, y, and z. Those who have the right to vote are p_1, p_2, and p_3, and they will make a choice based on values v_1, v_2, and v_3 respectively. The preference orders for candidates are shown in Table 3:

Table 3. List of voter values and their preference order

Voter	The value of voters	The order of preference for candidates
p_1	Academic performance v_1	$x > y > z$
p_2	Working ability v_2	$y > z > x$
p_3	Social skills v_3	$z > x > y$

What needs to be explained here is that p_1, p_2, and p_3 votes based on v_1, v_2 and v_3 does not mean that voters only vote based on a single value in Example 2. In fact, this implies that each of the voters makes a preference order for candidates after sorting the three values. For example, if p_1 believes that x is better than y, y is better than z based on v_1, then p_1's preference order is $x > y > z$. This shows that no matter how much better y or z is better than x based on value v_2 or v_3, p_1's preference order will not change. Similarly, the preference order of p_2 and p_3 is the same.

From the Definition 18, We can convert Example 2 to Fig. 1:

Argument A claims that x is better than y and z, that is, $Con(A) = x$. similarly, $Con(B) = y$ and $Con(C) = z$. It is obvious that arguments A, B, and C show the different values v_1 (blue), v_2 (yellow), and v_3 (green) for the candidates x, y, and z respectively.

Suppose the total value N is 10, we can assume the assignment for the candidates as shown in Table 4:

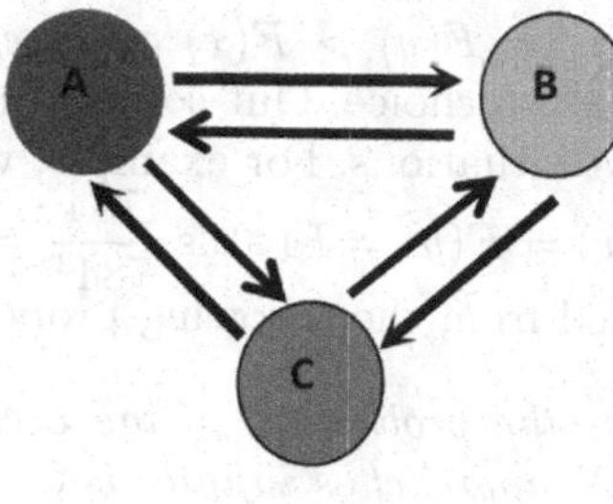

Fig. 1. The directed graph for Example 2

Table 4. List of voter values and their assignment for candidates

Voter	The value of voters	Assignment for candidates
p_1	Academic performance v_1	$f_{v1}^{p1}(x) = 5 > f_{v1}^{p1}(y) = 3 > f_{v1}^{p1}(z) = 2$
p_2	Working ability v_2	$f_{v2}^{p2}(y) = 6 > f_{v2}^{p2}(z) = 3 > f_{v2}^{p2}(x) = 1$
p_3	Social skills v_3	$f_{v3}^{p3}(z) = 7 > f_{v3}^{p3}(x) = 2 > f_{v3}^{p3}(y) = 1$

From the Definition 16 we know that $F(x) = 8$, $F(y) = 10$, $F(z) = 12$. According to the Definition 18, we can retain the original successful attack relation and delete the unsuccessful attack relation. For example, because $F(z) = 12 > F(y) = 10$, B's attack on C is unsuccessful, that is, C is not defeated by B, and C's attack on B is successful, that is, B is defeated by C. The same can be obtained: A is defeated by C, and A is defeated by B. Therefore, the conversion process of Fig. 1 is shown in Fig. 2:

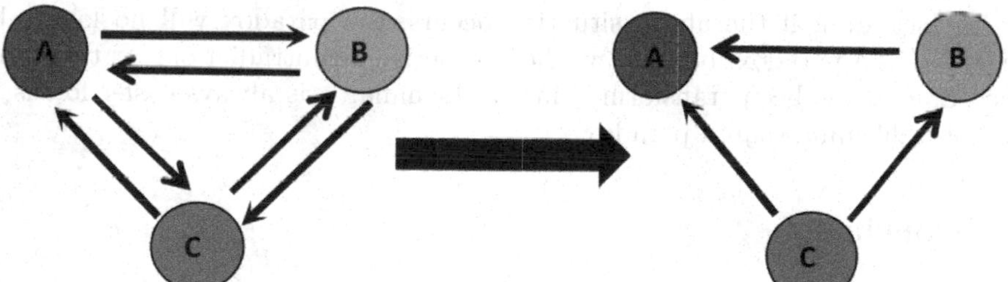

Fig. 2. The conversion process of Fig. 1

According to the Definition 18, if there is a situation similar to $F(x) > F(y)$ and $F(x) > F(z)$, then from the Definitions 18, 5 and Definition 6, we can get that the preferred extension and stable extension in the *VPAAF* is $\{C\}$, that is, the conclusion z of argument C should be elected as the president of the student union.

However, it should be noted that, similar to Example 1, the above assignment is only one of many assignments. It is very likely that $F(x) = F(y) = F(z)$,

$F(x) = F(y) > F(z)$, $F(z) = F(y) > F(x)$ or $F(x) = F(z) > F(y)$, which makes us still unable to make a choice. Our solution is to increase N to reduce the probability of the above situations. For example, when N is increased to by 100, the probability of $F(x) = F(y) = F(z)$ is $\frac{784}{784^3} = \frac{1}{784^2}$. The rationality of this method can be obtained from the following propositions:

Proposition 2. *Let P'' be the probability of the occurrence of the above four situations, limit of P'' as N approaches infinity is* 0, *that is,* $\lim_{N\to\infty} P'' = 0$

Proof. let S be the number of methods in which a positive integer N^+ is split into the sum of three different positive integers; then, based on three agents and three candidates, the number of each agent' alternatives is S, so, there will be S^3 situations in total. Let's consider the upper bound of the cases where $F(x) = F(y) = F(z)$ and $F(x) = F(y) > F(z)$ or $F(z) = F(y) > F(x)$ or $F(x) = F(z) > F(y)$ in different S^3 methods:

(1) the number of the case where $F(x) = F(y) = F(z)$ is S;
(2) the upper bound of the cases where $F(x) = F(y) > F(z)$ or $F(z) = F(y) > F(x)$ or $F(x) = F(z) > F(y)$ is $[C_3^2 \times S \times S \times \frac{N}{6}]$.

Therefore, the probability of the occurence of the above four cases must satisfy the inequality: $0 \le P'' < \frac{S + C_3^2 \times S \times S \times \frac{N}{6}}{S^3} = \frac{5S^2 + S}{S^3} = \frac{(2 + S \times N)}{2S^2}$, S is higher order infinity as N approach infinity, that is, $\lim_{N\to\infty} \frac{N}{S} = 0$;

Thus, $\lim_{N\to\infty} \frac{(2 + S \times N)}{2S^2} = 0$, then $0 \le P'' \le \lim_{N\to\infty} \frac{(2 + S \times N)}{2S^2} = 0$;

So, we have $\lim_{N\to\infty} P'' = 0$ by The Sandwich Theorem.

In fact, even if the above situation occurs, the paradox will no longer be regarded as a paradox, because we did not derive a contradiction. At this time, the paradox has been transformed into a dilemma, it is always easier for us to accept a dilemma than a paradox.

5 Conclusion

Argumentation plays two different roles when we are making decisions and discussing critical issues in our daily life. It helps people to choose one or several alternatives or to explain and prove the choices adopted so that people can make retrospective reasoning about the reasons for the decision results. This paper focused on the hidden values behind the decision-making results and proposed an extended argument framework $VPAAF$. It not only achieves the goal of exploring some problems in decision theory, which include resolving the Voting paradox to a certain extent and expanding the application field of VAF, but also explicitly reveals the value factors implied in the decision results, making it possible to do retrospective reasoning of the results with the assistance of arguments. At the same time, although considering that daily life is mostly related to decision making problems involving 2 or 3 types of values, the number of agents, values, and alternatives involved in the cases examined in this paper are equal, and the corresponding values are relatively small. Extending the research results of this paper to a larger number of values and alternatives, etc. is another focus of interest in future work.

References

1. Amgoud, L., Prade, H.: Using arguments for making and explaining decisions. Artif. Intell. **173**(3–4), 413–436 (2009)
2. Arrow, K.: Social Choice and Individual Values. Yale University Press, New Haven (1951)
3. Atkinson, K., Bench-Capon, T.: Addressing moral problems through practical reasoning. J. Appl. Log. **6**(2), 135–151 (2008)
4. Baroni, P., Giacomin, M.: On principle-based evaluation of extension-based argumentation semantics. Artif. Intell. **171**(10–15), 675–700 (2007)
5. Bench-capon, T.: Value based argumentation frameworks. Artif. Intell. **171**(1), 444–453 (2002)
6. Bench-capon, T., Atkinson, K.: Abstract Argumentation and Values. In: Simari, G., Rahwan, I. (eds.) Argumentation in Artificial Intelligence, pp. 45–64. Springer, Heidelberg (2009). https://doi.org/10.1007/978-0-387-98197-0_3
7. Bench-Capon, T., Prakken, H., Sartor, G.: Argumentation in legal reasoning. In: Simari, G., Rahwan, I. (eds.) Argumentation in Artificial Intelligence, pp. 363–382. Springer, Boston (2009). https://doi.org/10.1007/978-0-387-98197-0_18
8. Besnard, P., Hunter, A.: Elements of Argumentation. MIT Press, Cambridge (2008)
9. Bench-Capon, T., George, C.: Christie, the notion of an ideal audience in legal argument. Artif. Intell. Law **9**(1), 59–71 (2001)
10. Davenport, A.J., Kalagnanam, J.A.: Computational study of the Kemeny rule for preference aggregation. In: AAAI, vol. 4, pp. 697–702 (2004)
11. Dimopoulos, Y., Mailly, J.G., Moraitis, P.: Argumentation-based negotiation with incomplete opponent profiles//13èmes Journées d'Intelligence Artificielle Fondamentale (JIAF 2019), pp. 91–100 (2019)
12. Duncan, B.: The Theory of Committees and Elections. Econometrica, pp. 1341–1356 (1958)
13. Dung, P.M.: On the acceptability of arguments and its fundamental role in nonmonotonic reasoning, logic programming and n-person games. Artif. Intell. **77**(2), 321–357 (1995)
14. McBurney, P., Parsons, S., Rahwan, I. (eds.): Proceedings of the Eighth International Workshop on Argumentation in Multi-Agent Systems. LNCS, vol. 7543. Springer, Heidelberg (2012). https://doi.org/10.1007/978-3-642-33152-7
15. Perelman, C., Berman, H.J.: Justice, Law, and Argument. Essays on Moral and Legal Reasoning. Springer, Heidelberg (1980). https://doi.org/10.1007/978-94-009-9010-4
16. Saari, D.G.: The Borda dictionary. Soc. Choice Welf. **7**(4), 279–317 (1990). https://doi.org/10.1007/BF01376279
17. Sen, A.K.: A possibility theorem on majority decisions. Econometrica: J. Econ. Soc. 491–499 (1966)
18. Zhao, W.: An axiomatization of range voting with qualitative judgement. In: Studies in Logic, vol. 14, no. 3, pp. 01–15 (2021)

19. Dvoák, W., Woltran, S.: Complexity of abstract argumentation under a claim-centric view. Artif. Intell. **285**, 103290 (2020)
20. Dvoák, W., Rapberger, A., Woltran, S.: Argumentation semantics under a claim-centric view: properties, expressiveness and relation to SETAFs. In: Proceedings of the International Conference on Principles of Knowledge Representation and Reasoning, vol. 17, no. 1, pp. 341–350 (2020)

A New Fuzzy Analytic Hierarchy Process Method for Software Trustworthiness Measurement

Zelong Yue[1], Xinghua Yao[2], and Yixiang Chen[1](✉)

[1] National Engineering Research Center of Trustworthy Embedded Software, Software Engineering Institute, East China Normal University, Shanghai 200062, China
51205902148@stu.ecnu.edu.cn, yxchen@sei.ecnu.edu.cn

[2] School of Basic Medicine, Shanghai University of Traditional Chinese Medicine, Shanghai 201203, China
xhyao@shutcm.edu.cn

Abstract. Software trustworthiness measurement becomes a focus in software companies. In software trustworthiness measurement, traditional Analytic Hierarchy Process (shortly, AHP) is usually utilized to estimate software attributes' weights. However, the traditional AHP method only supports using definite numerical values and cannot quantify well decision makers' opinions on software attributes. By using interval-valued intuitionistic fuzzy set, a new method is proposed in this study based on the traditional AHP for software trustworthiness measurement. In the proposed method, an equation for calculating correlation coefficients between interval-valued intuitionistic fuzzy matrices is designed in order to characterize similarities among decision makers' opinions, and a parameter of threshold is introduced to select correlation coefficients for decision makers' weights calculation. Besides, an equation for calculating attribute weights is designed based on harmonic mean in order to stand out levels of attribute importance. The proposed method is experimented in a task of evaluating the resilience of an operating system, and it is compared to the other two classical methods. Our experimental results show that the proposed method produces attribute weights with great differences, and its ability is stronger in the aspect of describing decision makers' opinions.

Keywords: Interval-valued intuitionistic fuzzy set · Analytic hierarchy process · Attribute weight · Decision maker's weight · Software trustworthiness measurement

1 Introduction

Software plays an important role in a lot of industries and is ubiquitous in our daily lives. In wide applications, software errors are cared for by users. Software

Y. Chen and S. Zhang (Eds.): AILA 2022, CCIS 1657, pp. 239–253, 2022.
https://doi.org/10.1007/978-981-19-7510-3_18

failures not only lead to the losses of money and time but also endanger the lives of people [1]. It becomes a crucial problem whether the software is trustworthy or not. And software trustworthiness measurement has been attracting attention from researchers [2,3].

In software trustworthiness measurement, usually experts or decision makers (shortly, DMs) are called together to collect opinions on software attributes, and then the collected opinions are used to calculate decision makers' weights and software attributes' weights. Such a process is adopted in Multi-Criteria Decision-Making (shortly, MCDM). Since the early 1970s, many MCDM methods have been developed. As a famous MCDM method, the Analytic Hierarchy Process (shortly, AHP), which is proposed by Saaty [4], has been widely applied [5,6]. Currently, the AHP is utilized to measure software trustworthiness [7,8]. A main advantage of the AHP is that it is easy to be understood by users. The AHP provides linguistic terms for DMs and requires few mathematical calculations. On the other side, a disadvantage of the traditional AHP method is that it only supports using definite numerical values to describe DMs' opinions [9,10]. Such a way of using definite value is unable to handle well inherent uncertainty and imprecision in a pair-wise comparison process [11]. To address such shortcomings, fuzzy AHP was developed. Instead of definite values, numerical intervals are used to describe DMs' opinions in fuzzy AHP.

Fuzzy AHP methods were developed [12,13] and applied widely [14,15]. Van Laarhoven and Pedrycz in [12] developed the first fuzzy AHP method, which utilized fuzzy rations by means of triangular fuzzy numbers. Chang in [13] leveraged triangular fuzzy numbers for pair-wise comparison scale in a fuzzy AHP method.

Atanassov extended the concept of fuzzy sets to intuitionistic fuzzy set (shortly, IFS) in [16]. The concept of IFS introduces the degree of membership and degree of non-membership to describe uncertainty in a decision-making procedure. In [17], Atanassov and Gargov extended IFS to interval-valued intuitionistic fuzzy set (shortly, IVIFS). In the concept of IVIFS, intervals are used to represent membership degree, non-membership degree, and hesitation in a decision-making process. Theories related to IVIFS have been developed by many researchers. Atanassov established relations and operations for IVIFS in [18]. Hong in [19] presented concepts of correlation coefficient. Xu and Chen in [20] developed distance measurement and similarity measurement in IVIFS. Xu presented aggregation operators for IVIFS in [21]. Such complete theories provide sufficient preconditions for combinations of IVIFS and AHP. Abdullah presented a preference scale by using IVIFS in [22]. Büyüközkan in [23] proposed an approach by integrating IVIFS, AHP and a method of Additive Ratio Assessment. However, the methods in [22] and [23] need judgements on DMs. It is not easy to collect judgements on DMs in the case of large number of decision makers.

To address the above limitations, judgements on DMs are quantified in a way of similarity among DMs' opinions on software attributes. Inspired by [24], DMs' opinions on attributes are described by interval-valued intuitionistic fuzzy matrices, which are transformed from linguistic-term matrices, and the similarities among DMs' opinions are quantified with correlation coefficients between interval-valued intuitionistic fuzzy matrices. An equation for calculating correlation coefficients among interval-valued intuitionistic fuzzy matrices in [24] sums correlation coefficients of each pair of corresponding entries in the two matrices. Because of that, all the diagonal entries in a linguistic-term matrix are the linguistic term of Equally Important, correlation coefficients of corresponding diagonal entries could not reflect true similarities among DMs' opinions on software attributes. Furthermore, such an equation in [24] could not reveal accurately similarities among DMs' opinions. One of the equations for calculating DM's weight in [12] sums all the correlation coefficients of interval-valued intuitionistic fuzzy matrices between a DM and each of the remaindering DMs. Such a calculating-DM-weight way could not show clearly differences among DMs' weights. Additionally, equations for calculating attributes' weights in [23] could not manifest clearly levels of attribute importance. In terms of the above three considerations, a new fuzzy AHP method is designed for software trustworthiness measurement by integrating interval-valued intuitionistic fuzzy set and traditional AHP method at the bases of methods in [24] and [23]. In the proposed method, an equation is designed for calculating correlation coefficients between DM's interval-valued intuitionistic fuzzy matrices by only taking into account correlation coefficients among non-diagonal elements. A parameter of threshold is introduced to select IVIFS matrices correlation coefficients for calculating DMs' weights. The threshold could help manifest differences among DMs' weights. Besides, equations for calculating attribute's weights are designed by leveraging harmonic mean in order to show clearly levels of attribute weights.

The rest of this paper is organized as follows. Section 2 gives basic concepts which are needed in our proposed method. In Sect. 3, the proposed method is presented in detail. An example is given in Sect. 4 to show the advantages of the proposed method with respect to resilience trustworthiness measurement. Section 5 discusses the proposed method. Conclusions and future work are described in Sect. 6.

2 Basic Concepts and Operations

This section introduces basic concepts related to fuzzy sets and interval-valued intuitionistic fuzzy sets, and also gives related operations in our method.

Definition 1. (Fuzzy Set [25]) *A fuzzy set A defined on the universe of discourse* $X = \{x_1, x_2, \ldots, x_n\}$ *is given by*

$$A = \{\langle x, \mu_A(x)\rangle \mid x \in X\}, \tag{1}$$

where μ_A *denotes the membership function of the fuzzy set* A, $\mu_A\colon X \mapsto [0,1]$, *for every* $x \in X$, $\mu_A(x)$ *denotes the membership degree of* x *in* A.

Anatassov extended the fuzzy set and define an intuitionistic fuzzy set in 1986 [16].

Definition 2. (Intuitionistic Fuzzy Set, IFS for abbreviation [16]) *An intuitionistic fuzzy set A defined on the universe of discourse* $X = \{x_1, x_2, \ldots, x_n\}$ *is given by*

$$A = \{\langle x, \mu_A(x), \nu_A(x)\rangle \mid x \in X\}, \tag{2}$$

where $\mu_A, \nu_A\colon X \mapsto [0,1]$, *for every* $x \in X$, $\mu_A(x)$ *denotes the membership degree and* $\nu_A(x)$ *denotes the non-membership degree of x in A, respectively. This interpretation entails a natural restriction, i.e.,*

$$0 \le \nu_A(x) + \mu_A(x) \le 1.$$

$\pi_A(x)\colon X \mapsto [0,1]$ *is defined as follows:*

$$\pi_A(x) = 1 - \nu_A(x) - \mu_A(x). \tag{3}$$

$\pi_A(x)$ *denotes hesitate degree or intuitionistic index of x in A.*

An interval-valued fuzzy set is a generalization of the notion of fuzzy set [26].

Definition 3. (Interval-Valued Fuzzy Set, IVFS for abbreviation [26]) *An interval-valued fuzzy set A defined on the universe of discourse* $X = \{x_1, x_2, \ldots, x_n\}$ *is given by*

$$A = \{\langle x, \bar{\mu}_A(x)\rangle \mid x \in X\}, \tag{4}$$

where $\bar{\mu}_A\colon X \mapsto D[0,1]$, $D[0,1]$ *is the set of all subintervals of the unit interval* $[0,1]$, *i.e. for every* $x \in X$, $\bar{\mu}_A(x) = [\mu_A^L(x), \mu_A^U(x)]$ *is an interval within* $[0,1]$, *where* $\mu_A^L(x)$ *is the lower bound of membership degree of x in A, and* $\mu_A^U(x)$ *is the upper bound of the membership degree of x in A.*

Definition 4. (Mappings between interval-valued fuzzy set and intuitionistic fuzzy set [17])

(1) Let $A = \{\langle x, [\mu_A^L(x), \mu_A^U(x)]\rangle \mid x \in X\}$ *be an interval-valued fuzzy set, where* $\mu_A^L(x)$ *and* $\mu_A^U(x)$ *separately represent the lower bound and the upper bound of the membership degree of x in A. A map f assigns an intuitionistic fuzzy set B to the IVFS A, i.e.,*

$$B = f(A) = \{\langle x, \mu_A^L(x), 1 - \mu_A^U(x)\rangle \mid x \in X\},$$

(2) Let $B = \{\langle x, \mu_B(x), \nu_B(x)\rangle \mid x \in X\}$ *be an intuitionistic fuzzy set, where* $\mu_B(x)$ *and* $\nu_B(x)$ *separately represent the membership degree and the non-membership degree of x in A. A map g assigns an interval-valued fuzzy set A to the IFS B, i.e.,*

$$A = g(B) = \{\langle x, [\mu_B(x), 1 - \nu_B(x)]\rangle \mid x \in X\}.$$

Combining the IFS and the IVFS, Anatassov defined the interval-valued intuitionistic fuzzy set [17].

Definition 5. (Interval-Valued Intuitionistic Fuzzy Set, IVIFS for abbreviation [17]) *An interval-valued intuitionistic fuzzy set* $\tilde{A}$ *defined on the universe of discourse* $X = \{x_1, x_2, \ldots, x_n\}$ *is given by*

$$\tilde{A} = \{\langle x, \tilde{\mu}_{\tilde{A}}(x), \tilde{\nu}_{\tilde{A}}(x)\rangle \mid x \in X\}, \tag{5}$$

where $\tilde{\mu}_{\tilde{A}}, \tilde{\nu}_{\tilde{A}}: X \mapsto D[0,1]$. *For every* $x \in X$,

$$0 \leq \mu^L_{\tilde{A}}(x) + \nu^L_{\tilde{A}}(x) \leq \mu^U_{\tilde{A}}(x) + \nu^U_{\tilde{A}}(x) \leq 1,$$

$\tilde{\pi}_{\tilde{A}}(x) = [\pi^L_{\tilde{A}}(x), \pi^U_{\tilde{A}}(x)]$, *where* $\pi^L_{\tilde{A}}(x) = 1 - \mu^U_{\tilde{A}}(x) - \nu^U_{\tilde{A}}(x), \pi^U_{\tilde{A}}(x) = 1 - \mu^L_{\tilde{A}}(x) - \nu^L_{\tilde{A}}(x)$. $\tilde{\mu}_{\tilde{A}}(x), \tilde{\nu}_{\tilde{A}}(x), \tilde{\pi}_{\tilde{A}}(x)$ *denote the membership degree, non-membership degree and the intuitionistic index of* x *in* $\tilde{A}$*, respectively.*

Remark 1. For convenience, let $\tilde{\mu}_{\tilde{A}}(x_i) = [a_i, b_i], \tilde{\nu}_{\tilde{A}}(x_i) = [c_i, d_i]$, then $\tilde{\alpha}_i = ([a_i, b_i], [c_i, d_i])$ is called an interval-valued intuitionistic fuzzy number (shortly, IVIFN).

Park defined correlation coefficient between two interval-valued intuitionistic fuzzy sets $\tilde{A}$ and $\tilde{B}$ in [27].

Definition 6. (Correlation coefficients between interval-valued intuitionistic fuzzy sets [27]) *Let* $X = \{x_1, x_2, \ldots, x_n\}$ *be a finite set and* $\tilde{A} = \{\langle x_i, \tilde{\mu}_{\tilde{A}}(x_i), \tilde{\nu}_{\tilde{A}}(x_i)\rangle \mid x_i \in X\}, \tilde{B} = \{\langle x_i, \tilde{\mu}_{\tilde{B}}(x_i), \tilde{\nu}_{\tilde{B}}(x_i)\rangle \mid x_i \in X\}$ *be two interval-valued intuitionistic fuzzy sets over the set of* X. *A correlation coefficient between* $\tilde{A}$ *and* $\tilde{B}$ *is defined to be*

$$c(\tilde{A}, \tilde{B}) = \frac{\gamma(\tilde{A}, \tilde{B})}{(\gamma(\tilde{A}, \tilde{A}) \cdot \gamma(\tilde{B}, \tilde{B}))^{\frac{1}{2}}}, \tag{6}$$

where

$$\begin{aligned}\gamma(\tilde{A}, \tilde{B}) = \frac{1}{2}\sum_{i=1}^{n}[&\mu^L_{\tilde{A}}(x_i) \cdot \mu^L_{\tilde{B}}(x_i) + \mu^U_{\tilde{A}}(x_i) \cdot \mu^U_{\tilde{B}}(x_i) + \nu^L_{\tilde{A}}(x_i) \cdot \nu^L_{\tilde{B}}(x_i) \\ &+ \nu^U_{\tilde{A}}(x_i) \cdot \nu^U_{\tilde{B}}(x_i) + \pi^L_{\tilde{A}}(x_i) \cdot \pi^L_{\tilde{B}}(x_i) + \pi^U_{\tilde{A}}(x_i) \cdot \pi^U_{\tilde{B}}(x_i)],\end{aligned} \tag{7}$$

$$\begin{aligned}\gamma(\tilde{A}, \tilde{A}) = \frac{1}{2}\sum_{i=1}^{n}[&(\mu^L_{\tilde{A}}(x_i))^2 + (\mu^U_{\tilde{A}}(x_i))^2 + (\nu^L_{\tilde{A}}(x_i))^2 \\ &+ (\nu^U_{\tilde{A}}(x_i))^2 + (\pi^L_{\tilde{A}}(x_i))^2 + (\pi^U_{\tilde{A}}(x_i))^2].\end{aligned} \tag{8}$$

The calculation of $\gamma(\tilde{B}, \tilde{B})$ *is in the same manner as* $\gamma(\tilde{A}, \tilde{A})$.

Following Definition 6, Jia and Zhang extended the definition of correlation coefficient between IVIFSs to a correlation coefficient between interval-valued intuitionistic fuzzy matrices [24].

Definition 7. (Correlation coefficients between interval-valued intuitionistic fuzzy matrices [24]) *Let* $D_1 = [\alpha_{jk}]_{J\times K}$ *and* $D_2 = [\beta_{jk}]_{J\times K}$ *be two interval-valued intuitionistic fuzzy matrices, in which each element is an interval-valued intuitionistic fuzzy number. A correlation coefficient between* D_1 *and* D_2 *is defined by*

$$C(D_1, D_2) = \frac{1}{JK}\sum_{j=1}^{J}\sum_{k=1}^{K} c(\alpha_{jk}, \beta_{jk}), \tag{9}$$

Equation 9 in Definition 7 sums correlation coefficients of each pair of corresponding entries in two interval-valued intuitionistic fuzzy matrices. As we know, diagonal entries in each pair-wise comparison matrix obtained by the AHP method represent comparison results between identical attributes, and they are all the element of Equally Important. The diagonal entries in the comparison matrices are independent of decision makers, and they are not DMs' key opinions. Such a calculation way in Eq. 9 sums the correlation coefficients between each pair of corresponding entry in two pair-wise comparison matrices, and its calculated results include measurement of similarity among DMs' non-key opinions.

In order to reveal similarities among DMs' key opinions on software attributes, a new equation is proposed in the following Definition 8 by deleting computations of corresponding diagonal entries in Eq. 9.

Definition 8. *Let* $D_1 = [\alpha_{jk}]_{m\times m}$ *and* $D_2 = [\beta_{jk}]_{m\times m}$ *be two interval-valued intuitionistic fuzzy matrices. A correlation coefficient between* D_1 *and* D_2 *is defined by*

$$\bar{C}(D_1, D_2) = \frac{1}{m^2 - m}\sum_{j=1}^{m}\sum_{k=1,k\neq j}^{m} c(\alpha_{jk}, \beta_{jk}) \tag{10}$$

To aggregate IVIFNs, Xu introduced an operator of interval-valued intuitionistic fuzzy weighted arithmetic in [21].

Definition 9. (Interval-Valued Intuitionistic Fuzzy Weighted Arithmetic, IVIFWA for abbreviation [21]) *Let* $\tilde{\alpha_i} = ([a_i, b_i], [c_i, d_i])$ *be interval-valued intuitionistic fuzzy number,* $i = 1, 2, \ldots, n$. *An operator of interval-valued intuitionistic fuzzy weighted arithmetic is defined by*

$$\begin{aligned} IVIFWA_\sigma(\tilde{\alpha_1}, \tilde{\alpha_2}, \ldots, \tilde{\alpha_n}) = \Bigg(\Bigg[1 - \prod_{i=1}^{n}(1 - a_i)^{\sigma_i}, 1 - \prod_{i=1}^{n}(1 - b_i)^{\sigma_i}\Bigg], \\ \Bigg[\prod_{i=1}^{n} c_i^{\sigma_i}, \prod_{i=1}^{n} d_i^{\sigma_i}\Bigg]\Bigg) \end{aligned} \tag{11}$$

where $\sigma = (\sigma_1, \sigma_2, \ldots, \sigma_n)^\top$ *is a vector of weights,* σ_i *is a weight of* $\tilde{\alpha_i}$, *satisfying* $\sigma_i \geq 0$ *and* $\sum_{i=1}^{n}\sigma_i = 1$, $i = 1, 2, \ldots, n$.

3 Methods

Büyüközkan proposed a method based on methodologies of additive ratio assessment and AHP to evaluate decision makers' weights and attributes' weights in

[23]. However, the method requires decision makers to make judgements on each other. Such a requirement is not easy to execute in the case of large number of decision makers. Also, the judgement results are undesirable in the case that decision makers have prejudices. Inspired by [24], a method in this study is designed to address the above disadvantage.

Our proposed method consists of eight steps. The steps are listed as follows.

Step 1. Call together n Decision Makers (DMs) as a committee.

Step 2. Determine m software attributes to evaluate and design linguistic scales which are used to describe DMs' opinions on software attributes.

Step 3. Collect DMs' opinions on software attributes.

Step 4. Construct a pair-wise comparison matrix by the way of establishing comparison matrices in the AHP method [4], then transform linguistic terms in the obtained matrix into IVIFNs according to Table 1 [22].

Table 1. Conversion preference scale of IVIFN

Preference on comparison	Acronym	IVIFN		
		$[\mu_{\tilde{A}}^L(x), \mu_{\tilde{A}}^U(x)]$	$[\nu_{\tilde{A}}^L(x), \nu_{\tilde{A}}^U(x)]$	$[\pi_{\tilde{A}}^L(x), \pi_{\tilde{A}}^U(x)]$
Equally Important	EI	[0.38, 0.42]	[0.22, 0.58]	[0.00, 0.40]
Equally Very Important	EVI	[0.29, 0.41]	[0.12, 0.58]	[0.01, 0.59]
Moderately Important	MI	[0.10, 0.43]	[0.03, 0.57]	[0.00, 0.87]
Moderately More Important	MMI	[0.03, 0.47]	[0.03, 0.53]	[0.00, 0.94]
Strongly Important	SI	[0.13, 0.53]	[0.07, 0.47]	[0.00, 0.80]
Strongly More Important	SMI	[0.32, 0.62]	[0.08, 0.38]	[0.00, 0.60]
Very Strongly More Important	VSMI	[0.52, 0.72]	[0.08, 0.28]	[0.00, 0.40]
Extremely Strong Important	ESI	[0.75, 0.85]	[0.05, 0.15]	[0.00, 0.20]
Extremely More Important	EMI	[1.00, 1.00]	[0.00, 0.00]	[0.00, 0.00]

For reciprocal preferences, related IVIFNs could be obtained by interchanging $[\mu_{\tilde{A}}^L(x), \mu_{\tilde{A}}^U(x)]$ and $[\nu_{\tilde{A}}^L(x), \nu_{\tilde{A}}^U(x)]$. For example, the IVIFN for EVI is ([0.29, 0.41], [0.12, 0.58]), then the IVIFN for 1/EVI is ([0.12, 0.58], [0.29, 0.41]).

Step 5. Calculate DMs' weights using correlation coefficients. The weight of DM_i is calculated according to the following Eq. 12 [24] and Eq. 13.

$$\lambda_i = \frac{\theta_i}{\sum_{j=1}^{n} \theta_j} \tag{12}$$

$$\theta_i = \sum_{i' \in S_i} \bar{C}(D_i, D_{i'}) \tag{13}$$

where $S_i = \{i' \mid \bar{C}(D_i, D_{i'}) \geq T, i' \neq i, i' \in \{1, 2, \ldots, n\}\}$, D_i denotes interval-valued intuitionistic fuzzy matrices of DM_i, $i = 1, 2, \ldots, n$.

The symbol T in Eq. 13 means a threshold, which is valued in the closed interval of [0, 1]. Such a setting is to present clearly differences among DMs' weights. The value of T could be set according to research needs. Specially, the obtained DMs' weights in the setting of $T = 0$ are the same as the weight results by the related calculation method in [24].

So, our method in Eq. 12 and Eq. 13 for calculating DMs' weights extends the related calculation method in [24].

Step 6. By using the IVIFWA operator [21] (i.e., Eq. 11), aggregate per-attributes' results of opinions on software attributes for each decision maker. In each aggregation, the used vector of weights is $\sigma = (\sigma_1, \sigma_2, \ldots, \sigma_m)^\top$, where $\sigma_i = \frac{1}{m}$, $i = 1, 2, \ldots, m$.

Step 7. Calculate consistent ratio about opinions of the decision maker DM_i according to Eq. 14 [22]

$$CR_{DM_i} = \frac{RI - \frac{\sum_{k=1}^{m} \pi^U_{\tilde{A}_{DM_i}}(x_k)}{m}}{m-1}, \tag{14}$$

where CR_{DM_i} means consistent ratio for the decision maker DM_i's opinions on software attributes, $i = 1, 2, \ldots, n$. $\tilde{A}_{DM_i} = \{\langle x_k, \tilde{\mu}_{\tilde{A}_{DM_i}}(x_k), \tilde{\nu}_{\tilde{A}_{DM_i}}(x_k)\rangle, k = 1, 2, \ldots, m\}$ denotes an interval-valued intuitionistic fuzzy set of aggregation results of the decision maker DM_i's opinions on software attributes, $i = 1, 2, \ldots, n$. RI is a random index which is valued according to table Table 2 [28]. If the absolute value of CR_{DM_i} is not more than 0.10, then the consistency of the judgement matrix is acceptable in this study. Otherwise, it is not acceptable, and go back to Step 1.

Table 2. Random Index (RI)

len	1	2	3	4	5	6	7	8	9	10
RI	0.00	0.00	0.58	0.90	1.12	1.24	1.32	1.41	1.45	1.49

Step 8. Calculate attributes' weights using Eq. 15 [23] and Eq. 16.

$$w_k = \frac{1 - \breve{w}_k}{m - \sum_{j=1}^{m} \breve{w}_j} \tag{15}$$

$$\breve{w}_k = 1 - \frac{\sum_{i=1}^{n} \lambda_i \cdot \frac{2}{\frac{1}{\mu^L_{\tilde{A}_{DM_i}}(x_k)} + \frac{1}{\mu^U_{\tilde{A}_{DM_i}}(x_k)}}}{\sum_{i=1}^{n} \lambda_i \cdot \left(\frac{2}{\frac{1}{\mu^L_{\tilde{A}_{DM_i}}(x_k)} + \frac{1}{\mu^U_{\tilde{A}_{DM_i}}(x_k)}} + N(i,k) \right)} \tag{16}$$

$$N(i,k) = \begin{cases} \frac{2}{\frac{1}{\nu^L_{\tilde{A}_{DM_i}}(x_k)} + \frac{1}{\nu^U_{\tilde{A}_{DM_i}}(x_k)}}, & \text{if } \nu^L_{\tilde{A}_{DM_i}}(x_k) \cdot \nu^U_{\tilde{A}_{DM_i}}(x_k) \neq 0 \\ 0, & \text{otherwise} \end{cases}$$

where x_k denotes the attribute of $attr_k$, $k = 1, 2, \ldots, m$. $\tilde{A}_{DM_i} = \{\langle x_k, [\mu^L_{\tilde{A}_{DM_i}}(x_k), \mu^U_{\tilde{A}_{DM_i}}(x_k)], [\nu^L_{\tilde{A}_{DM_i}}(x_k), \nu^U_{\tilde{A}_{DM_i}}(x_k)]\rangle, k = 1, 2, \ldots, m\}$ means an interval-valued intuitionistic fuzzy set of aggregation results of the decision maker DM_i's opinions on software attributes, $i = 1, 2, \ldots, n$. w_k denotes the weight of the k-th attribute, $\check{w}_k$ denotes the pre-weight of the k-th attribute, $k = 1, 2, \ldots, m$. λ_i means the weight of the decision maker DM_i, $i = 1, 2, \ldots, n$.

Based on the obtained attributes' weights, a software trustworthiness measurement value could be calculated by Eq. 17 [7,8].

$$R = ATTR_1^{w_1} \times ATTR_2^{w_2} \times \ldots \times ATTR_m^{w_m} \tag{17}$$

In Eq. 17, the w_k represents the weight of the k-th attribute and $\sum_{k=1}^{m} w_k = 1$, $ATTR_k$ represents the trustworthiness measurement value of the k-th attribute $attr_k$.

4 Examples

In this section, evaluating the resilience of an operating system is taken as an example. The resilience of an operating system is described by three attributes, including the attribute of survivability denoted by $attr_1$, the attribute of recoverability denoted by $attr_2$, and the attribute of adaptability denoted by $attr_3$. In the example, our proposed method is compared with methods in [23] and [8]. Eight experts (i.e., DMs) are called together to provide their opinions on the resilience attributes. Based on their opinions, DMs' weights and attributes' weights are calculated.

Step 1. Eight DMs are called together online to provide their opinions on resilience attributes of operating system.

Step 2. Questionnaires with respect to the resilience of an operating system are provided to DMs to ask for opinions on the three attributes, including survivability, recoverability, and adaptability. Our used linguistic scales in the questionnaires are the same as in Table 1.

Step 3. Opinions on the resilience attributes from eight DMs are collected. Because of space limitation, opinions from only two DMs are shown in Table 3[1].

Table 3. Two DMs' judgements on attributes

(a) DM_1's judgements

Attrs	$attr_1$	$attr_2$	$attr_3$
$attr_1$	EI	1/MI	VSMI
$attr_2$	MI	EI	SI
$attr_3$	1/VSMI	1/SI	EI

(b) DM_2's judgements

Attrs	$attr_1$	$attr_2$	$attr_3$
$attr_1$	EI	1/SMI	EI
$attr_2$	SMI	EI	SMI
$attr_3$	EI	1/SMI	EI

[1] Please check out https://jihulab.com/lukedyue/aila2022 for all results.

Step 4. Based on Table 1 and Table 3, pair-wise comparison matrices are constructed, in which all the elements are IVIFNs. Because of space limitation, only two matrices of our obtained comparison matrices are separately given in the form of table in Table 4 and Table 5.

Table 4. IVIFNs corresponding to DM_1's judgements

Attrs	$attr_1$	$attr_2$	$attr_3$
$attr_1$	([0.38, 0.42], [0.22, 0.58])	([0.03, 0.57], [0.10, 0.43])	([0.52, 0.72], [0.08, 0.28])
$attr_2$	([0.10, 0.43], [0.03, 0.57])	([0.38, 0.42], [0.22, 0.58])	([0.13, 0.53], [0.07, 0.47])
$attr_3$	([0.08, 0.28], [0.52, 0.72])	([0.07, 0.47], [0.13, 0.53])	([0.38, 0.42], [0.22, 0.58])

Step 5. The threshold of T in Eq. 12 is set to be 0.90. DMs' weights are calculated by using Eq. 12 and Eq. 13. The obtained DMs' weights are given in Table 6.

Step 6. Aggregate DMs' judgements with the IVIFWA operator by using the obtained matrices in Step 4 and the weight vector of $(\frac{1}{3}, \frac{1}{3}, \frac{1}{3})^{\top}$. For better understanding, a calculation process of aggregating DM_1's judgements on the attribute of survivability (i.e., $attr_1$) is given in Eq. 18. Because of the space limitation, aggregation results for only two DMs are shown in Table 7.

$$\begin{aligned}
&IVIFWA_{attr_1(DM_1)} \\
&= \left[\begin{pmatrix} (1-(1-0.38)^{\frac{1}{3}} \times (1-0.03)^{\frac{1}{3}} \times (1-0.52)^{\frac{1}{3}}), \\ (1-(1-0.42)^{\frac{1}{3}} \times (1-0.57)^{\frac{1}{3}} \times (1-0.72)^{\frac{1}{3}}) \end{pmatrix}, \right. \\
&\left. \begin{pmatrix} (0.22)^{\frac{1}{3}} \times (0.10)^{\frac{1}{3}} \times (0.08)^{\frac{1}{3}}, \\ (0.58)^{\frac{1}{3}} \times (0.43)^{\frac{1}{3}} \times (0.28)^{\frac{1}{3}} \end{pmatrix}\right] \\
&= [(0.339, 0.588), (0.121, 0.412)]
\end{aligned} \tag{18}$$

Table 5. IVIFNs corresponding to DM_2's judgements

Attrs	$attr_1$	$attr_2$	$attr_3$
$attr_1$	([0.38, 0.42], [0.22, 0.58])	([0.08, 0.38], [0.32, 0.62])	([0.38, 0.42], [0.22, 0.58])
$attr_2$	([0.32, 0.62], [0.08, 0.38])	([0.38, 0.42], [0.22, 0.58])	([0.32, 0.62], [0.08, 0.38])
$attr_3$	([0.22, 0.58], [0.38, 0.42])	([0.08, 0.38], [0.32, 0.62])	([0.38, 0.42], [0.22, 0.58])

Table 6. Results of DMs' weight

DM	DM_1	DM_2	DM_3	DM_4	DM_5	DM_6	DM_7	DM_8
Weight	0.1185	0.1418	0.1433	0.1073	0.1189	0.1088	0.1319	0.1295

Table 7. Aggregation results for two DMs

Attrs	DM_1	DM_2
$attr_1$	$([0.339, 0.588], [0.121, 0.412])$	$([0.293, 0.407], [0.249, 0.593])$
$attr_2$	$([0.214, 0.462], [0.077, 0.538])$	$([0.341, 0.562], [0.112, 0.438])$
$attr_3$	$([0.190, 0.395], [0.246, 0.605])$	$([0.237, 0.467], [0.299, 0.533])$

Step 7. The consistency ratio of DM_1's judgement matrix is calculated using Eq. 14.

$$CR_{DM_1} = \frac{0.58 - \left(\frac{1-0.339-0.121+1-0.214-0.077+1-0.190-0.246}{3}\right)}{3-1} = -0.012$$

The absolute value of CR_{DM_1} is no more than 0.10. So, consistency of DM_1's judgement matrix is considered to be acceptable. Similarly, consistency ratios for other DMs' judgement matrices are calculated. All the achieved consistency ratio results are listed in Table 8, and all the DMs' judgement matrices are acceptable.

Table 8. Results of consistency ratios

DMs	DM_1	DM_2	DM_3	DM_4	DM_5	DM_6	DM_7	DM_8
CR	−0.012	0.045	0.020	0.013	0.013	0.004	−0.008	−0.023

Step 8. Calculate attributes' weights using the Eq. 15 and Eq. 16 based on the obtained DMs' weights in Step 5 and the aggregated results in Step 6. Our obtained results of attribute weights are shown in Table 9.

Table 9. Results of attribute weight

	Survivability	Recoverability	Adaptability
w	0.322	0.410	0.268

Using methods in [7] and [8], attributes' trustworthiness measurement values in our example are calculated. For the three attributes of survivability, recoverability, and adaptability, our obtained trustworthiness measurement values are 9.240, 6.973, and 6.088, respectively. By Eq. 17, the resilience trustworthiness measurement value R is calculated as follows. The obtained value of R is 7.362.

$$R = 9.240^{0.322} \times 6.973^{0.410} \times 6.088^{0.268} = 7.362$$

For comparisons, an original method is applied to this example. In the original method, an equation in [23] (i.e., Eq. 19) is used to calculate attributes' pre-weights instead of using Eq. 16. Beyond that, the original method uses the same equations and steps as our proposed method.

$$\check{w}_k = 1 - \frac{\sum_{i=1}^{n} \lambda_i \cdot \frac{\mu^L_{\tilde{A}_{DM_i}}(x_k)+\mu^U_{\tilde{A}_{DM_i}}(x_k)}{2}}{\sqrt{\sum_{i=1}^{n} \lambda_i \cdot \frac{(\mu^L_{\tilde{A}_{DM_i}}(x_k))^2+(\mu^U_{\tilde{A}_{DM_i}}(x_k))^2+(\nu^L_{\tilde{A}_{DM_i}}(x_k))^2+(\nu^U_{\tilde{A}_{DM_i}}(x_k))^2}{2}}} \quad (19)$$

Also, the traditional AHP method in [8] is applied to the task in this example. For applying the traditional AHP method, the linguistic terms in our method are kept, and they are quantified with definite numeric values. The achieved results for the three methods are shown in Table 10.

Table 10. Evaluation results for three methods

Methods	Results			
	$attr_1$'s weight	$attr_2$'s weight	$attr_3$'s weight	Resilience Measurement Value
Our method	0.322	0.410	0.268	7.362
Original method [23]	0.327	0.382	0.291	7.349
Traditional AHP [8]	0.250	0.621	0.129	7.352

Evaluation results in Table 10 show that the orders of attribute weights are the same for the three methods, and differences between obtained attribute weights by our proposed method are greater than the attribute weight differences by using the original method in [23]. The greater the differences of attribute weights are, the more evident the levels of attribute importance are. The traditional AHP method only supports using definite numerical values to describe DMs' opinions, although its obtained attribute weight differences are the maximal. The way of the definite numerical values is not enough to quantify DMs' opinions. Our proposed method not only supports using interval-valued fuzzy set but also produces attribute weights with great differences.

5 Discussion

The traditional AHP method uses definite numerical values to describe DMs' opinions. Such a way could not characterize sufficiently linguistic terms which are DMs' opinions on software attributes. On the other side, DMs' opinions on attributes usually are assigned weights, which are called DMs' weights. DMs' weights may be collected from a manager or in a way that decision makers provide judgements on each other. Such a way of achieving DMs' weights is hard to execute in the case of large number of decision makers. Considering the above two disadvantages, we proposed a new method for software trustworthiness measurement by leveraging interval-valued intuitionistic fuzzy set and calculating DMs' weights based on correlation coefficients between DMs' opinions on software attributes.

As we know, pair-wise comparison matrices need to be constructed by the related means in the AHP method according to DMs' opinions on software attributes. In the obtained matrices, the diagonal entries are all the linguistic terms of Equally Important. The diagonal elements in the matrices could not show strong similarities among DMs' opinions. Our designed equation (i.e., Eq. 10 in Definition 8) throws away computing correlation coefficients of corresponding diagonal entries. Such a calculation reveals true similarities among DMs' opinions.

It is observed that each decision maker's weight obtained by a method in [24] is close to an average weight. Considering such a disadvantage of the method in [24], a parameter of threshold is introduced in our proposed method as we see in Eq. 12 and Eq. 13. Such a parameter is established in order to manifest differences among DMs' weights. In the case of the threshold being zero, the calculation of DMs' weights in our method is the same as the method in [24].

In our method, attributes' weights are calculated by using harmonic mean as we see in Eq. 15 and Eq. 16, instead of using calculating-attribute-weights equations in [23]. Our equations manifest more evident levels of attribute importance than the method in [23]. Also, the equations are more suitable for such a case where the lower bounds of IVIFN's membership degree and non-membership degree vary more significantly than the upper bounds.

6 Conclusions

This paper focuses on the software trustworthiness measurement. A new method is designed by combining the Analytic Hierarchy Process method with interval-valued intuitionistic fuzzy set. In our method, there are three improvements. The first one is that, an equation is improved for calculating correlation coefficients between interval-valued intuitionistic fuzzy matrices in order to quantify similarities among DMs' opinions on software attributes. The second one is that, a parameter of threshold is introduced in order to improve the calculation of DMs' weights. The third one is that, equations for the attribute weights calculation are improved by leveraging harmonic mean in order to manifest levels of attribute importance. In a task of evaluating the resilience of an operating system, our method is compared to the traditional AHP method and a method in [23]. Our experimental results show that, the proposed method achieves the same order of attribute weights as the above two methods. Furthermore, our method not only obtains greater differences among attribute weights than the method in [23], but also describes more sufficiently DMs' opinions than the traditional AHP method.

As we know, there are no definite indexes for evaluating calculating-attribute-weights methods in the research of software trustworthiness measurement. In the future, indexes for the evaluation of measuring-software-trustworthiness methods will be explored. Also, more applications will be practiced for the proposed method.

Acknowledgement. This work is supported by the East China Normal University - Huawei Trustworthiness Innovation Center and the Shanghai Trusted Industry Internet Software Collaborative Innovation Center.

References

1. Wong, W.E., Li, X., Laplante, P.A.: Be more familiar with our enemies and pave the way forward: a review of the roles bugs played in software failures. J. Syst. Softw. **133**, 68–94 (2017)
2. He, J., et al.: Review of the achievements of major research plan of trustworthy software. Sci. Found. China **32**(3), 291–296 (2018)
3. Maza, S., Megouas, O.: Framework for trustworthiness in software development. Int. J. Perform. Eng. **17**(2), 241–252 (2021)
4. Saaty, T.L.: A scaling method for priorities in hierarchical structures. J. Math. Psychol. **15**(3), 234–281 (1977)
5. Di Angelo, L., Di Stefano, P., Fratocchi, L., Marzola, A.: An AHP-based method for choosing the best 3D scanner for cultural heritage applications. J. Cult. Herit. **34**, 109–115 (2018)
6. Ishizaka, A., Pearman, C., Nemery, P.: AHPSort: an AHP-based method for sorting problems. Int. J. Prod. Res. **50**(17), 4767–4784 (2012)
7. Tao, H.: Research on the measurement models of software trustworthiness based on attributes. Doctor, East China Normal University, Shanghai, China, April 2011
8. Wang, B.: Research on trustworthiness measurement models based on software component. Doctor, East China Normal University, Shanghai, China, October 2019
9. Kahraman, C., Cebeci, U., Ulukan, Z.: Multi-criteria supplier selection using fuzzy AHP. Logist. Inf. Manag. **16**(6), 382–394 (2003)
10. Wang, T.C., Chen, Y.H.: Applying consistent fuzzy preference relations to partnership selection. Omega **35**(4), 384–388 (2007)
11. Deng, H.: Multicriteria analysis with fuzzy pairwise comparison. Int. J. Approx. Reason. **21**(3), 215–231 (1999)
12. van Laarhoven, P., Pedrycz, W.: A fuzzy extension of Saaty's priority theory. Fuzzy Sets Syst. **11**(1–3), 229–241 (1983)
13. Chang, D.Y.: Applications of the extent analysis method on fuzzy AHP. Eur. J. Oper. Res. **95**(3), 649–655 (1996)
14. Peng, G., Han, L., Liu, Z., Guo, Y., Yan, J., Jia, X.: An application of fuzzy analytic hierarchy process in risk evaluation model. Front. Psychol. **12**, 715003 (2021)
15. Tan, R., Aviso, K., Huelgas, A., Promentilla, M.: Fuzzy AHP approach to selection problems in process engineering involving quantitative and qualitative aspects. Process Saf. Environ. Prot. **92**(5), 467–475 (2014)
16. Atanassov, K.T.: Intuitionistic fuzzy sets. Fuzzy Sets Syst. **20**(1), 87–96 (1986)
17. Atanassov, K., Gargov, G.: Interval valued intuitionistic fuzzy sets. Fuzzy Sets Syst. **31**(3), 343–349 (1989). https://doi.org/10.1007/978-3-7908-1870-3_2
18. Atanassov, K.T.: Operators over interval valued intuitionistic fuzzy sets. Fuzzy Sets Syst. **64**(2), 159–174 (1994)
19. Hong, D.H.: A note on correlation of interval-valued intuitionistic fuzzy sets. Fuzzy Sets Syst. **95**(1), 113–117 (1998)
20. Xu, Z.S., Chen, J.: An overview of distance and similarity measures of intuitionistic fuzzy sets. Int. J. Uncertain. Fuzziness Knowl.-Based Syst. **16**(04), 529–555 (2008)
21. Xu, Z.: Methods for aggregating interval-valued intuitionistic fuzzy information and their application to decision making. Control Decis. **22**(2), 215–219 (2007)

22. Abdullah, L., Najib, L.: A new preference scale MCDM method based on interval-valued intuitionistic fuzzy sets and the analytic hierarchy process. Soft. Comput. **20**(2), 511–523 (2016). https://doi.org/10.1007/s00500-014-1519-y
23. Büyüközkan, G., Göçer, F.: An extension of ARAS methodology under interval valued intuitionistic fuzzy environment for digital supply chain. Appl. Soft Comput. **69**, 634–654 (2018)
24. Jia, Z., Zhang, Y.: Interval-valued intuitionistic fuzzy multiple attribute group decision making with uncertain weights. Math. Probl. Eng. **2019**, 1–9 (2019)
25. Zadeh, L.: Fuzzy sets. Inf. Control **8**(3), 338–353 (1965)
26. Gorzałczany, M.B.: A method of inference in approximate reasoning based on interval-valued fuzzy sets. Fuzzy Sets Syst. **21**(1), 1–17 (1987)
27. Park, D.G., Kwun, Y.C., Park, J.H., Park, I.Y.: Correlation coefficient of interval-valued intuitionistic fuzzy sets and its application to multiple attribute group decision making problems. Math. Comput. Model. **50**(9–10), 1279–1293 (2009)
28. Saaty, T.L.: The Analytic Hierarchy Process: Planning, Priority Setting, Resource Allocation. McGraw-Hill International Book Co, New York (1980)

A Novel Trustworthiness Measurement Method for Software System Based on Fuzzy Set

Qilong Nie[1], Yixiang Chen[1(✉)], and Hongwei Tao[2]

[1] National Engineering Research Center of Trustworthy Embedded Software, East China Normal University, Shanghai 200062, China
51205902018@stu.ecnu.edu.cn, yxchen@sei.ecnu.edu.cn

[2] College of Computer and Communication Engineering, Zhengzhou University of Light Industry, Zhengzhou 450001, China
hongweitao@zzuli.edu.cn

Abstract. The software has been a part of our daily life. However, software systems are becoming more and more complex, with many uncertainties, unavoidable software bugs, failures, and even disasters. The measurement of software trustworthiness has already attracted attention from both academia and industry. Today, component-based software systems (CBSS) have become mainstream due to their high reusability and low development cost. How to accurately measure the trustworthiness of CBSS has become an urgent problem to be solved. In this paper, to overcome this problem, we calculate the importance of components according to the fault propagation impact and function importance and propose a fuzzy criticality model to distinguish critical components from non-critical components. Finally, we propose a hierarchical trustworthiness computing model to measure the trustworthiness of software systems. Through the results of experiments, the necessity of determining critical components and the rationality of the hierarchical trustworthiness measurement model are verified.

Keywords: Component-based software system · Fuzzy criticality model · Hierarchical trustworthiness computing model

1 Introduction

Software systems are becoming a crucial part of today's society and playing an essential role in everything from military, cultural, economic, and political to people's daily lives. There will be many problems in the software system because the software becomes bigger and more complex, such as uncertainty, abnormal vulnerability, failure, bugs, failure, and other problems [11]. The trustworthiness of software has become the focus of attention.

The component-based software systems (CBSS) can reduce the coupling degree of the software system, improve the development efficiency of the software

Y. Chen and S. Zhang (Eds.): AILA 2022, CCIS 1657, pp. 254–269, 2022.
https://doi.org/10.1007/978-981-19-7510-3_19

system, which have gradually become mainstream. In the CBSS, a component is a unit that can be deployed independently, which has an interface is easily assembled and used by third parties [5]. As a module runs independently, components improve the reuse rate and shorten the development time. But, it also increases the risk of system failure because once a component fails, the failure will be propagated, causing the system not to run correctly. How to measure the trustworthiness of the CBSS has become an urgent issue to be solved.

At present, there has been some research on the trustworthiness measurement of the CBSS. Chinnaiyan et al. used the Markov process to evaluate the reliability of component-based software system, which considered the impact of failures within the system [9]. Mao et al. proposed a general reliability model based on component-based software systems to measure the reliability of common software systems on the market [15]. Zhang et al. proposed a component dynamic transfer graph, which considers the dependencies between the components [24]. Krishnamurthy et al. combined failure rate with software architecture to compute path reliability for dynamic operation [12]. Chen et al. proposed a reliability assessment method based on the effects of components [7]. Zheng et al. proposed autonomic trust management for a component-based software system to measure the trustworthiness of system automatically [22].

In this paper, we propose a hierarchical trustworthiness measurement method for CBSS to help developers improve systems' trustworthiness during the software life cycle. Compared to the other existing methods, our approach pays more attention to the importance of components of software systems. We compute the importance of the components according to the impact of fault propagation and function importance and distinguish the critical components and the non-critical components according to the fuzzy criticality model. Finally, we compute the trustworthiness of CBSS according to the hierarchical trustworthiness measurement method.

2 Background and Related Work

2.1 Trustworthiness

Software trustworthiness is a relatively new concept introduced to software engineering in the 1980s. In recent years, various kinds of literature have given different definitions of trustworthiness. The U.S. Department of Defense proposed in 1985 that a system is considered trustworthy if it uses sufficient hardware and software integrity metrics to ensure that it can handle a range of sensitive or classified information at the same time [16]. The ISO/IEC15408 standard proposed that in a system if the behavior of the components, operations, or processes involved in the calculation is predictable under any operating conditions and can be well resistant to application software, bugs, and specific physical disturbances if it is destroyed, the system is trustworthy [1]. Liu et al. proposed a trustworthy software as a software system whose dynamic behavior and results always meet people's expectations and do not fail when disturbed [11].

Many scholars and institutions also divided trustworthiness into sub-attributes. National Institute of Standards and Technology (NIST) established a trustworthiness attributes model including accuracy, usability, reliability, availability, resilience, testability, maintainability, performance, safety, security, precision, and conformance [4]. Chen and Tao et al. presented trustworthy attributes based on the whole life cycle, including functionality, reliability, maintainability, survivability, real-time, and safety [6,19,20]. Commonly, trustworthiness is a comprehensive, multidimensional, multidisciplinary concept. It is a challenge to measure the trustworthiness of a software system.

2.2 Software System Trustworthiness Measurement

The software quality mainly reflects the objective state of the software itself, and the trustworthiness measurement of the software system is a quantitative evaluation of the quality of the software system. At present, there has been a lot of research related to software trustworthiness measurement, and Tao et al. proposed an axiom-based measurement model to evaluate the improvement properties [18,21]. Zheng et al. introduced a statistical analysis method of the dynamics of software trustworthiness [25,27]. Ding et al. proposed a software trustworthiness measurement method based on evidence theory [10]. Malaskas et al. proposed a trustworthiness measurement method based on a questionnaire and multivariate statistical analysis, which focused more on the behavior of the software at runtime and compared these behaviors to normal behaviors [17]. The above-mentioned methods mainly regard the software system as a whole by obtaining the attributes of the system and correctly measuring the trustworthiness of the system. Different from these methods, Wang et al. proposed a hierarchical software trustworthiness rating model to calculate the level of a software system [3]. Chen et al. proposed a trustworthiness measurement method for CBSS based on the importance and the trustworthiness of components [7].

3 Methodology

In this section, we propose a trustworthiness computing method for component-based software systems, including a fuzzy criticality model and hierarchical trustworthiness computing mode, which can effectively compute the value of the software system's trustworthiness according to the value of the components' trustworthiness.

3.1 Fuzzy Criticality Model

Different elements provide different effects on the software system and their faults and errors will bring different impacts on the software system's trustworthiness [13]. A critical element's fault may lead to the system's error. In order to calculate the importance of elements, we refer to the method proposed by Chen et al. We comprehensively considered and calculated the importance of each element from the three aspects of self-influence, fault influence, and fault propagation, and then selected some critical elements for system trustworthiness evaluation [7].

There are two kinds of ways for an element to affect the software system's trustworthiness: self-influence and fault propagation influence. Self-influence means that the element is responsible for important functions in the system. Once a fault occurs, it will directly cause some functions to fail, and affect the system's trustworthiness. The impact of fault propagation means that after an element fails, the element will propagate the error to other related elements, resulting in a decrease in the trustworthiness of the system.

In order to get the fault propagation influence of elements, we build an element trustworthiness dependency model, as shown in Fig. 1. The dependence of element B on element A represents the extent to which element B's dependability is (or would be) affected by that of element A [2]. In actual scenarios, the dependency between elements is a function call relationship. The value of dependency between element A and element B represents the intensity of dependency, which can be determined by the ratio of function calls $w_{AB} \in [0, 1]$ [26].

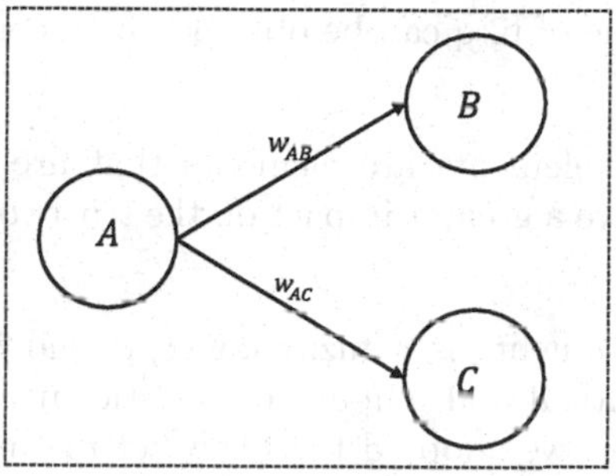

Fig. 1. Elements trustworthiness dependency model

If many other elements depend on the element e_i at the same time, that is, data can flow from element e_i to other elements. And it means that errors of element e_i have more paths and a wider range to propagate, which will have a large impact on the system's trustworthiness. So we use the out-degree importance of elements to represent the fault propagation influence of elements.

For the computing of the element's out-degree importance, it is necessary to comprehensively consider the number of elements' out-degree and the weight of dependencies. The out-degree of the element indicates the number of other elements depending on it, and the weight of dependencies indicates the degree of dependency between elements. Therefore, we use three aspects of the out-degree of edge to describe the influence importance of the elements, as shown in formula (1).

$$IMout(e_i) = \max_{r_{ij} \in R_{ee}} \{w_{ij}\} \times \sum_{r_{ij} \in R_{ee}} \{w_{ij}\} \times outdegree \quad (1)$$

where e_i indicates the element. R_{ee} indicates an element set which contains all elements with dependencies, and $r_{ij} \in R_{ee}$ indicates element r_j depends on element r_i. The symbol $\max_{r_{ij} \in R_{ee}} \{w_{ij}\}$ means the maximum of the edge value starting from this element e_i, which is used to represent the greatest degree of influence on other elements. The symbol $\sum_{r_{ij} \in R_{ee}} w_{ij}$ means the sum of the edge value starting from this element e_i, which is used to represent the sum of influence degree on other elements. The symbol *outdegree* means the number of the edge starting from element e_i, which is used to represent the number of elements can be affected by element e_i. Then, the self-importance $S(e_i)$ of the element e_i represents the importance of its function. we used $S(e_i) \in [1, 10]$ to describe the self-importance of the element e_i, which can be obtained by a comprehensive evaluation of multiple experts.

Herein, the importance of the element e_i can be calculated using formula (2).

$$IM(e_i) = p \cdot IMout(e_i) + q \cdot S(e_i) \tag{2}$$

where p, q is the weight of the influence importance and the self-importance, and $p + q = 1$. The value of p, q can be obtained by a comprehensive evaluation of multiple experts.

Definition 1. Critical elements are elements that are highly important in a software system and have a greater impact on the trustworthiness of the software system.

The criticality of elements is a fuzzy concept, and it is difficult to clearly distinguish between critical and non-critical entities using the general method. Therefore, in this paper, we adopted the fuzzy set method created by Zadeh to describe the phenomenon of fuzzy and distinguish the critical elements [23]. This method takes the object to be investigated, and the fuzzy concept reflecting it as a certain fuzzy set establishes an appropriate membership function and analyzes the fuzzy object through the relevant operations and transformations of the fuzzy set.

In our fuzzy set method, we set domain $U \in [1, 10]$ indicates the importance of the elements, and the value of U indicates the criticality of the elements.

Before determining the degree of membership function, we need to find a function that satisfies the following properties: 1. The value of the function should be between (0, 1], because when the function value is 0, the weight of the element is 0, and the trustworthiness value of the element to the power of 0 is 1, which makes no sense; 2. The function should conform to the characteristics of sparse at both ends and dense at the middle, because in actual scenarios, the importance of most elements should be about 7, and only a small amount of elements appear to be very critical or non-critical; 3. The function is convenient for derivation, which is convenient for subsequent sensitivity analysis.

Above all, we use the Gaussian probability distribution function as the degree of membership function, which is shown as formula (3). μ indicates the position of the symmetry axis, and σ indicates the standard deviation. The Gaussian function satisfies the value range $f(x) \in (0, 1]$, it is a monotonically increasing function, and the function meets the characteristic of sparse at both ends and dense in the middle while $x \leq u$.

$$f(x) = \frac{1}{\sqrt{2\pi\sigma}} \exp\left(-\frac{(x-\mu)^2)}{2\sigma^2}\right) \tag{3}$$

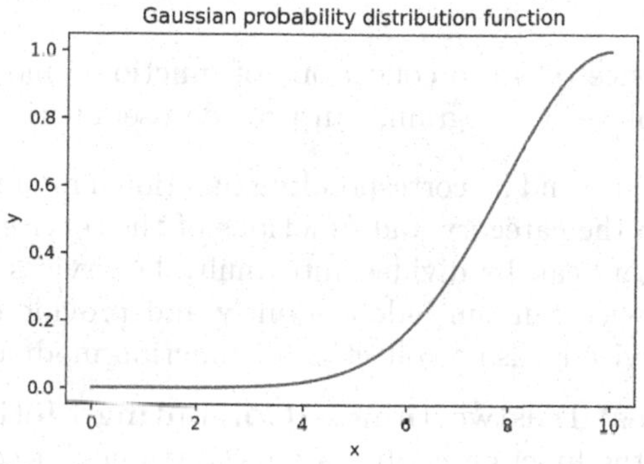

Fig. 2. Gaussian probability distribution function

Therefore, we make some modifications and simplifications to the Gaussian function and set a partial large membership function represented as formula (4) and shown in Fig. 2. It can be seen that its value is between (0, 1], and the fast growth rate in the middle section is in line with most Elements' importance is between 7 and 8.

$$A(x) = e^{-\left(\frac{x-10}{\sigma}\right)^2} \tag{4}$$

where A(x) indicates the criticality of the element, x indicates the importance of the elements, which can be calculated by the formula (2). σ indicates the standard deviation of the element importance set.

Through the fuzzy criticality model, we can calculate the value of criticality A(x) of each element. We set a threshold t and identify the elements with a critical degree greater than threshold t as critical components. This enables the identification of critical and non-critical elements.

3.2 Hierarchical Trustworthiness Computing Model

A component-based software system contains a large number of components, which can be divided into different modules according to different functions. In order to be able to calculate the trustworthiness of the software system effectively, we establish a hierarchical model to calculate the trustworthiness of each function module and service so that the calculation of the trustworthiness of the complex software system is more reliable. Therefore, we divide the software system into three layers, including the services layer, function modules layer, and components layer, respectively.

Definition 2. Functional module refers to collections of components required to complete a specific function in a software system.

Therefore, the system can be divided into many function modules based on functions, and which components are in the function module can be determined by functions.

Definition 3. Service refers to collections of functional modules in a software system, like an application program, can provide users with multiple functions.

Herein, each service and its corresponding functional module set can be determined according to the category and functions of the service.

A software system can be divided into multiple services according to functions, and each service can run independently and provide different functions. In other words, service is also a collection of function modules.

Function Modules' Trustworthiness Computing Method

The computing of the function module's trustworthiness needs to use the trustworthiness of each component, and there are already many methods for calculating the trustworthiness of the component. Herein, we assume that the trustworthiness of each component has been obtained.

We assume there is n components in this function module, and m critical components have been selected through the fuzzy criticality model, which component $cp_i(1 \leq i \leq m)$ is a critical component and component $cp_j(m+1 \leq j \leq n)$ is a non-critical component. The weight of each critical component in the critical components set and the corresponding weight of each non-critical component can be obtained by the normalization method which is given in formula (5), where $IM(cp_i)$ indicates the importance of component cp_i, which is given by the component output degree and the importance of the component in formula (2). $A(IM(cp_i))$ indicates the criticality of the component cp_i, which is given in formula (4).

The weights α_i of the critical component cp_i are given as formula (5).

$$\begin{cases} \alpha_{\mathrm{i}} = \dfrac{A(IM(cp_i))}{\sum\limits_{A(IM(cp_k)) \geq t} A(IM(cp_k))} \\ \sum\limits_{i=1}^{m} \alpha_i = 1 \end{cases} \tag{5}$$

The weights β_j of the non-critical component cp_j are given as formula (6).

$$\begin{cases} \beta_{\mathrm{j}} = \dfrac{A(IM(cp_j))}{\sum\limits_{A(IM(cp_l)) < t} A(IM(cp_l))} \\ \sum\limits_{j=m+1}^{n} \beta_j = 1 \end{cases} \tag{6}$$

where t indicates the criticality threshold, which is used to distinguish critical components from non-critical components. $A(IM(cp_k)) \geq t$ indicates component cp_k is the critical component.

Definition 4. The trustworthiness of functional module is given by the trustworthiness of critical components and non-critical components, as shown in the formula (7).

$$\begin{cases} FM = Q * \prod_{i=1}^{m} CP_i^{\alpha_i} + (1-Q) * \prod_{j=m+1}^{n} CP_j^{\beta_j} \\ \sum_{i=1}^{m} \alpha_i = 1 \ , \ \sum_{j=m+1}^{n} \beta_j = 1 \\ 0 \leq \alpha_i, \beta_j \leq 1 \end{cases} \tag{7}$$

where CP_i indicates the trustworthiness of the component cp_i. And FM indicates the trustworthiness of the function module fm. Q indicates the weight of critical components, which can be obtained by a comprehensive evaluation of multiple experts.

Services' Trustworthiness Computing Method

The calculation of the services' trustworthiness also need to use the trustworthiness of each function module, and the trustworthiness of each function module can be calculated by formula (7).

First of all, we also need to distinguish critical function modules from non-critical function modules according to the fuzzy criticality model. We also assume there are y function modules in the service, and x critical function modules have been selected through the fuzzy criticality model, in which $fm_i(1 \leq i \leq x)$ indicates a critical function module and $fm_j(x+1 \leq j \leq y)$ indicates a non-critical function module. The weight of each critical function module in the set of critical function modules and the corresponding weight of each non-critical function module is obtained by the following normalization method, where $IM(fm_i)$ indicates the importance of function module fm_i, which is given by the function module output degree and the importance of the function module in formula (2).

The weights α_i of the critical function module fm_i are given as formula (8).

$$\begin{cases} \alpha_{\mathrm{i}} = \frac{A(IM(fm_i))}{\sum\limits_{A(IM(fm_k)) \geq t} A(IM(fm_k))} \\ \sum_{i=1}^{x} \alpha_i = 1 \end{cases} \tag{8}$$

The weights β_j of the non-critical function module fm_j are given as formula (9).

$$\begin{cases} \beta_{\mathrm{j}} = \frac{A(IM(fm_j))}{\sum\limits_{A(IM(fm_l)) < t} A(IM(fm_l))} \\ \sum_{j=x+1}^{y} \beta_j = 1 \end{cases} \tag{9}$$

where $A(IM(fm_i))$ indicates the criticality of the function module fm_i, which can be calculated by formula (4).

Definition 5. The trustworthiness of server is given by the trustworthiness of critical function module and non-critical function module, as shown in the formula (10).

$$\begin{cases} SE = R * \prod_{i=1}^{x} FM_i^{\alpha_i} + (1-R) * \prod_{j=x+1}^{y} FM_j^{\beta_j} \\ \sum_{i=1}^{x} \alpha_i = 1 \ , \ \sum_{j=x+1}^{y} \beta_j = 1 \\ 0 \leq \alpha_i, \beta_j \leq 1 \end{cases} \tag{10}$$

where FM_i indicates the value of the function module fm_i's trustworthiness. SE indicates the value of the severs' trustworthiness. R indicates the weight of critical function module, which can be obtained by a comprehensive evaluation of multiple experts.

Considering that each service is integral and essential to the system, so that each services is critical. And we assume there is k services in the CBSS. We can get the weight of each service by the formula (11).

The weights T_i of each service se_i are given as formula (11).

$$\begin{cases} T_i = \frac{A(IM(se_i))}{\sum_{l=1}^{k} A(IM(se_l))} \\ \sum_{i=1}^{k} T_i = 1 \end{cases} \tag{11}$$

Based on the trustworthiness of each service, we can compute the trustworthiness of the CBSS as shown in formula (12).

$$\begin{cases} SYS = T_1 * SE_1 + T_2 * SE_2 + \cdots + T_k * SE_k \\ \sum_{i=1}^{k} T_i = 1 \end{cases} \tag{12}$$

where SYS indicates the trustworthiness of the software system. SE_i indicates the trustworthiness of the service se_i. T_i indicates the weight of each service, which can be calculated by normalization method.

4 Experiments and Analysis

4.1 Case Studies of Two Simple Systems

Without loss of generality, two general case studies adapted from literature [7,8,13,14] are used to demonstrate the effectiveness of the proposed method, a simple multi-input/single-output (MISO) software system and a simple multiple-input/multi-output (MIMO) software system, as shown in Fig. 3, which are consisting of 10 components. A directed graph is used to describe the dependency relations of components in the MISO/MIMO system. Considering a trustworthiness simulation case of a practice software system [7], a simple MISO system is shown in Fig. 3(a). And C_1, C_2 are the input components, C_{10} is the output component. But component C_7 is changed to another output component in the MIMO system, which is shown in Fig. 3(b).

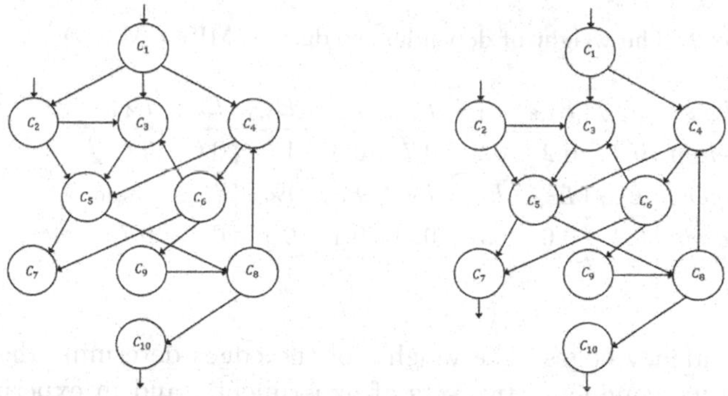

(a) The MISO software system. (b) The MIMO software system.

Fig. 3. MISO/MIMO software systems

The initial trustworthiness and self-importance of each component in the MISO/MIMO system are shown in Table 1. The dependency degree of components of the MISO/MIMO system is demonstrated in Table 2 [14]. The weight of edge $E_{i,j}$, which indicates that component C_j depends on component C_i, and the degree of dependence is $w_{i,j}$.

According to the data in Table 1 and Table 2, we computed the importance and criticality of each component and selected the critical components according to the fuzzy criticality model. Then according to the selected results of critical components, the corresponding weights of each component were calculated, and finally, the trustworthiness of the system is calculated. Computed the trustworthiness of the MISO system is 9.8209, and the trustworthiness of MIMO is 9.7711. Due to the limited space, all the computing results of the MIMO system are shown in Table 3.

Results and Analysis. During the computing of the MIMO system, we set the threshold of criticality $t = 0.6$, and got the critical components set as $C_1, C_2, C_5, C_7, C_8, C_{10}$. Finally, we got the trustworthiness of MIMO is 9.7711. The results of selected critical components set and trustworthiness of the MIMO system are similar to the computing results 9.6845 by Lo et al. [14] proposed model, which shows that our results are reasonable and thus our method is reasonable.

To measure the impact of components on the system trustworthiness, assume a simulated attack experiment on the MIMO system. We suppose a component is attacked and the trustworthiness was dropped by 2. And the attack is propagated

Table 1. The trustworthiness and self importance of each component

Components	C_1	C_2	C_3	C_4	C_5	C_6	C_7	C_8	C_9	C_{10}
Trustworthiness	9.9	9.8	9.9	9.6	9.8	9.5	9.8	9.6	9.7	9.9
Self Importance (MISO)	9	5	5.2	4.8	7.2	3.6	4.6	5.1	2	9
Self Importance (MIMO)	9	9	5.2	4.8	7.2	3.6	9	5.1	2	9

Table 2. The weight of dependency edges in MISO/MIMO

Edges	$E_{1,2}$	$E_{1,3}$	$E_{1,4}$	$E_{2,3}$	$E_{2,5}$	$E_{3,5}$	$E_{4,5}$	$E_{4,6}$
Weight	0.6	0.2	0.2	0.7	0.3	1.0	0.6	0.4
Edges	$E_{5,7}$	$E_{5,8}$	$E_{6,3}$	$E_{6,7}$	$E_{6,9}$	$E_{9,8}$	$E_{8,4}$	$E_{8,10}$
Weight	0.4	0.6	0.3	0.3	0.1	0.1	0.25	0.75

through the dependency edges. The weights of the edges determine the degree of influence. And we conducted two sets of experiments, and in experiment 1, we assume that the critical component C_1 is attacked and the trustworthiness is dropped by 2. At this time, the trustworthiness of other components and the trustworthiness of the system are computed, and the variation δ_1 of each component is shown in Table 4. In experiment 2, we assume that the non-critical component C_6 was attacked and the trustworthiness is also dropped by 2. At this time, the variation δ_2 of each component is also shown in Table 4. In experiment 1, the trustworthiness of the MIMO system was decreased by **9.3562**. But the trustworthiness of the MIMO system in experiment 2 is reduced by **9.4433**.

It is not difficult to see from the above experiments that the more essential components have a more significant impact on the system, which also shows the necessity of dividing the components into critical components and non-critical components.

4.2 Case Study of a Simple Game Software System

To illustrate that the hierarchical trustworthiness computing model is effective, we carry out a study case on a game system on Github (It can be available on https://github.com/jzyong/game-server). First, we used Structure 101 Studio 5

Table 3. The computing results of MIMO

Components	C_1	C_2	C_3	C_4	C_5	C_6	C_7	C_8	C_9	C_{10}
Importance	6.89	9.12	5.74	5.82	7.5	3.95	6.6	6.57	1.718	6.6
Criticality	0.66	0.97	0.45	0.47	0.76	0.21	0.61	0.6	0.05	0.61
Critical component	T	T	F	F	T	F	T	T	F	T
System Trustworthiness	**9.7711**									

Table 4. The variation of components' trustworthiness in MIMO

Components	C_1	C_2	C_3	C_4	C_5	C_6	C_7	C_8	C_9	C_{10}
δ_1	2	0	0.4	0.4	0.56	0.24	0	0	0	0
δ_2	0	0	0.6	0	0.6	2	0.6	0.02	0.2	0

for Java to analyze the project. We divided the system into three services: Server Computing Service, Web Service, and Client Service, and the server computing service is the core service. Each service can be divided into multiple functional modules according to specific functions. The architecture of the game system is shown in Fig. 4.

We further divide functional modules into 43 components, including the registered cluster, public, tool, configure, core-engine, message, etc. To simplify the process of computing, we number the components from 1 to 43. And components $C_1 \sim C_8$ belong to the playability function module, components $C_9 \sim C_{17}$ belong to the management function module, components $C_{18} \sim C_{24}$ belong to the network function module, components $C_{25} \sim C_{35}$ belong to the login function module and components $C_{36} \sim C_{43}$ belong to the register function module. And we use the Structure 101 Studio 5 for Java to get the usage relationship (including returns, calls, parameter, reference, is type, etc.) and usage times. We obtain the dependency according to the usage relationships by reversing the usage relationships firstly. A recalls B means A depends on B, and there is an edge from B to A in the dependency model. There are 68 dependency relations between components, which are shown in Table 5.

According to the weight of dependency edges and self-importance of components in Table 5 and Table 6. We compute the importance and criticality of each component through the fuzzy criticality model. The value of σ is obtained by computing 2.8. We set t = 0.6 because about 1/4 of the components are identified as critical components at this time, which is in line with the proportion of critical components in the actual scene. The criticality of critical components

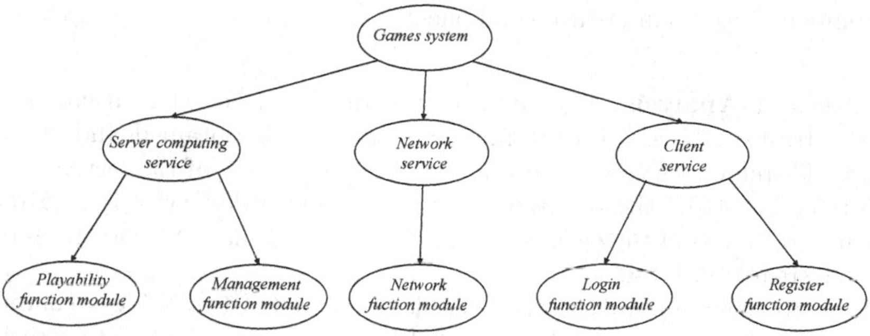

Fig. 4. Elements trustworthiness dependency model

Table 5. The dependency edges' weight of game system

Edges	$E_{1,2}$	$E_{1,3}$	$E_{2,3}$	$E_{4,3}$	$E_{4,5}$	$E_{5,3}$	$E_{6,3}$	$E_{7,4}$	$E_{7,5}$	$E_{8,2}$	$E_{8,3}$	$E_{8,5}$	$E_{9,16}$	$E_{10,16}$
Weights	0.61	0.63	0.21	0.19	0.43	0.11	0.61	0.52	0.65	0.1	0.1	0.69	0.2	0.39
Edges	$E_{12,17}$	$E_{13,10}$	$E_{13,14}$	$E_{14,16}$	$E_{15,10}$	$E_{15,14}$	$E_{15,16}$	$E_{15,17}$	$E_{17,10}$	$E_{18,21}$	$E_{18,24}$	$E_{18,25}$	$E_{18,26}$	$E_{19,21}$
Weights	0.23	0.31	0.57	0.3	0.38	0.43	0.45	0.03	0.4	0.28	0.6	0.6	0.54	0.37
Edges	$E_{21,24}$	$E_{22,19}$	$E_{22,24}$	$E_{23,19}$	$E_{23,20}$	$E_{23,24}$	$E_{25,35}$	$E_{26,25}$	$E_{26,27}$	$E_{26,30}$	$E_{27,29}$	$E_{27,30}$	$E_{27,32}$	$E_{28,25}$
Weights	0.46	0.68	0.61	0.31	0.47	0.49	0.57	0.25	0.5	0.6	0.56	0.38	0.4	0.23
Edges	$E_{31,29}$	$E_{33,27}$	$E_{33,35}$	$E_{34,27}$	$E_{35,30}$	$E_{37,36}$	$E_{37,39}$	$E_{37,41}$	$E_{37,42}$	$E_{38,37}$	$E_{38,41}$	$E_{39,36}$	$E_{40,36}$	$E_{40,39}$
Weights	0.19	0.22	0.44	0.28	0.2	0.15	0.27	0.41	0.33	0.27	0.51	0.49	0.23	0.24
Edges	$E_{8,3}$	$E_{8,5}$	$E_{9,16}$	$E_{19,24}$	$E_{20,24}$	$E_{21,20}$	$E_{28,32}$	$E_{29,25}$	$E_{29,30}$	$E_{40,42}$	$E_{43,36}$	$E_{44,38}$		
Weights	0.54	0.56	0.04	0.27	0.4	0.39	0.36	0.12	0.39	0.84	0.3	0.28		

Table 6. The trustworthiness and self importance of components

Components	C_1	C_2	C_3	C_4	C_5	C_6	C_7	C_8	C_9	C_{10}	C_{11}
Trustworthiness	9.15	9.77	9.07	9.36	9.04	9.84	9.05	9.61	9.91	9.23	9.74
Self Importance	6	5	5	4	3	3	5	6	7	4	6
Components	C_{12}	C_{13}	C_{14}	C_{15}	C_{16}	C_{17}	C_{18}	C_{19}	C_{20}	C_{21}	C_{22}
Trustworthiness	9.33	9.41	9.06	9.33	9.86	9.28	9.14	9.67	9.78	9.02	9.78
Self Importance	5	7	8	6	6	5	9	8	6	6	7
Components	C_{23}	C_{24}	C_{25}	C_{26}	C_{27}	C_{28}	C_{29}	C_{30}	C_{31}	C_{32}	C_{33}
Trustworthiness	9.86	9.14	9.9	9.53	9.25	9.56	9.74	9.07	9.94	9.51	9.79
Self Importance	8	9	6	4	3	4	3	5	5	6	7
Components	C_{34}	C_{35}	C_{36}	C_{37}	C_{38}	C_{39}	C_{40}	C_{41}	C_{42}	C_{43}	
Trustworthiness	9.46	9.9	9.07	9.44	9.8	9.67	9.34	9.47	9.84	9.9	
Self Importance	4	6	5	3	5	5	6	7	4	6	

and non-critical components is quite different, so the distinction is noticeable. The results of computing are shown in Table 7. We determine the critical components in each function module through the fuzzy criticality model. As shown in Table 7, T indicates the component is a critical component, while F indicates the component is a non-critical component.

Results and Analysis. According to the results of the critical components set, the trustworthiness of each functional module is obtained and shown in Table 8. Therefore, we calculated that the trustworthiness of the server computing service is 9.4107, the trustworthiness of the network service is 9.4510, and the trustworthiness of the client service is 9.5993. And the trustworthiness of the Game System is **9.4794**.

As a comparison, we use the non-hierarchical trustworthiness computing method and the model which Chen et al. [7] proposed to calculate the trustworthiness of the system and get the results as below. In the non-hierarchical trustworthiness computing method, due to there is no distinguishing module, all com-

Table 7. The importance and criticality of components

Components	C_1	C_2	C_3	C_4	C_5	C_6	C_7	C_8	C_9	C_{10}	C_{11}
Importance	5.37	3.82	3.8	3.39	2.41	2.61	4.65	5.53	5.22	3.19	5.19
Critical	T	F	F	F	F	F	F	T	T	F	F
Components	C_{12}	C_{13}	C_{14}	C_{15}	C_{16}	C_{17}	C_{18}	C_{19}	C_{20}	C_{21}	C_{22}
Importance	3.87	5.76	5.95	5.79	4.5	3.89	9.3	6.16	4.59	4.94	6.18
Critical	F	T	T	T	F	F	T	T	F	F	T
Components	C_{23}	C_{24}	C_{25}	C_{26}	C_{27}	C_{28}	C_{29}	C_{30}	C_{31}	C_{32}	C_{33}
Importance	6.94	6.6	4.68	4.45	3.65	3.34	2.62	3.8	3.82	4.5	5.52
Critical	T	T	F	F	F	F	F	F	F	F	T
Components	C_{34}	C_{35}	C_{36}	C_{37}	C_{38}	C_{39}	C_{40}	C_{41}	C_{42}	C_{43}	
Importance	3.14	4.52	3.8	3.46	4.24	3.93	6.34	5.2	3.1	4.69	
Critical	F	F	F	F	F	F	T	T	F	F	

ponents were regarded as a set, and the value of the variance σ of the Gaussian function changed so that the selection results of critical components were also different. Different from the results in Table 7, these components $c_1, c_8, c_9, c_{33}, c_{41}$ changed to the non-critical components, this is because the importance of other components in the system is higher, which makes the importance of these components is no longer ranked in the forefront of the system. So they were no longer divided into critical components. At this time, we computed the trustworthiness of the system is **9.2128**. And the result of the system's trustworthiness calculated by the model of Chen et al. [7] is **9.2627**.

According to the results, the system's trustworthiness obtained by the hierarchical trustworthiness computing method is higher than the other two methods. It is because, different from the other two methods, our method considers that most component-based software systems are developed using component-based software development techniques nowadays, and developers modularise the systems, with different components playing different roles in the module. Our method focuses on the importance of each component in the module and obtains the trustworthiness of critical components and non-critical components, and obtains the trustworthiness of the module and the trustworthiness of the system. In the other two methods, they do not consider the modularization of the system, which will cause some components that play an essential role in the module to be ignored, resulting in the wrong division of critical components. As a result, the computer result of the system's trustworthiness is low. So here, we think the results of our experiments are reliable.

Table 8. The trustworthiness of function modules

Function modules	*playability*	*management*	*network*	*login*	*register*
Trustworthiness	9.3643	9.4417	9.4510	9.7319	9.4666

5 Conclusion and Future Work

As software systems become more complex, how correctly measuring the trustworthiness of a software system has become an urgent problem to be solved. In this paper, we proposed a hierarchical trustworthiness measure method for component-based software systems (CBSS) based on the fuzzy criticality model. Experiments show that our approach can obtain reasonable measurements that can guide designers to work effectively in the design phase of development, helping maintainers maintain software more accessible.

In the future work, we will focus on using this trustworthiness measurement method to calculate the trustworthiness of CBSS dynamically and develop a set of components fault detective and components fault isolation methods accordingly.

Acknowledgements. This paper is funded by East China Normal University-Huawei Trustworthiness Innovation Center and Shanghai Trusted Industry Internet Software Collaborative Innovation Center.

References

1. ISO/IEC 15408-1-2005. Information Technology-Security Techniques Evaluation Criteria for IT Security, Part 1: Introduction and General Model (2005)
2. Avizienis, A., Laprie, J.C., Randell, B., Landwehr, C.: Basic concepts and taxonomy of dependable and secure computing. IEEE Trans. Dependable Secure Comput. **1**(1), 11–33 (2004)
3. Bo, L., Xudong, L., Huaimin, W., Bing, X., Xiaoguang, M.: A software trust rating model. Comput. Sci. Explor. **4**(3), 231–239 (2010)
4. Boland, T., Cleraux, C., Fong, E.: Toward a preliminary framework for assessing the trustworthiness of software. National Institute of Standards and Technology (2010)
5. Chen, H., Wang, J., Dong, W.: High confidence software engineering technology. Electron. J. (z1), 6 (2003)
6. Chen, Y., Tao, H.: Software Trustworthiness Measurement and Enhancement Specification. China Science Publishing Media Ltd., Beijing (2019)
7. Chen, Y., Yan, X., Khan, A.A.: A novel reliability assessment method based on the effects of components. In: 2019 IEEE 19th International Conference on Software Quality, Reliability and Security (QRS), pp. 69–76. IEEE (2019)
8. Cheung, R.C.: A user-oriented software reliability model. IEEE Trans. Softw. Eng. **2**, 118–125 (1980)
9. Chinnaiyan, R., Somasundaram, S.: Evaluating the reliability of component-based software systems. Int. J. Qual. Reliab. Manag. **27**, 78–88 (2010)

10. Ding, S., Yang, S.L., Fu, C.: A novel evidential reasoning based method for software trustworthiness evaluation under the uncertain and unreliable environment. Expert Syst. Appl. **39**(3), 2700–2709 (2012)
11. Ke, L., Zhiguang, S., Ji, W., Jifeng, H., Zhaotian, Z., Yuwen, Q.: "Basic research on trusted software" overview of major research programs. Chin. Sci. Found. **22**(3), 145–151 (2008)
12. Kristiansen, M., Winther, R., Natvig, B.: On component dependencies in compound software. Int. J. Reliab. Qual. Saf. Eng. **17**(05), 465–493 (2010)
13. Li, K., Yu, M., Liu, L., Zhai, J., Liu, W.: A novel reliability analysis approach for component-based software based on the complex network theory. Softw. Test. Verif. Reliab. **28**(6), e1674 (2018)
14. Lo, J.H., Kuo, S.Y., Lyu, M.R., Huang, C.Y.: Optimal resource allocation and reliability analysis for component-based software applications. In: Proceedings 26th Annual International Computer Software and Applications, pp. 7–12. IEEE (2002)
15. Mao, X., Deng, Y.: General model of reliability based on component software. J. Softw. **15**, 27–32 (2004)
16. Qiu, L., Zhang, Y., Wang, F., Kyung, M., Mahajan, H.R.: Trusted computer system evaluation criteria. National Computer Security Center, Citeseer (1985)
17. Taibi, D., del Bianco, V., Carbonare, D.D., Lavazza, L., Morasca, S.: Towards the evaluation of OSS trustworthiness: lessons learned from the observation of relevant OSS projects. In: Russo, B., Damiani, E., Hissam, S., Lundell, B., Succi, G. (eds.) OSS 2008. ITIFIP, vol. 275, pp. 389–395. Springer, Boston, MA (2008). https://doi.org/10.1007/978-0-387-09684-1_37
18. Tao, H., Chen, Y., Wu, H.: A reallocation approach for software trustworthiness based on trustworthy attributes. Mathematics **8**(1), 14 (2019)
19. Tao, H., Chen, Y., Wu, H.: Theoretical and empirical validation of software trustworthiness measure based on the decomposition of attributes. Connect. Sci. **34**(1), 1181–1200 (2022). https://doi.org/10.1080/09540091.2022.2061424
20. Tao, H., Fu, L., Chen, Y., Han, L., Wang, X.: Improved allocation and reallocation approaches for software trustworthiness based on mathematical programming. Symmetry **14**(3), 628 (2022). https://doi.org/10.3390/sym14030628
21. Tao, H., Wu, H., Chen, Y.: An approach of trustworthy measurement allocation based on sub-attributes of software. Mathematics **7**(3), 237 (2019)
22. Yan, Z., Prehofer, C.: Autonomic trust management for a component-based software system. IEEE Trans. Dependable Secure Comput. **8**(6), 810–823 (2010)
23. Zadeh, L.A.: Fuzzy sets. In: Fuzzy Sets, Fuzzy Logic, and Fuzzy Systems: Selected Papers by Lotfi A Zadeh, pp. 394–432. World Scientific (1996)
24. Zhang, W., Zhang, W.: Research on an improved path-based component software reliability model. Comput. Sci. **38**(2), 148–151 (2011)
25. Zhang, X., Li, W., Zheng, Z., Guo, B.: Optimized statistical analysis of software trustworthiness attributes. Science China Inf. Sci. **55**(11), 2508–2520 (2012). https://doi.org/10.1007/s11432-012-4646-z
26. Zhang, X., Jiang, S., Qiao, X., Cao, Z., Zhang, L.: Critical components identification for service-oriented systems. Symmetry **11**(3), 427 (2019)
27. Zheng, Z., Ma, S., Li, W., Wei, W., Jiang, X., Zhang, Z., Guo, B.: Dynamical characteristics of software trustworthiness and their evolutionary complexity. Sci. China Ser. F Inf. Sci. **52**(8), 1328–1334 (2009). https://doi.org/10.1007/s11432-009-0137-2

The Trustworthiness Measurement Model of Component-Based Software Based on Combination Weight

Yanfang Ma[1,2](✉), Xiaotong Gao[2], and Wei Zhou[2]

[1] Changzhou Institute of Technology, Changzhou, China
clmyf@163.com
[2] Huaibei Normal University, Huaibei, China

Abstract. Software trustworthiness is an important indicator to assess the quality of software and can be portrayed by the attributes of the software. The different attribute produces different influence on the software quality. Therefore, it is important to study the weight allocation for different attributes to measure software trustworthiness reasonably. Usually, the weight of trustworthy attributes is affected by two aspects, one is the assessment of experts and the other is the hidden information in attributes. The component-based software has become popular in the field of software engineering due to its advantages. So a trustworthiness measurement model of component-based software is proposed by combining the weights. Firstly, a new method of weight allocation for trustworthy attributes is proposed based on Fuzzy Analytical Hierarchy Process and the grey correlation method. Secondly, the trustworthiness measurement model of component-based software will be established based on the combination structures of components. Finally, the station ticketing system is used to illustrate the rationality of the model.

Keywords: Trustworthiness · Component-based software · Measurement · Weight

1 Introduction

With the rapid development of computer technology, the software is widely used in various industries such as aerospace, finance, medical, automotive, and shopping. A large number of software products have penetrated people's daily life [1]. However, as the scale of software continues to expand, failures and faults of software systems are inevitable and cause many negative impacts on people, which even threaten human lives and property safety [2]. Software is not always completely trustworthy and does not perform exactly as expected, which produces the software trustworthiness problem [3].

Supported by the Natural Science Foundation of Anhui Province (No. 2108085MF204), the National Natural Science Foundation of China (No. 62162014, 62077029), the Abroad Visiting of Excellent Young Talents of Universities in Anhui Province (No. GXGWFX2019022).

Y. Chen and S. Zhang (Eds.): AILA 2022, CCIS 1657, pp. 270–285, 2022.
https://doi.org/10.1007/978-981-19-7510-3_20

Trustworthiness, as the focus of international industry and academia, is a key technology foundation for the development of software technology and the information industry [4]. There exists a large number of excellent results. Prof. J.F.He proposed that "trustworthy software" means that the operation behavior and results of software systems always meet people's expectations and can provide continuous services even when they are disturbed [5]. The team of Prof. Y.X. Chen, from the perspective of attributes, has achieved many significant research during the software trustworthiness [6–10]. The team of Prof. D.X.Wang, from the perspective of evidence, has made great contributions to the trustworthiness model [11,12].

Due to the increasing complexity of the software, more requirements for the quality and productivity of software products have been put forward, and the development method of component-based software has become one of the important development methods in the field of software engineering due to its advantages of reusability, saving development cost, effort, and time [13,14], and so on. Considering that different combination structures of components will produce different software systems, their trustworthiness will be influenced by the combination structures [15,16]. A large number of researches have existed to study component-based software from some perspectives, such as formal languages [17], framework structures [18], etc. However, the quality of components plays an important role in the quality of component-based software. The component is usually developed by a third party and has some important attributes. These attributes also play a critical role in the quality of the component.

Different trustworthy attributes have different influences on component trustworthiness, for example, security is an important attribute for the component which is designed to ensure the security of the system, and the weight of security needs to be increased to reflect its importance. Many methods already exist for allocating weights to trustworthy attributes. For example, the article [19] allocated weights according to the AHP method. The paper [20] obtained stable expert two-way weights by calculating the consistency of individual experts' decisions and the experts' knowledge. The authors in [21] gave an embedded software trustworthiness measurement method based on the entropy weight method. Prof. S.L.Yang et al. proposed a weight allocation method by combining TOPSIS method [22]. The authors in [23] studied the sensitivity of attributes and combined it with experts' evaluation to establish a subjective and objective weight allocation model. The authors in [24] analyzed the relationships between attributes and used directed graphs to describe the interactions between attributes, and assigned weights for attributes according to the degree of the directed graph and the weights on each edge.

From the aspect of the experts' evaluation, experts have produced many excellent research results on weight allocation. However, weight allocation methods based on experts' evaluation often influence the results due to the ambiguity and personal preference of expert evaluation. At the same time, some hidden information in attributes can also reflect the quality of software. Therefore, it is necessary to consider the experts' assessment of the attributes and the hidden information of the trustworthy attributes together when allocating weights to the attributes. In this paper, we will use the FAHP (Fuzzy Analytical Hierarchy Process) method and the grey correlation degree to combine the experts' evaluation and the hidden information in attributes to establish a new weight allocation method. Furthermore, the trustworthiness measurement model of the component-based software is established based on the combined weight allocation method and different structures of components.

In Sect. 2, the combination weight allocation method is proposed based on the FAHP method and grey correlation method. Section 3 investigates the component-based software trustworthiness measurement model. Section 4 takes the station ticketing system as an example to study the concrete implementation, and Sect. 5 concludes the paper.

2 Combination Weight Allocation Method

2.1 Subjective Weight

FAHP theory is built based on AHP, incorporating fuzzy theory and taking into account the fuzziness in the human judgment process [25]. In the software system, the weight allocation of attributes is affected by many factors, such as fuzziness and arbitrariness in the experts' judgment. However, FAHP has the advantage of defuzzification. Therefore, this paper adopts the FAHP method to calculate the subjective weights of attributes. The FAHP method judges the importance of the attributes by using triangular fuzzy numbers.

The general expression of the triangular fuzzy number is $N = (l, m, u)$. In the domain U, if $u_N(x) : U \to [0, 1]$ exists, $u_N(x)$ is the degree of affiliation of $x \in N$, $l \leq m \leq u$, where $x = m$ is the median of the N affiliation degree of 1 and l, u are the lower and upper bound values of N. The values of l, u determine the degree of fuzziness, and the larger $u - l$ is, the greater the degree of fuzziness.

$$u_N(x) \begin{cases} x - l/m - l, x \in [l, m] \\ u - x/u - x, x \in [m, u] \\ 0, other \end{cases} \tag{1}$$

The triangular fuzzy numbers $N_1(l_1, m_1, u_1)$, $N_2(l_2, m_2, u_2)$ satisfy the following rules:

$$\begin{aligned} N_1 + N_2 &= (l_1 + l_2, m_1 + m_2, u_1 + u_2) \\ 1/N_1 &= (1/u_1, 1/m_1, 1/l_1) \end{aligned} \tag{2}$$

Let $h_{ij} = (l_{ij}, m_{ij}, u_{ij})$ be the relative importance of attribute i to attribute j, where $h_{ji} = (h_{ij})^{-1}$, $i \neq j$, $i, j = 1, \cdots q$. Referring to the [1,9] evaluation scale method proposed in the AHP method [19], the evaluation standards of the triangular fuzzy numbers is given as Table 1 shown. Experts compare the importance of trustworthy attributes two-by-two through their experience and knowledge, and give the value of h_{ij}.

Table 1. Triangular fuzzy number evaluation standards

Compare standards	Triangular fuzzy number	Countdown
Both attributes are equally important	$(1, 1, 1)$	$(1, 1, 1)$
Between equally important and slightly important	$(1, 2, 3)$	$(1/3, 1/2, 1)$
The former is slightly more important than the latter	$(2, 3, 4)$	$(1/4, 1/3, 1/2)$
Between slightly important and more important	$(3, 4, 5)$	$(1/5, 1/4, 1/3)$
The former is more important than the latter	$(4, 5, 6)$	$(1/6, 1/5, 1/4)$
Between more important and strongly important	$(5, 6, 7)$	$(1/7, 1/6, 1/5)$
The former is more strongly important than the latter	$(6, 7, 8)$	$(1/8, 1/7, 1/6)$
Between strongly important and extremely important	$(7, 8, 9)$	$(1/9, 1/8, 1/7)$
The former is extremely more important than the latter	$(8, 9, 9)$	$(1/9, 1/9, 1/8)$

The calculation process of the FAHP method is described as follows, the detail can refer to [25].

1. Construct the fuzzy judgment matrix, compare the importance of trustworthy attributes, and give triangular fuzzy numbers to construct the fuzzy judgment matrix $H = (h_{ij})_{q \times q}$.
2. According to the fuzzy judgment matrix, the fuzzy subjective weights are calculated.
$$\tilde{w}_j = \frac{\sum_{i=1}^{q} h_{ij}}{\sum_{i=1}^{q} \sum_{j=1}^{q} h_{ij}} = \left(\tilde{w}_j^l, \tilde{w}_j^m, \tilde{w}_j^u\right), \tag{3}$$
where $\tilde{w}_j$ $(j = 1 \cdots q)$ is the fuzzy subjective weight of the jth attribute.
3. To defuzzify the fuzzy subjective weights, the affiliation limit element averaging method is used, which cuts the data flat according to the value taken by the affiliation degree, and averages all the elements that are greater than or equal to this affiliation degree after the cut, the affiliation degree can be selected in [0,1], and this paper determines the affiliation degree $a = 0.5$, followed by the calculation and simplification to obtain the defuzzification formula as follow:
$$w_j^s = \frac{\tilde{w}_j^l + 4\tilde{w}_j^m + \tilde{w}_j^u}{6}, \tag{4}$$
where w_j^s $(j = 1 \cdots q)$ is the subjective weight of the jth attribute.

2.2 Objective Weight

Various factors can lead to defects in the design and development process of components, such as the development environment of components, the development language used, and the compilation environment. There is a negative correlation between defects and trustworthiness. If the correlation between defect data and a trustworthy attribute is high, then the software is more prone to defects when the value of that trustworthy attribute is low, and more weight is assigned to that attribute. By inverting the defect data to obtain the reference sequence, and comparing the sequence curve of the reference sequence and the trustworthy attribute, the correlation between the trustworthy attribute and the defect can be obtained, and the weights are allocated according to the correlation.

Suppose that component has q trustworthy attributes, each component is tested and the defect data is collected, and the number of defects of the component is num. [26] gives the detailed method to collect the defect data and the counting standards. The x_0 is the reference data obtained from the defect data. The evaluation object can be described as x_j, which is the value of the jth trustworthy attribute, and the correlation of each trustworthy attribute with the defect data can be obtained by comparing the reference data and the evaluation object, and the weights are allocated according to the correlation, and the detail process is as follows:

1. From the defect perspective, the defect data are collected, and considering the negative correlation between the defect data and trustworthy attributes, the defect data are inverted and the defect data is calculated by referring to the method of processing defect data in [26].
$$x_0 = e^{-num/f}, \tag{5}$$
where f is the control parameter of the component, generally f is the number of lines of the code of the component.
2. Calculation of correlation coefficient. Calculate the absolute difference between comparison data of component evaluation objects and the defect reference data.
$$\Delta_j = |x_j - x_0|, j = 1, 2, \cdots q. \tag{6}$$
Furthermore, the two-level maximum difference $\Delta\,(\max)$ and the two-level minimum $\Delta\,(\min)$ can then be obtained.
$$\Delta\,(\max) = \max_{1 \le j \le q} (\Delta_j)\,, \Delta\,(\min) = \min_{1 \le j \le q} (\Delta_j)\,. \tag{7}$$
And subsequently, the correlation between each trustworthy attribute and the defect data is calculated as follows:
$$\xi_j = \Delta\,(\min) + \rho\Delta\,(\max)/\Delta_j + \rho\Delta\,(\max)\,, j = 1, 2, \cdots, q \tag{8}$$
where ξ_j is the correlation coefficient between x_j and x_0, ξ_j represents the correlation between the jth trustworthy attribute and the defect reference data of the component, and ρ is the discrimination coefficient to weaken the distortion of the two-level maximum difference $\Delta\,(\max)$ by too large, and ρ is generally taken as 0.5.
3. Calculating objective weights. The higher the correlation degree, the higher the correlation between this trustworthy attribute and the component defect data, then the weight assigned to this trustworthy attribute needs to be increased. Let w_j^o be the objective weight of the jth trustworthy attribute, can be given by the following formula:
$$w_j^o = \xi_j \Big/ \sum_{j=1}^{q} \xi_j. \tag{9}$$

2.3 Combination Weight

The subjective weight method is simple, based on expert experience entirely, while the objective weight is too dependent on the sample and the data. Both methods suffer from information loss. Therefore, when allocating weights to attributes, a combination of the subjective allocation method (FAHP) and the objective allocation method (grey correlation) is used to make up for the deficiencies brought by a single assignment. The combination of allocation minimizes the loss of information and makes the result of allocation as close as possible to the actual result. Game theory is the study of how decision makers make decisions to maximize their utility and the equilibrium of decisions between different decision-makers. By applying the idea of game theory, the subjective weights and objective weights are regarded as the two sides of the game, and the optimal combination of weights can be regarded as the two sides of the game reaching the equilibrium state [27]. The steps are as follows:

1. Combining the subjective weight $W^s = (w_1^s \cdots w_q^s)$ and objective weight $W^o = (w_1^o \cdots w_q^o)$ to form combination weight $W^* = (w_1^* \cdots w_q^*)$,
$$W^* = \lambda_1^* W^s + \lambda_2^* W^o, \tag{10}$$
where, λ_1^* and λ_2^* are linear combination coefficients.
2. According to the idea of game theory, the objective function is established, and the optimal linear combination coefficient λ_1, λ_2 are sought with the objective of minimizing the sum of the deviation of indicator combination weight $W^* = (w_1^* \cdots w_q^*)$ from $W^o = (w_1^o \cdots w_q^o)$ and $W^s = (w_1^s \cdots w_q^s)$. At this time, the indicator combination weight is the optimal combination weight. The objective function and constraints are as follows.
$$\min \left(\|W^* - W^s\|_2 + \|W^* - W^o\|_2 \right), \tag{11}$$
$$s.t.\ \lambda_1^* + \lambda_2^* = 1, \lambda_1^* \geq 0, \lambda_2^* \geq 0.$$
3. According to the principle of differentiation, the following conditions need to be satisfied when the above model takes the smallest value.
$$\begin{cases} \lambda_1^* W^s (W^s)^T + \lambda_2^* W^s (W^o)^T = W^s (W^s)^T, \\ \lambda_1^* W^o (W^s)^T + \lambda_2^* W^o (W^o)^T = W^o (W^o)^T. \end{cases} \tag{12}$$
4. The combination coefficients are normalized and the combination weights are calculated,
$$\lambda_1 = \frac{|\lambda_1^*|}{|\lambda_1^*| + |\lambda_2^*|}, \lambda_2 = \frac{|\lambda_2^*|}{|\lambda_1^*| + |\lambda_2^*|}. \tag{13}$$

The combination weights are a linear combination of subjective and objective weights, and according to the above model, the linear combination coefficients λ_1 and λ_2 are obtained. Thus the combination weights W can be obtained,
$$W = \lambda_1 W^s + \lambda_2 W^o. \tag{14}$$

3 The Trustworthiness Measurement Model of Component-Based Software

3.1 Trustworthiness Measurement Model of Component

Measuring component trustworthiness is important for constructing highly trustworthy component-based software. Component trustworthiness can be portrayed by the trustworthy attributes of the components, so we can build a trustworthiness measurement model of a single component based on the model in [28]. Assuming that component C_i has q trustworthy attributes, the trustworthiness measurement model of a single component is obtained as follows:

$$T_{C_i} = \prod_{j=1}^{q} y_{ij}^{\lambda_1^i w_{ij}^s + \lambda_2^i w_{ij}^o}, \tag{15}$$

where T_{C_i} denotes the trustworthiness of component C_i, y_{ij} denotes the value of the jth trustworthy attribute of component C_i. w_{ij}^s and w_{ij}^o denote the subjective weight and objective weight of the jth trustworthy attribute of component C_i. λ_1^i and λ_2^i are linear combination coefficients of component C_i.

3.2 The Trustworthiness Measurement Models of Component-Based Software

The trustworthiness of component-based software is influenced by the trustworthiness of the components and the combination structure of the components. The component trustworthiness can be obtained through Subsect. 3.1. This section delves into the inter-component relationships and proposes corresponding trustworthiness measurement models for four different combination structures.

Suppose that there are n components to constitute structure and each component has q trustworthy attributes, where T_{C_i} represents the trustworthiness of component C_i, the weight of C_i is denoted as α_i. Then the four combination structures will be introduced in detail.

Sequence Structure. The structure requires that all of these components execute successfully, and only if the previous component executes successfully can the subsequent components execute in sequence. Under the sequence structure, the components conform to the cascade rule, and the trustworthiness of each component directly affects the trustworthiness of the whole component-based software. Figure 1 shows how the sequence structure is combined. Only when component C_1 executes successfully will component C_2 and subsequent component be executed sequentially.

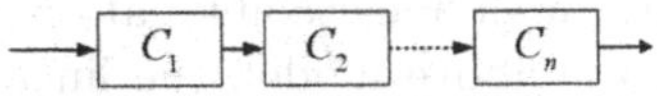

Fig. 1. Combination structure of sequence structure.

Let T_S represent the trustworthiness of the sequence structure,

$$\begin{cases} T_S = \prod_{i=1}^{n} T_{C_i}^{\alpha_i} \\ T_{C_i} = \prod_{j=1}^{q} y_{ij}^{\lambda_1^i w_{ij}^S + \lambda_2^i w_{ij}^O} \end{cases} \tag{16}$$

Branch Structure. In the branch structure, the program selects one branch at a time to execute, and each branch has a corresponding probability of being selected. In this structure, the selected components have a direct influence on the whole structure, but each component has a different probability of being selected. Figure 2 shows the branch structure. In this structure, the components $C_1 ... C_n$ have the probability of being selected, and if the component is selected, then execute it.

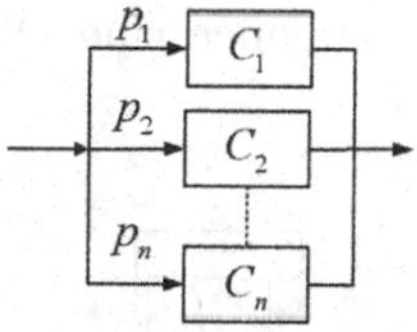

Fig. 2. Combination structure of branch structure.

Suppose that the probability of component C_i being selected to execute is denoted as p_i. T_B is used to represent the trustworthiness of the branch structure.

$$\begin{cases} T_B = \sum_{i=1}^{n} p_i T_{C_i}, \sum_{i=1}^{n} p_i = 1, \\ T_{C_i} = \prod_{j=1}^{q} y_{ij}^{\lambda_1^i w_{ij}^s + \lambda_2^i w_{ij}^o} . \end{cases} \tag{17}$$

Parallel Structure. Parallel structure is usually divided into "and parallel" and "or parallel". "And parallel" requires all components are successfully executed. The calculation of the trustworthiness is similar to sequence structure. Next, we only focus on "or parallel", which is mostly used in software with high trustworthiness requirements, where multiple components complete the same service. If the execution of the master component is successful, the execution of the subsequent component is executed, and if the execution of the master component fails, the redundant component continues to complete the service. The parallel structure runs successfully as long as one of the components in that structure runs successfully. Figure 3 shows how the parallel structure is combined. The structure requires one of the components $C_1 ... C_n$ to execute successfully, only then can the subsequent components be executed. Thus the structure is more trustworthy than any of the components.

Using T_P to represent the trustworthiness of the parallel structure, so that

$$\begin{cases} T_P = 1 - \min\limits_{1 \le i \le n} \left(1 - T_{C_i}^{\alpha_i}\right) \\ T_{C_i} = \prod_{j=1}^{q} y_{ij}^{\lambda_1^i w_{ij}^S + \lambda_2^i w_{ij}^O} \end{cases} \tag{18}$$

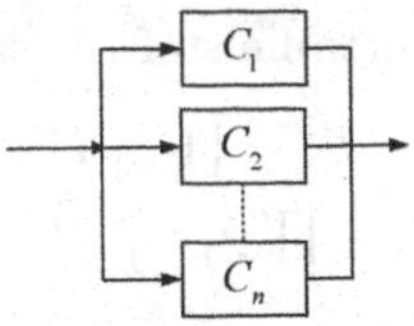

Fig. 3. Combination structure of parallel structure.

Loop Structure. If a condition is satisfied, the loop is started, and if not, the loop is jumped out, where the loop can be a single component or a composite component made of components connected by one or more of the above. Figure 4 shows how the loop structure is combined. Suppose the trustworthiness of the loop A denoted as T_A, and the loop body A is executed t times, which is equivalent to A loops t satisfies the sequence rule. The trustworthiness of the loop structure is noted as T_L.

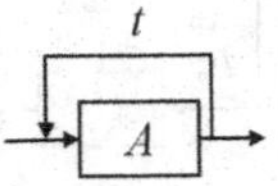

Fig. 4. Combination structure of loop structure.

Propose the trustworthiness measurement model for loop structure:

$$\begin{cases} T_L = T_A^t, \\ T_{C_i} = \prod_{j=1}^{q} y_{ij}^{\lambda_1^i w_{ij}^s + \lambda_2^i w_{ij}^o}. \end{cases} \tag{19}$$

4 Case Study

Assuming a station ticketing system, the system consists of nine components, as shown in Fig. 5. Component C_1 completes the account login function, component C_2 realizes the third-party login function such as QQ, WeChat and Alipay, component C_3 completes the tour ticket inquiry function, component C_4 realizes the user refund function, C_5 completes the tour ticket reservation function, component C_6 completes the bank card payment function, component C_7 completes the WeChat payment function, component C_8 completes the Alipay payment function, and component C_9 completes the account withdrawal function.

Component C_1 and component C_2 are connected by or parallel structure to form composite component C_A, which completes the user login function, and the users can enter the system for operation as long as one of component C_1 and C_2 is running successfully. Component C_6, C_7, and C_8 are connected by or parallel to form the composite component C_B, and only one component needs to run successfully for the users to complete the payment function. Component C_3, C_5, and C_B are connected sequentially, and each component needs to be successfully executed to constitute the composite component C_D. When the user chooses to return the ticket, the software executes component C_4 to achieve the task. After returning the ticket, the probability of choosing to rebook the ticket is $p_3 = 0.7$, and the probability of choosing to exit the system is $p_4 = 0.3$. Component C_D and C_4 are connected through the branch structure to form the composite component C_E, and according to the statistics of the ticketing system, the probability of the user choosing to purchase the ticket is $p_1 = 0.9$, and the probability of the user choosing to return the ticket is $p_2 = 0.1$, component C_E and component C_D form the composite component C_F through the branch structure. The component C_9 completes the refund and booking functions. Finally, component C_F completes the exit function. The whole system will loop t times.

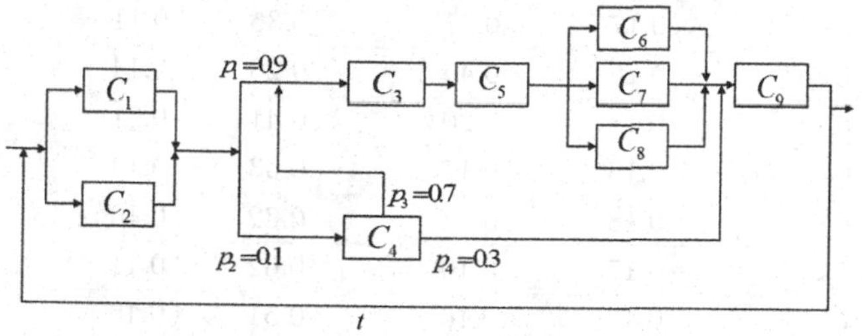

Fig. 5. Structure of station ticketing system.

1. Calculate the combination weights of trustworthy attributes.

When the incremental model is used for component-based software development, each functional module and component is retained and developers need to review it independently before developing the components to ensure that it is trustworthy in terms of single components before moving on to the next development. Component evidence is collected during the testing phase to obtain the values of trustworthy attributes. It is assumed that each component of this station ticketing system possesses four attributes: reliability, correctness, security, and availability. The values of trustworthy attributes are shown in Table 2.

Table 2. The values of trustworthy attributes

Component	Reliability	Correctness	Security	Availability
C_1	0.87	0.84	0.88	0.89
C_2	0.90	0.86	0.90	0.85
C_3	0.80	0.87	0.82	0.92
C_4	0.81	0.96	0.85	0.87
C_5	0.82	0.84	0.86	0.88
C_6	0.96	0.92	0.93	0.95
C_7	0.98	0.92	0.90	0.88
C_8	0.97	0.91	0.94	0.97
C_9	0.86	0.88	0.90	0.84

(1) Calculate the subjective weights of trustworthy attributes.
The experts use the FAHP method to calculate the subjective weights of the nine components in turn.

Table 3. The subjective weights of trustworthy attributes

Component	Reliability	Correctness	Security	Availability
C_1	0.06	0.33	0.50	0.11
C_2	0.20	0.25	0.40	0.15
C_3	0.17	0.37	0.35	0.11
C_4	0.15	0.45	0.26	0.14
C_5	0.18	0.20	0.41	0.21
C_6	0.16	0.11	0.63	0.14
C_7	0.18	0.10	0.62	0.10
C_8	0.17	0.10	0.62	0.11
C_9	0.17	0.16	0.51	0.16

The subjective weights presented in Table 3, which are calculated through formula (3)–(4) based on the fuzzy judgment matrix, the detail can be viewed in Subsect. 2.1.

(2) Calculate objective weights. Collect defect data, count the number of defects for each component and then obtain the reference data x_0, and $x_0 = (0.83, 0.84, 0.85, 0.82, 0.80, 0.88, 0.86, 0.90, 0.82)$. According to the formula (5)–(8) the objective weights of trustworthy attributes can be calculated in Table 4.

Table 4. The objective weights of trustworthy attributes

Component	Reliability	Correctness	Security	Availability
C_1	0.22	0.39	0.22	0.17
C_2	0.16	0.30	0.16	0.38
C_3	0.22	0.33	0.28	0.17
C_4	0.33	0.19	0.26	0.22
C_5	0.35	0.26	0.21	0.17
C_6	0.20	0.31	0.27	0.22
C_7	0.15	0.23	0.27	0.34
C_8	0.18	0.42	0.22	0.18
C_9	0.27	0.27	0.18	0.27

(3) Calculation of combination weights. According to the game theory problem, the subjective and objective weights are linearly programmed to find the

linear combination coefficients λ_1^i, λ_2^i as Table 5 shows. According to the formula (14), the combination weights are calculated in Table 6 shown by combining subjective weights and objective weights in Table 3 and Table 4.

Table 5. Linear combination coefficients

Linear combination coefficient	C_1	C_2	C_3	C_4	C_5	C_6	C_7	C_8	C_9
λ_1^i	0.92	0.51	0.63	0.70	0.60	0.91	0.87	0.69	0.75
λ_2^i	0.08	0.49	0.37	0.30	0.40	0.09	0.13	0.31	0.25

Table 6. The combination weights of trustworthy attributes

Component	Reliability	Correctness	Security	Availability
C_1	0.07	0.33	0.48	0.12
C_2	0.18	0.27	0.29	0.26
C_3	0.19	0.36	0.32	0.13
C_4	0.21	0.37	0.26	0.16
C_5	0.25	0.22	0.33	0.20
C_6	0.16	0.12	0.59	0.13
C_7	0.18	0.12	0.57	0.13
C_8	0.17	0.20	0.50	0.13
C_9	0.19	0.19	0.43	0.19

2. Calculating the trustworthiness of components.

The components trustworthiness can be portrayed by the trustworthy attributes of the components. Through the values of trustworthy attributes in Table 2 and the combination weights in Table 6, the trustworthiness of components can be obtained as Table 7.

Table 7. Trustworthiness of components

Component	C_1	C_2	C_3	C_4	C_5	C_6	C_7	C_8	C_9
Trustworthiness	0.867	0.876	0.867	0.884	0.849	0.934	0.913	0.943	0.877

3. Calculation of component-based software trustworthiness.

The weights of the components are obtained through expert evaluation, as shown in Table 8.

Table 8. The weights of components

Component	C_1	C_2	C_3	C_4	C_5	C_6	C_7	C_8	C_9
Weight	0.08	0.07	0.15	0.13	0.18	0.10	0.10	0.10	0.09

In component C_A, component C_1, C_2 are connected by or in parallel to form a composite component, and the weights occupied by component C_1, C_2 in component C_A are shown in Table 9.

Table 9. The ratio of component C_A

Component C_i	C_1	C_2
Weight α_A^i	0.533	0.467

where α_A^i denotes the weight ratio of component C_i in component C_A, $\alpha_A^i = \alpha_i \Big/ \sum_{i=1}^{2} \alpha_i$, T_{C_A} denotes the trustworthiness of component C_A, $T_{C_A} = 1 - \min\left(1 - T_{C_1}^{\alpha_A^1}, 1 - T_{C_2}^{\alpha_A^2}\right) = 0.940$, and α_A denotes the weight of component C_A in the whole system, $\alpha_A = \alpha_1 + \alpha_2 = 0.15$.

In component C_B, components C_6, C_7 and C_8 are connected by or parallel to form a composite component. It is easy to get $T_{C_B} = 0.980$, and $\alpha_B = \alpha_6 + \alpha_7 + \alpha_8 = 0.30$.

In component C_D, components C_3, C_5 and C_B are connected by sequence structure to form a composite component, and the weights occupied by components C_3, C_5 and C_B in the component C_D are shown in Table 10.

Table 10. The ratio of component C_D

Component C_i	C_3	C_5	C_B
Weight α_D^i	0.238	0.286	0.476

Where α_D^i denotes the weight ratio of component C_i in component C_D, T_{C_D} denotes the trustworthiness of component C_D, $T_{C_D} = T_{C_3}^{\alpha_D^3} \times T_{C_5}^{\alpha_D^5} \times T_{C_B}^{\alpha_D^B} = 0.937$, and $\alpha_D = \alpha_3 + \alpha_5 + \alpha_B = 0.63$.

In component C_E, the weights that component C_4 and C_D occupy in the component C_E are shown in Table 11.

Table 11. The ratio of component C_E

Component C_i	C_4	C_D
Weight α_E^i	0.171	0.829

Where α_E^i denotes the weight ratio of component C_i in component C_E, $\alpha_E^i = \alpha_i / \alpha_4 + \alpha_D$, T_{C_E} denotes the trustworthiness of component C_E. Component C_4

has $p_3 = 0.7$ probability of executing component C_D and $p_4 = 0.3$ probability of exiting directly. When component C_4 chooses to execute component C_D, it can be regarded as a sequence of components C_4, C_D, and when the component C_D does not choose to execute, then the trustworthiness of the branch is equal to the trustworthiness of component C_4. $T_{C_E} = 0.7\left(T_{C_4}^{\alpha_E^4} \times T_{C_D}^{\alpha_E^D}\right) + 0.3T_{C_4} = 0.915$.

Component C_E and component C_4 are connected by a branch structure to form a composite component C_F, and T_{C_F} denotes the trustworthiness of component C_F. $T_{C_F} = 0.1T_{C_E} + 0.9T_{C_D} = 0.935$, where $\alpha_F = \alpha_4 + \alpha_D = 0.76$.

The entire station ticketing system can be regarded as components C_A, and C_F, C_9 connected sequence, then the trustworthiness of the entire software system $T = T_{C_A}^{\alpha_A} \times T_{C_F}^{\alpha_F} \times T_{C_9}^{\alpha_9} = 0.933$ (Fig. 6).

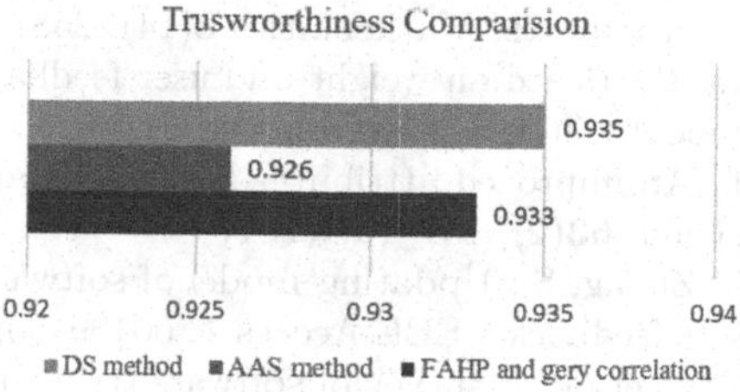

Fig. 6. Trustworthiness comparision.

Comparing the DS method [7] and the weight allocation base on AHP and sensitivity (AAS) method [23], the data in this paper are centered, where the DS method completely relies on the experience and knowledge of experts and ignores the ambiguity of expert evaluation, and the AAS method have focused on the sensitivity of the attributes in addition to the expert evaluation. This method focuses on the effect of dynamic changes in attributes on trustworthiness, ignoring the information contained inside the attributes under static. And the weight assignment method based on FAHP and grey correlation integrate the subjective characteristics of experts and the objective characteristics of data.

5 Conclusion

A reasonable weight allocation method needs to consider the amount of information from both subjective and objective aspects. On the one hand, it needs to consider the requirements and evaluation of experts, and on the other hand, it needs to pay attention to the connection and mutual influence between objective data. In this paper, the FAHP method is used to calculate the subjective weights, while the grey correlation degree is used to calculate the degree of influence of the attributes on the defect data, which reduces the information loss as much as possible. On the other hand, the connection of components in the component-based software is the connection between interfaces, and it is important to study the rules of the connection between interfaces and then construct a reasonable model to measure the trustworthiness of the component-based software.

References

1. Cho, J.H., Xu, S., Hurley, P.M., et al.: STRAM: measuring the trustworthiness of computer-based systems. ACM Comput. Surv. **51**(6), 1–47 (2019)
2. Wang, D.X., Wang, Q.: Trustworthiness evidence supporting evaluation of software process trustworthiness. J. Softw. **29**(11), 3412–3434 (2018)
3. Neumann, P.G.: Trustworthiness and truthfulness are essential. Commun. ACM. **60**(6), 26–28 (2017)
4. Martinez-Gil, F., Lozano, M., García, F.I., et al.: Using inverse reinforcement learning with real trajectories to get more trustworthy pedestrian simulations. Mathematics **8**, 1479 (2020)
5. He, J.F., Dan, Z.G., Wang, J.: Review of the achievement of major research plan on "trustworthy software". Bull. Natl. Nat. Sci. Found. China **32**(3), 291–296 (2018)
6. Tao, H., Chen, Y., Wu, H.A.: A reallocation approach for software trustworthiness based on trustworthy attributes. Mathematics **8**(14), 25829–25835 (2020)
7. Wei, Z., Ma, Y., Pan, H.: Based on weight and user feedback: a novel trustworthiness measurement model. Chin. J. Electron. **31**, 612–625 (2022)
8. Tao, H.W., Zhao, J.: An improved attributes-based software trustworthiness metrics model. Wuhan Univ. **63**(2), 151–157 (2017)
9. Wang, B., Chen, Y., Zhang, S.: Updating model of software component trustworthiness based on users feedback. IEEE Access **7**, 60199–60205 (2019)
10. Tao, H.W., Chen, Y.X., et al.: A survey of software trustworthiness measurements. Int. J. Performability Eng. **15**(9), 2364–2371 (2019)
11. Wang, D.X., Wang, Q., He, J.F.: Evidence-based software process trustworthiness model and evaluation method. J. Softw. **28**(7), 1713–1731 (2017). (in Chinese)
12. Wang, D.X., Wang, Q.: Trustworthiness evidence supporting evaluation of software process trustworthiness. J. Softw. **29**(11), 3412–3434 (2018). (in Chinese)
13. Tiwari, U.K., Kumar, S., Matta, P.: Execution-history based reliability estimation for component-based software: considering reusability-ratio and interaction-ratio. Int. J. Syst. Assur. Eng. Manag. **11**, 1003–1019 (2020)
14. Malik, P., Nautiyal, L., Ram, M.: A method for considering error propagation in reliability estimation of component-based software systems. Int. J. Math. Eng. Manag. Sci. **4**, 635–653 (2019)
15. Dong, G.L.: Trusted software design and evaluation based on component relationship. Hebei University (2012). (in Chinese)
16. Wang, B.H.: Research on Trustworthiness Measurement Models Based on Software Component. East China Normal University, Shanghai (2019). (in Chinese)
17. Wang, B.H., Liu, D.N., Zhang, S.: The performance quantitative model based on the specification and relation of the component. Mathematics **7**, 730 (2019)
18. Nie, F., Wu, D., Wang, R., et al.: Truncated robust principle component analysis with a general optimization framework. IEEE Tran. Softw. Eng. **44**(2), 1081–1097 (2022)
19. Huang, D.J.: Component-based software trustworthiness measurement and allocation model. Huaibei Normal University (2019). (in Chinese)
20. Cheng, S.L., Wang, Y.: Determination of objective weight of expert based on interval-valued intuitionistic 2-tuple linguistic setting. Comput. Eng. Des. **2**, 282–287 (2018)
21. Gu, H.H., Li, M.Z.: Objective weighting approach based embedded software trustworthiness evaluation method. Appl. Res. Comput. **29**(5), 1761–1763 (2010)

22. Shi, L., Yang, S.L., Li, K., et al.: Developing an evaluation approach for software trustworthiness using combination weights and TOPSIS. J. Softw. **7**(3), 532–542 (2012)
23. Wang, B.H., Zhang, S.: A subjective and objective integration approach of determining weights for trustworthy measurement. IEEE Access **6**, 25829–25835 (2018)
24. Zhang, J., Zhou, Y.: New software trustworthiness attribute weight distribution method based on attribute affection and importance. Appl. Res. Comput. **33**(5), 1390–1394 (2016). (in Chinese)
25. Xiong, G., Lan, J.L., Hu, Y.X., Liu, S.R.: Evaluation approach for network components performance using trustworthiness measurement. J. Commun. **37**(3), 117–128 (2016)
26. Liu, H., Tao, H.W., Chen, Y.X.: An approach for trustworthy evidence of source code oriented aerospace software trustworthiness measurement. Aerosp. Control Appl. **47**(2), 32–41 (2021). (in Chinese)
27. Kong, L.N.: Optimum selection of bridge reinforcement scheme based on game theory and evidence theory. J. Railway Sci. Eng. **17**(3), 556–562 (2020). (in Chinese)
28. Tao, H.W.: Research on attribute-based software trustworthy measurement model. East China Normal University, Shang Hai (2011). (in Chinese)

Author Index

Printed in the United States
by Baker & Taylor Publisher Services